Linear Systems

Optimal and Robust Control

Linear Systems
Optimal and Robust Control

Alok Sinha

CRC Press
Taylor & Francis Group
Boca Raton   London   New York

CRC Press is an imprint of the
Taylor & Francis Group, an informa business

Learning Resources
Centre

13082329

CRC Press
Taylor & Francis Group
6000 Broken Sound Parkway NW, Suite 300
Boca Raton, FL 33487-2742

International Standard Book Number-10: 0-8493-9217-9 (Hardcover)
International Standard Book Number-13: 978-0-8493-9217-7 (Hardcover)

Library of Congress Cataloging-in-Publication Data

Sinha, Alok. K.
    Linear systems : optimal and robust control / by Alok Sinha.
        p. cm.
    Includes bibliographical references and index.
    ISBN 0-8493-9217-9 (alk. paper)
    1. Control theory. 2. Linear system. I. Title.

QA402.3.S553 2007
629.8'32--dc22
                                           2006049289

Visit the Taylor & Francis Web site at
http://www.taylorandfrancis.com

and the CRC Press Web site at
http://www.crcpress.com

To
The Loving Memory of My Mother, Grandparents,
and Uncle
And
My Father

A COMMON MODELING ERROR

Prakrieh kriyamanani
Gunaih karmani sarvasah
Ahamkararavimudhatma
Karta 'ham iti manyate

While all kinds of work are done by the modes of nature, he whose soul is bewildered by the self-sense thinks "I am the doer." (*The Bhagavad Gita: An English Translation*, by S. Radhakrishnan, George Allen and Unwin, 1971, page 143.)

A ROBUST AND OPTIMAL CONTROL ALGORITHM

Karmay eva 'dhikaras te
Ma phalesu kadacana
Ma karmaphalahetur bhur
Ma te sango 'stv akarmani

To action alone hast thou a right and never at all to its fruits; let not the fruits of action be thy motive; neither let there be in thee any attachment to inaction. (*The Bhagavad Gita: An English Translation*, by S. Radhakrishnan, George Allen and Unwin, 1971, page 119.)

Contents

1 Introduction

1.1 OVERVIEW

This book contains materials on linear systems, and optimal and robust control, and is an outgrowth of two graduate level courses I have taught for many years at The Pennsylvania State University, University Park. The first course is on linear systems and optimal control, whereas the second course is on robust control. The unique feature of this book is that it presents the materials in a theoretically rigorous way while keeping the applications to practical problems in mind. Also, this is the first book containing H_∞ and sliding mode methods together.

The materials on linear systems include controllability, observability, and matrix fraction description. First, the concepts of state feedback control and observers are developed. Next, the optimal control is presented along with stochastic optimal control. Then, the lack of robustness of LQG control is discussed. This is followed by the presentation of robust control techniques. The derivation of H_∞ control theory is developed from the first principle. The sliding mode control of a linear system is presented. Then, it is shown how a blend of sliding mode control and H_∞ methods can enhance the robustness of the system.

One of the objectives is to make the book self contained as much as possible. For example, all the required concepts for stochastic processes are presented so that a student can understand LQG control without much prior background in stochastic processes. The book contains the presentation of theory with practical examples to illustrate the key theoretical concepts and to show their applications to practical problems. At the end of each chapter, exercise problems are included. The use of MATLAB software has been highlighted.

For my course on linear systems and optimal control, I have used the textbook by T. Kailath, *Linear Systems* (Prentice-Hall, 1980). This is a great book, and contains extensive amount of information on linear systems. For my course on robust control, I have used the textbook by K. Zhou and J. C. Doyle, *Essentials of Robust Control* (Prentice-Hall, 1998). This is also an excellent book. However, a typical engineering graduate student in most universities may find the materials in both these textbooks to be highly mathematical and, as a result, find them difficult to follow. Therefore, I have developed mathematical analyses in this book by keeping in view the background of a typical engineering student with a bachelor's degree. For example, the derivation of H_∞ does not require students to learn additional mathematical tools.

I have learned the linear system theory from the textbook of T. Kailath. Therefore, even though I have never met him, I would like to recognize T. Kailath as my virtual teacher. During my sabbatical at MIT, I was fortunate to interact with M. Athans and attend his course on multivariable control systems, for which G. Stein presented excellent lectures on H_∞ control. I would also like to thank E. F. Crawley for arranging my sabbatical and providing me an opportunity to do research on sliding mode control at the MIT Space Engineering Research Center, where I was lucky to find David Miller who helped me implement my controller on the development model of the Middeck Active Control Experiment (MACE). Lastly, I would like to thank my wife, Hansa, and daughters, Divya and Swarna, for their support.

1.2 CONTENTS OF THE BOOK

In Chapter 2, methods to develop a state space realization from a SISO transfer function are presented, along with the concepts of controllability and observability. The connection between a minimal order of the SISO state space realization and simultaneous controllability and observability is presented. This is followed by the matrix fraction description of the MIMO system, and a method is developed to find a state space realization with the minimal order. Lastly, poles and zeros of a MIMO system are defined.

In Chapter 3, the design of a full state feedback control system is presented for a SISO system along with its impact on poles and zeros of the closed-loop system. Next, the full state feedback control system is presented for a MIMO system. The necessary conditions for the optimal control are then derived and used to develop the linear quadratic (LQ) control theory and the minimum time control.

In Chapter 4, methods to estimate states are developed on the basis of inputs and outputs of a deterministic and a stochastic system. Then, theories and examples of optimal state estimation and linear quadratic Gaussian control are presented.

In Chapter 5, the fundamental concepts of robust control are developed. The robustness of LQ and LQG control techniques developed in Chapter 3 and Chapter 4 are examined. Lastly, theories for H_2, H_∞, and μ techniques are presented along with Bode's sensitivity integrals and illustrative examples.

In Chapter 6, basic concepts of sliding modes are presented along with the sliding mode control of a linear system with full state feedback. Then, it is shown how H_∞ and sliding mode theories can be blended to control an uncertain linear system with full state feedback. Next, the sliding mode control of a deterministic linear system is developed with the feedback of estimated states. Lastly, the optimal sliding Gaussian (OSG) control theory is presented for a stochastic system.

2 State Space Description of a Linear System

First, methods to develop a state space realization from a SISO transfer function are presented, as well as the concepts of controllability and observability. The connection between a minimal order of the SISO state space realization and the simultaneous controllability and observability is presented. This is followed by the matrix fraction description (MFD) of the MIMO system, and a method is developed to find a state space realization with the minimal order. Lastly, poles and zeros of a MIMO system are defined.

2.1 TRANSFER FUNCTION OF A SINGLE INPUT/SINGLE OUTPUT (SISO) SYSTEM

The dynamics of a single input/single output (SISO) linear system can be represented in general by the following nth order differential equation:

$$\frac{d^n y}{dt^n} + a_1 \frac{d^{n-1}y}{dt^{n-1}} + a_2 \frac{d^{n-2}y}{dt^{n-2}} + \ldots + a_n y =$$
$$b_0 \frac{d^n u}{dt^n} + b_1 \frac{d^{n-1}u}{dt^{n-1}} + b_2 \frac{d^{n-2}u}{dt^{n-2}} + \ldots + b_n u \tag{2.1.1}$$

where $y(t)$ is the output and $u(t)$ is the input. The coefficients $a_1, a_2, \ldots, a_n$ and b_0, $b_1, \ldots, b_n$ are system parameters. These parameters are constants for a time-invariant system.

Taking the Laplace transformation of (2.1.1) and setting all initial conditions to be zero,

$$\frac{y(s)}{u(s)} = g(s) \tag{2.1.2}$$

where

$$g(s) = \frac{b_0 s^n + b_1 s^{n-1} + \ldots + b_{n-1} s + b_n}{s^n + a_1 s^{n-1} + a_2 s^{n-2} + \ldots a_{n-1} s + a_n} \tag{2.1.3}$$

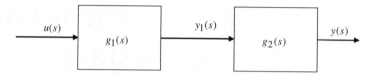

FIGURE 2.1 Cascaded linear systems.

The expression $g(s)$ is described as the transfer function with which the following facts can be attributed:

1. The transfer function is the ratio of Laplace transforms of the output and the input.
2. All initial conditions associated with the input and the output are taken to be zero.
3. The transfer function is only defined for a linear and time-invariant system.

Taking $u(t) = \delta(t)$, the unit impulse function, it can be seen that

$$g(t) = L^{-1}(g(s)) \tag{2.1.4}$$

is the unit impulse response function of the system. In other words, the transfer function of a linear time-invariant system is the Laplace transformation of the unit impulse response of the system.

To appreciate the usefulness of the transfer function approach, consider the system shown in Figure 2.1 in which the output of the first subsystem $y_1(s)$ is the input to the next subsystem. This is a typical situation found in the study or design of a control system. The output $y(s)$ and the input $u(s)$ are related as follows:

$$\frac{y(s)}{u(s)} = g_1(s)g_2(s) \tag{2.1.5}$$

In time domain, the output $y(t)$ is related to the input $u(t)$ via the convolution integral (Kuo, 1995). More specifically,

$$y(t) = \int_0^t g_2(t - \tau)y_1(\tau)d\tau \tag{2.1.6a}$$

and

$$y_1(t) = \int_0^t g_1(t - \tau)u(\tau)d\tau \tag{2.1.6b}$$

Therefore,

$$y(t) = \int_0^t g_2(t-\tau) \int_0^\tau g_1(\tau-v)u(v)dvd\tau \qquad (2.1.7)$$

Comparing Equation 2.1.5 and Equation 2.1.7, it is obvious that the input–output relationship in s-domain is much simpler than that in time domain.

2.2　STATE SPACE REALIZATIONS OF A SISO SYSTEM

Method I

Define $\xi(t)$ such that

$$\frac{d^n\xi}{dt^n} + a_1 \frac{d^{n-1}\xi}{dt^{n-1}} + a_2 \frac{d^{n-2}\xi}{dt^{n-2}} + \ldots + a_n\xi = u(t) \qquad (2.2.1)$$

Assuming that all initial conditions on $y(t)$ are zero and using the principle of superposition (Kailath, 1980), Equation 2.1.1 leads to

$$y(t) = b_0 \frac{d^n\xi}{dt^n} + b_1 \frac{d^{n-1}\xi}{dt^{n-1}} + b_2 \frac{d^{n-2}\xi}{dt^{n-2}} + \ldots + b_n\xi \qquad (2.2.2)$$

Although initial conditions on $y(t)$ and its higher derivatives have been taken to be zero, the Equation 2.2.2 is valid for nonzero initial conditions on $y(t)$ and its higher derivatives. The treatment of nonzero $y(0)$, $\dot{y}(0),\ldots,$ and $\frac{d^n y}{dt^n}(0)$ is related to the **observability** issue and will be discussed later.

Define

$$x_1 = \xi$$

$$x_2 = \frac{d\xi}{dt}$$

$$x_3 = \frac{d^2\xi}{dt^2}$$

$$\vdots \qquad\qquad (2.2.3)$$

$$x_n = \frac{d^{n-1}\xi}{dt^{n-1}}$$

Using (2.2.1) and (2.2.3), the following n first-order differential equations are obtained as follows:

$$\frac{dx_1}{dt} = x_2$$

$$\frac{dx_2}{dt} = x_3$$

$$\vdots$$

$$\frac{dx_{n-1}}{dt} = x_n$$

$$\frac{dx_n}{dt} = -a_n x_1 - a_{n-1} x_2 - \ldots - a_1 x_n + u(t) \qquad (2.2.4)$$

Using Equation 2.2.2, the output $y(t)$ is related to variables defined in (2.2.3) as follows:

$$y(t) = b_0 \frac{dx_n}{dt} + b_1 x_n + b_2 x_{n-1} + \ldots + b_n x_1 \qquad (2.2.5)$$

Using (2.2.4),

$$y(t) = (b_n - b_0 a_n)x_1 + (b_{n-1} - b_0 a_{n-1})x_2 + \ldots + (b_1 - b_0 a_1)x_n + b_0 u(t) \quad (2.2.6)$$

The system represented by (2.2.4) and (2.2.6) can be realized using n analog integrators as shown in Figure 2.2. The variables $x_1, x_2, x_3, \ldots, x_n$ turn out to be outputs of integrators and are described as **state variables**. In matrix form, Equation 2.2.4 and Equation 2.2.6 are described as follows:

$$\frac{d\mathbf{x}}{dt} = A_c \mathbf{x}(t) + \mathbf{b}_c u(t) \qquad (2.2.7)$$

and

$$y(t) = \mathbf{c}_c \mathbf{x}(t) + b_0 u(t) \qquad (2.2.8)$$

where

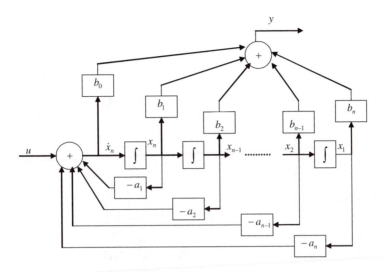

FIGURE 2.2 Analog computer simulation diagram (Method I).

$$A_c = \begin{bmatrix} 0 & 1 & 0 & 0 & . & . & . & 0 \\ 0 & 0 & 1 & . & . & . & . & 0 \\ . & . & . & . & . & . & . & . \\ -a_n & . & . & . & . & . & -a_2 & -a_1 \end{bmatrix} \qquad (2.2.9)$$

$$\mathbf{b}_c = \begin{bmatrix} 0 \\ 0 \\ . \\ . \\ . \\ 1 \end{bmatrix} \qquad (2.2.10)$$

and

$$\mathbf{c}_c = [b_n - b_0 a_n \quad b_{n-1} - b_0 a_{n-1} \ldots b_1 - b_0 a_1] \qquad (2.2.11)$$

METHOD II

Defining $p = \dfrac{d}{dt}$, Equation 2.1.1 can be written as

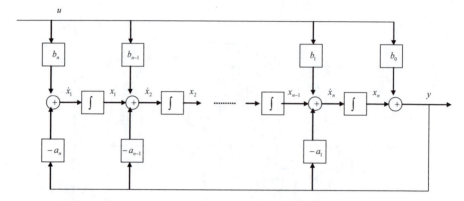

FIGURE 2.3 Analog computer simulation diagram (Method II).

$$p^n(y(t)-b_0u(t))+p^{n-1}(a_1y(t)-b_1u(t))+p^{n-2}(a_2y(t)-b_2u(t))+\ldots+$$
$$(a_ny(t)-b_nu(t))=0 \tag{2.2.12}$$

Dividing (2.2.12) by p^n (Wiberg, 1971),

$$y(t)=b_0u(t)-\frac{a_1y(t)-b_1u(t)}{p}-\frac{a_2y(t)-b_2u(t)}{p^2}-\ldots-\frac{a_ny(t)-b_nu(t)}{p^n} \tag{2.2.13}$$

The analog computer simulation diagram corresponding to Equation 2.2.13 is shown in Figure 2.3.

Defining the outputs of integrators as state variables $x_1, x_2, x_3,\ldots, x_n$, the following relationships are obtained:

$$y(t)=x_n(t)+b_0u(t)$$

and

$$\frac{dx_1}{dt}=-a_ny(t)+b_nu(t)=-a_nx_n+(b_n-b_0a_n)u$$

$$\frac{dx_2}{dt}=-a_{n-1}y(t)+x_1(t)+b_{n-1}u(t)=-a_{n-1}x_n+x_1+(b_{n-1}-b_0a_{n-1})u$$

$$\vdots \tag{2.2.14}$$

$$\frac{dx_n}{dt}=-a_1y(t)+x_{n-1}(t)+b_1u(t)=-a_1x_n+x_{n-1}+(b_1-b_0a_1)u$$

Equations 2.2.14 can be put in the matrix form as follows:

$$\frac{d\mathbf{x}}{dt} = A_o \mathbf{x}(t) + \mathbf{b}_o u(t) \tag{2.2.15a}$$

and

$$y(t) = \mathbf{c}_o x(t) + b_0 u(t) \tag{2.2.15b}$$

where

$$A_o = \begin{bmatrix} 0 & 0 & . & . & . & -a_n \\ 1 & 0 & . & . & 0 & -a_{n-1} \\ 0 & 1 & . & . & 0 & -a_{n-2} \\ . & . & . & . & . & . \\ 0 & 0 & . & . & 1 & -a_1 \end{bmatrix} \tag{2.2.16}$$

$$\mathbf{b}_o = \begin{bmatrix} b_n - b_0 a_n \\ b_{n-1} - b_0 a_{n-1} \\ . \\ . \\ . \\ b_1 - b_o a_1 \end{bmatrix} \tag{2.2.17}$$

and

$$\mathbf{c}_o = \begin{bmatrix} 0 & 0 & . & . & 0 & 1 \end{bmatrix} \tag{2.2.18}$$

Notation

Often, a state space realization is represented by the symbol $\{A, \mathbf{b}, \mathbf{c}\}$ by which we mean the following:

$$\dot{\mathbf{x}} = A\mathbf{x}(t) + \mathbf{b}u(t) \tag{2.2.19a}$$

$$y(t) = \mathbf{c}\mathbf{x}(t) \tag{2.2.19b}$$

EXAMPLE 2.1

$$g(s) = \frac{s^3 + 2s + 5}{s^3 + 2s^2 + 4s + 3} \tag{2.2.20}$$

Here, $b_0 = 1$. From Method I:

$$A_c = \begin{bmatrix} 0 & 1 & 0 \\ 0 & 0 & 1 \\ -3 & -4 & -2 \end{bmatrix} \quad \mathbf{b}_c = \begin{bmatrix} 0 \\ 0 \\ 1 \end{bmatrix} \quad \mathbf{c}_c = \begin{bmatrix} 2 & 2 & 2 \end{bmatrix} \tag{2.2.21}$$

From Method II:

$$A_o = \begin{bmatrix} 0 & 0 & -3 \\ 1 & 0 & -4 \\ 0 & 1 & -2 \end{bmatrix} \quad \mathbf{b}_o = \begin{bmatrix} 2 \\ 2 \\ 2 \end{bmatrix} \quad \mathbf{c}_o = \begin{bmatrix} 0 & 0 & 1 \end{bmatrix} \tag{2.2.22}$$

PROPERTIES OF STATE SPACE MODELS

1. Duality

For any given realization $\{A, \mathbf{b}, \mathbf{c}\}$ of a system transfer function, there is a dual (Kailath, 1980) realization $\{A^T, \mathbf{c}^T, \mathbf{b}^T\}$.

It is interesting to note that

$$A_o = A_c^T, \quad \mathbf{b}_o = \mathbf{c}_c^T, \text{ and } \mathbf{c}_o = \mathbf{b}_c^T \tag{2.2.23}$$

Hence, the state space realization obtained by Method I is dual to that obtained by Method II, and *vice versa*.

2. Nonuniqueness of State Space Realization

Consider the following transformation:

$$\mathbf{x}(t) = T\overline{\mathbf{x}}(t) \tag{2.2.24}$$

where T is any nonsingular matrix. For the realization $\{A, \mathbf{b}, \mathbf{c}\}$,

$$\frac{d\mathbf{x}}{dt} = A\mathbf{x} + \mathbf{b}u(t) \tag{2.2.25}$$

$$y(t) = \mathbf{cx}(t) \tag{2.2.26}$$

Substituting (2.2.24) into (2.2.25) and (2.2.26),

$$\frac{d\overline{\mathbf{x}}}{dt} = \overline{A}\overline{\mathbf{x}} + \overline{\mathbf{b}}u(t) \tag{2.2.27}$$

$$y(t) = \overline{\mathbf{c}}\overline{x}(t) \tag{2.2.28}$$

where

$$\overline{A} = T^{-1}AT \ , \quad \overline{\mathbf{b}} = T^{-1}\mathbf{b} \ , \quad \text{and} \quad \overline{\mathbf{c}} = \mathbf{c}T \tag{2.2.29}$$

Hence, a new realization, $\{\overline{A}, \overline{\mathbf{b}}, \overline{\mathbf{c}}\}$, has been obtained. As the choice of the nonsingular matrix T is arbitrary, there are clearly many realizations or nonunique state space realizations corresponding to a given transfer function (Kailath, 1980). In matrix theory, the transformation (2.2.24) is known as similarity transformation, and A and $\overline{A}$ are called **similar matrices**.

2.3 SISO TRANSFER FUNCTION FROM A STATE SPACE REALIZATION

Taking the Laplace transformation of equations (2.2.19),

$$y(s) = \mathbf{c}(sI - A)^{-1}\mathbf{x}(0) + \mathbf{c}(sI - A)^{-1}\mathbf{b}u(s) \tag{2.3.1}$$

For the definition of a transfer function, $\mathbf{x}(0) = 0$. Therefore,

$$g(s) = \frac{y(s)}{u(s)} = \mathbf{c}(sI - A)^{-1}\mathbf{b} \tag{2.3.2}$$

Equation 2.3.2 can be expressed as

$$g(s) = \frac{\mathbf{c}Adj(sI - A)\mathbf{b}}{\det(sI - A)} \tag{2.3.3}$$

where

$$Adj(sI - A) = A^{n-1} + (s + a_1)A^{n-2} + \ldots + (s^{n-1} + a_1 s^{n-2} + \ldots + a_{n-1})I_n \tag{2.3.4}$$

and

$$\det(sI_n - A) = s^n + a_1 s^{n-1} + a_2 s^{n-2} + \ldots + a_{n-1} s + a_n \qquad (2.3.5)$$

Further, it can be shown that

$$\mathbf{c} Adj(sI - A)\mathbf{b} = b_1 s^{n-1} + b_2 s^{n-2} + \ldots + b_{n-1} s + b_n \qquad (2.3.6)$$

The results (2.3.5) and (2.3.6) are also true for $\bar{A}$, $\bar{\mathbf{b}}$, and $\bar{\mathbf{c}}$. Hence, the transfer function is unique for all **similar** state space realizations.

2.4 SOLUTION OF STATE SPACE EQUATIONS

2.4.1 HOMOGENEOUS EQUATION

Consider the solution of Equation 2.2.19 with $u(t) = 0$; i.e.,

$$\frac{d\mathbf{x}}{dt} = A\mathbf{x} ; \quad \mathbf{x}(0) = \mathbf{x}_0 \qquad (2.4.1)$$

Taking the Laplace transformation of (2.4.1),

$$\mathbf{x}(s) = (sI - A)^{-1}\mathbf{x}_0 \qquad (2.4.2)$$

Now,

$$(sI - A)^{-1} = \frac{1}{s}\left(I - \frac{A}{s}\right)^{-1} = \frac{1}{s}\left(I + \frac{A}{s} + \frac{A^2}{s^2} + \ldots\right) \qquad (2.4.3)$$

Taking the inverse Laplace transformation of (2.4.3),

$$L^{-1}(sI - A)^{-1} = I + At + A^2 \frac{t^2}{2} + \ldots; \quad t \geq 0 \qquad (2.4.4)$$

The matrix exponential is defined as follows:

$$e^{At} = L^{-1}[(sI - A)^{-1}] \qquad (2.4.5)$$

Lastly, taking the inverse Laplace transformation of (2.4.2),

$$\mathbf{x}(t) = e^{At}\mathbf{x}_0 ; \quad t \geq 0 \tag{2.4.6}$$

The matrix e^{At} is also known as the **state transition matrix** because it takes the initial state $\mathbf{x}_0$ to a state $\mathbf{x}(t)$ in time t.

Properties of e^{At}

1. $\dfrac{d}{dt}(e^{At}) = Ae^{At} = e^{At}A$ $\hspace{4cm}$ (2.4.7)

2. $e^{A(t_1+t_2)} = e^{At_1}e^{At_2}$ $\hspace{4cm}$ (2.4.8)

3. If e^{At} is nonsingular, $(e^{At})^{-1} = e^{-At}$ $\hspace{2.5cm}$ (2.4.9)

EXAMPLE 2.2

$$A = \begin{bmatrix} -4 & 3 \\ -1 & 0 \end{bmatrix} \quad \mathbf{x}(0) = \begin{bmatrix} 0 \\ 1 \end{bmatrix} \tag{2.4.10}$$

$$(sI - A)^{-1} = \begin{bmatrix} \dfrac{s}{(s+1)(s+3)} & \dfrac{3}{(s+1)(s+3)} \\ -\dfrac{1}{(s+1)(s+3)} & \dfrac{s+4}{(s+1)(s+3)} \end{bmatrix} \tag{2.4.11}$$

$$L^{-1}[(sI - A)^{-1}] = e^{At} = \begin{bmatrix} 1.5e^{-3t} - 0.5e^{-t} & -1.5e^{-3t} + 1.5e^{-t} \\ 0.5e^{-3t} - 0.5e^{-t} & -0.5e^{-3t} + 1.5e^{-t} \end{bmatrix} \tag{2.4.12}$$

Then, Equation 2.4.6 yields

$$\mathbf{x}(t) = \begin{bmatrix} -1.5e^{-3t} + 1.5e^{-t} \\ -0.5e^{-3t} + 1.5e^{-t} \end{bmatrix} ; \quad t \geq 0 \tag{2.4.13}$$

2.4.2 INHOMOGENEOUS EQUATION

Taking the Laplace transformation of (2.2.19),

$$\mathbf{x}(s) = (sI - A)^{-1}\mathbf{x}(0) + (sI - A)^{-1}\mathbf{b}u(s) \tag{2.4.14}$$

Utilizing the definition (2.4.5), the inverse Laplace transformation of (2.4.14) yields

$$\mathbf{x}(t) = e^{At}\mathbf{x}(0) + \int_0^t e^{A(t-\tau)}\mathbf{b}u(\tau)d\tau \; ; \; t \geq 0 \qquad (2.4.15)$$

EXAMPLE 2.3

Let

$$A = \begin{bmatrix} -4 & 3 \\ -1 & 0 \end{bmatrix} \quad \mathbf{b} = \begin{bmatrix} 1 \\ 0 \end{bmatrix} \quad \mathbf{x}(0) = \begin{bmatrix} 0 \\ 1 \end{bmatrix}; \text{ and } u(t) = 1 \text{ for } t \geq 0 \qquad (2.4.16)$$

$$e^{A(t-\tau)}\mathbf{b} = \begin{bmatrix} 1.5e^{-3(t-\tau)} - 0.5e^{-(t-\tau)} \\ 0.5e^{-3(t-\tau)} - 0.5e^{-(t-\tau)} \end{bmatrix} \qquad (2.4.17)$$

Then

$$\int_0^t e^{A(t-\tau)}\mathbf{b}u(\tau)d\tau = \begin{bmatrix} 1.5e^{-3t}\int_0^{t} e^{3\tau}d\tau - 0.5e^{-t}\int_0^{t} e^{\tau}d\tau \\ 0.5e^{-3t}\int_0^{t} e^{3\tau}d\tau - 0.5e^{-t}\int_0^{t} e^{\tau}d\tau \end{bmatrix} = $$

$$\begin{bmatrix} -0.5e^{-3t} + 0.5e^{-t} \\ -\dfrac{1}{3} - \dfrac{0.5}{3}e^{-3t} + 0.5e^{-t} \end{bmatrix} \qquad (2.4.18)$$

Hence, from (2.4.15),

$$\mathbf{x}(t) = \begin{bmatrix} -2e^{-3t} + 2e^{-t} \\ -\dfrac{1}{3} - \dfrac{2}{3}e^{-3t} + 2e^{-t} \end{bmatrix}; \quad t \geq 0 \qquad (2.4.19)$$

2.5 OBSERVABILITY AND CONTROLLABILITY OF A SISO SYSTEM

2.5.1 OBSERVABILITY

The state space model $\{A, \mathbf{b}, \mathbf{c}\}$ can be developed on the basis of the governing differential Equation 2.1.1 or the transfer function (2.1.3). If the initial conditions on the output and its derivatives are nonzero, how would one get correct values of initial states? To answer this question, consider the output Equation 2.2.19b,

$$y(t) = \mathbf{c}\mathbf{x}(t)$$

Differentiating it and using (2.2.19a),

$$\frac{dy}{dt} = \mathbf{c}A\mathbf{x}(t) + \mathbf{c}\mathbf{b}u(t) \tag{2.5.1}$$

Continuing this differentiation process,

$$\frac{d^2y}{dt^2} = \mathbf{c}A^2\mathbf{x}(t) + \mathbf{c}A\mathbf{b}u(t) + \mathbf{c}\mathbf{b}\frac{du}{dt}$$

$$\vdots \tag{2.5.2}$$

$$\frac{d^{n-1}y}{dt^{n-1}} = \mathbf{c}A^{n-1}\mathbf{x}(t) + \mathbf{c}A^{n-2}\mathbf{b}u(t) + \mathbf{c}A^{n-3}\mathbf{b}\frac{du}{dt} + \ldots + \mathbf{c}\mathbf{b}\frac{d^{n-1}u}{dt^{n-1}}$$

Representing (2.5.1) and (2.5.2) in matrix form,

$$\mathbf{y}(t) = \Theta\mathbf{x}(t) + T\mathbf{u}(t) \tag{2.5.3}$$

where

$$\mathbf{y}(t) = \begin{bmatrix} y & \dfrac{dy}{dt} & . & . & \dfrac{d^{n-1}y}{dt^{n-1}} \end{bmatrix}^T \tag{2.5.4}$$

$$\Theta = \begin{bmatrix} \mathbf{c} \\ \mathbf{c}A \\ \mathbf{c}A^2 \\ . \\ . \\ . \\ \mathbf{c}A^{n-1} \end{bmatrix} \tag{2.5.5}$$

$$\mathbf{u}(t) = \begin{bmatrix} u & \dfrac{du}{dt} & . & . & \dfrac{d^{n-1}u}{dt^{n-1}} \end{bmatrix}^T \tag{2.5.6}$$

$$T = \begin{bmatrix} 0 & 0 & 0 & . & . & 0 \\ cb & 0 & 0 & . & . & 0 \\ cAb & cb & 0 & . & . & 0 \\ . & . & . & . & . & . \\ . & . & . & . & . & . \\ cA^{n-2}b & . & . & . & cb & 0 \end{bmatrix} \qquad (2.5.7)$$

From (2.5.3),

$$\Theta x(0) = y(0) - Tu(0) \qquad (2.5.8)$$

Therefore,

$$x(0) = \Theta^{-1}y(0) - \Theta^{-1}Tu(0) \qquad (2.5.9)$$

provided the matrix Θ is nonsingular. Hence, the condition for initial states to be calculated from the initial values of input, output, and their derivatives is that the matrix Θ should be nonsingular. The matrix Θ is known as the *observability matrix*.

The solution (2.4.15) of the state space equation indicates that $x(t)$ can be calculated for any $u(t)$ if the initial value $x(0)$ is known.

Observability of State Space Realization Using Method I

Using (2.5.5),

$$\Theta = \begin{bmatrix} b_n - b_0 a_n & . & . & . & . & b_1 - b_0 a_1 \\ -a_n(b_1 - b_0 a_1) & . & . & . & . & . \\ . & . & . & . & . & . \\ . & . & . & . & . & . \\ . & . & . & . & . & . \end{bmatrix} \qquad (2.5.10)$$

The determinant and hence the singularity of the observability matrix will depend on system parameters. Hence, the realization may or may not be observable.

Observability of State Space Realization Using Method II

Using (2.5.5),

$$\Theta = \begin{bmatrix} 0 & 0 & 0 & . & . & 1 \\ 0 & 0 & . & . & 1 & 0 \\ . & . & . & . & . & . \\ . & . & . & . & . & . \\ 1 & 0 & 0 & . & . & 0 \end{bmatrix} \qquad (2.5.11)$$

It can be easily seen that

$$\det \Theta = -1 \text{ or } +1 \qquad (2.5.12)$$

Therefore, Θ is nonsingular irrespective of the system parameter values. Hence, this state space realization is always observable.

EXAMPLE 2.4

Consider a spring-mass-damper system Figure 2.4, with the following differential equation:

$$m\ddot{x} + \alpha\dot{x} + \beta x = f(t) \qquad (2.5.13)$$

where $f(t)$ is the applied force. Dividing (2.5.13) by m,

$$\ddot{x} + \frac{\alpha}{m}\dot{x} + \frac{\beta}{m}x = \frac{f(t)}{m} = u(t) \qquad (2.5.14)$$

Defining states as

$$x_1 = x \quad \text{and} \quad x_2 = \dot{x} \qquad (2.5.15a,b)$$

state equations are

$$\dot{\mathbf{x}} = A\mathbf{x}(t) + \mathbf{b}u(t) \qquad (2.5.16)$$

where

$$\mathbf{x}(t) = \begin{bmatrix} x_1(t) \\ x_2(t) \end{bmatrix}; \quad A = \begin{bmatrix} 0 & 1 \\ -\beta/m & -\alpha/m \end{bmatrix}; \quad \mathbf{b} = \begin{bmatrix} 0 \\ 1 \end{bmatrix} \qquad (2.5.17)$$

Case I: Position Output

If the position of the mass, x, is measured by a sensor,

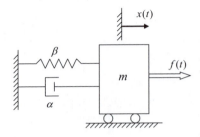

FIGURE 2.4 A spring-mass-damper system.

$$y(t) = \mathbf{c}\mathbf{x}(t) \qquad (2.5.18)$$

where

$$\mathbf{c} = [1 \quad 0] \qquad (2.5.19)$$

Then the observability matrix is

$$\Theta = \begin{bmatrix} 1 & 0 \\ 0 & 1 \end{bmatrix} \qquad (2.5.20)$$

Hence, the state space realization is observable for all values of spring stiffness and damping constant.

Case II: Velocity Output

If the velocity of the mass, $\dot{x}$, is measured by a sensor,

$$y(t) = \mathbf{c}\mathbf{x}(t) \qquad (2.5.21)$$

where

$$\mathbf{c} = [0 \quad 1] \qquad (2.5.22)$$

Then the observability matrix is

$$\Theta = \begin{bmatrix} 0 & 1 \\ -\beta/m & -\alpha/m \end{bmatrix} \qquad (2.5.23)$$

Therefore,

$$\det \Theta = \beta / m \tag{2.5.24}$$

Hence, the state space realization is not observable if the spring stiffness is zero, i.e., if there is no spring in the system. This is an interesting result because the displacement can be obtained by integrating velocity:

$$x_1(t) = x_1(0) + \int_0^t x_2(t)dt \tag{2.5.25}$$

However, the initial condition $x_1(0)$ is needed. In the absence of a spring, the loss of observability implies that $x_1(0)$ cannot be obtained, and the position of the system cannot be observed.

2.5.2 Controllability

The linear differential equation

$$\frac{d\mathbf{x}}{dt} = A\mathbf{x} + \mathbf{b}u(t) \tag{2.5.26}$$

is controllable if and only if it can be transferred from *any* initial state to *any* final state in a *finite* time.

Therefore, for the linear and time-invariant system (2.5.26), the issue is to find $u(t); t_0 \le t \le t_f$, which will take the system from any $\mathbf{x}(t_0)$ to any $\mathbf{x}(t_f)$ in a finite time $t_f - t_0$.

Using the solution of (2.5.26),

$$\mathbf{x}(t_f) = e^{A(t_f - t_0)}\mathbf{x}(t_0) + \int_{t_0}^{t_f} e^{A(t_f - \tau)}\mathbf{b}u(\tau)d\tau \tag{2.5.27}$$

This is an integral equation because the unknown function $u(.)$ appears under an integral sign. Define a matrix $P(t_f, t_0)$ as follows:

$$P(t_f, t_0) = \int_{t_0}^{t_f} e^{A(t_f - \tau)}\mathbf{b}\mathbf{b}^T e^{A^T(t_f - \tau)}d\tau \tag{2.5.28}$$

This matrix $P(t_f, t_0)$ is known as the *controllability Gramian*.

The solution of (2.5.27) is found (Friedland, 1985) to be

$$u(\tau) = \mathbf{b}^T e^{A^T(t_f-\tau)} P^{-1}(t_f,t_0)[\mathbf{x}(t_f) - e^{A(t_f-t_0)}\mathbf{x}(t_0)] \tag{2.5.29}$$

This result can be verified by substituting (2.5.29) into (2.5.27). Hence, the control input $u(.)$ can be found if and only if the matrix $P(t_f,t_0)$ is nonsingular. Defining a new variable $v = t_f - t_0$,

$$P(t_f,t_0) = \int_0^{t_f-t_0} e^{Av}\mathbf{bb}^T e^{A^T v} dv \tag{2.5.30}$$

Note that $P(t_f,t_0) = P(t_f - t_0)$; i.e., $P(t_f,t_0)$ only depends on $t_f - t_0$ for a linear and time-invariant system.

A test for linear independence of any functions $\{\ell_i(\tau); 0 \le \tau \le t_f - t_0, i = 1,2,....,n\}$ is that their Gramian matrix G (Appendix A) is non-singular where

$$G = \int_0^{t_f-t_0} \underline{\ell}(\tau)\underline{\ell}^T(\tau)d\tau \quad \text{and} \quad \underline{\ell}^T(\tau) = \begin{bmatrix} \ell_1(\tau) & . & . & . & \ell_n(\tau) \end{bmatrix} \tag{2.5.31}$$

Equation 2.5.30 can be written as

$$P(t_f,t_0) = \int_0^{t_f-t_0} e^{Av}\mathbf{b}(e^{Av}\mathbf{b})^T dv \tag{2.5.32}$$

Comparing (2.5.31) and (2.5.32), nonsingularity of the matrix P implies that elements of the vector $e^{Av}\mathbf{b}$ are linearly independent over $(0,t_f - t_0)$. Therefore, the pair $\{A,\mathbf{b}\}$ is controllable over $(0,t_f - t_0)$ if and only if the elements of the vector $e^{Av}\mathbf{b}$ are linearly independent.

EXAMPLE 2.5

$$A = \begin{bmatrix} 0 & 1 \\ -4 & 0 \end{bmatrix} \quad \text{and} \quad \mathbf{b} = \begin{bmatrix} 0 \\ 1 \end{bmatrix} \tag{2.5.33}$$

Find e^{At} :

$$sI - A = \begin{bmatrix} s & -1 \\ 4 & s \end{bmatrix} \tag{2.5.34}$$

$$(sI - A)^{-1} = \begin{bmatrix} \dfrac{s}{s^2+4} & \dfrac{1}{s^2+4} \\ \dfrac{-4}{s^2+4} & \dfrac{s}{s^2+4} \end{bmatrix} \tag{2.5.35}$$

$$e^{At} = L^{-1}[(sI - A)^{-1}] = \begin{bmatrix} \cos 2t & 0.5\sin 2t \\ -2\sin 2t & \cos 2t \end{bmatrix} \tag{2.5.36}$$

Find $e^{Av}\mathbf{b}$:

$$e^{Av}\mathbf{b} = \begin{bmatrix} \cos 2v & \dfrac{\sin 2v}{2} \\ -2\sin 2v & \cos 2v \end{bmatrix}\begin{bmatrix} 0 \\ 1 \end{bmatrix} = \begin{bmatrix} \dfrac{\sin 2v}{2} \\ \cos 2v \end{bmatrix} \tag{2.5.37}$$

From the controllability Gramian,

$$e^{Av}\mathbf{bb}^T e^{A^T v} = \begin{bmatrix} \dfrac{\sin^2 2v}{4} & \dfrac{\sin 2v \cos 2v}{2} \\ \dfrac{\sin 2v \cos 2v}{2} & \cos^2 2v \end{bmatrix} \tag{2.5.38}$$

Let

$$t_0 - t_f = \Delta t \tag{2.5.39}$$

Hence,

$$P(t_f, t_0) = \int_0^{\Delta t} e^{Av}\mathbf{bb}^T e^{A^T v}\,dv$$

$$= \begin{bmatrix} \dfrac{\Delta t - (\sin(4\Delta t))/4}{8} & \dfrac{-\cos(4\Delta t)+1}{16} \\ \dfrac{-\cos(4\Delta t)+1}{16} & \dfrac{\Delta t + (\sin(4\Delta t))/4}{2} \end{bmatrix} \tag{2.5.40}$$

$$\det(P(\Delta t)) = \frac{(\Delta t)^2 - 0.25\sin^2(2\Delta t)}{16} \tag{2.5.41}$$

Theorem

The system is completely controllable if and only if the **controllability matrix**

$$C = [\mathbf{b} \quad A\mathbf{b} \quad A^2\mathbf{b} \quad . \quad . \quad A^{n-1}\mathbf{b}] \tag{2.5.42}$$

is nonsingular.

Proof

Step I: Consider that the matrix P, Equation 2.5.28, is singular. In this case, elements of the vector $e^{Av}\mathbf{b}$ are linearly dependent. As a result, there exists a nonzero vector q such that

$$z(t) = \mathbf{b}^T e^{A^T t} \mathbf{q} = 0 \quad \text{for } 0 \le t \le t_f - t_0 \tag{2.5.43}$$

Differentiating (2.5.43),

$$\dot{z}(t) = \mathbf{b}^T A^T e^{A^T t} \mathbf{q} = 0$$

$$\ddot{z}(t) = \mathbf{b}^T (A^T)^2 e^{A^T t} \mathbf{q} = 0$$

$$\vdots \tag{2.5.44}$$

$$\frac{d^{n-1}}{dt^{n-1}} z(t) = \mathbf{b}^T (A^T)^{n-1} e^{A^T t} \mathbf{q} = 0$$

Putting (2.5.43) and (2.5.44) in matrix form,

$$\begin{bmatrix} \mathbf{b}^T \\ \mathbf{b}^T A^T \\ . \\ . \\ \mathbf{b}^T (A^T)^{n-1} \end{bmatrix} e^{A^T t} \mathbf{q} = 0 \tag{2.5.45}$$

Using the definition (2.5.42),

$$C^T e^{A^T t} \mathbf{q} = 0 \tag{2.5.46}$$

Let

$$C^T = \begin{bmatrix} \mathbf{r}_1 & \mathbf{r}_2 & \cdot & \cdot & \mathbf{r}_n \end{bmatrix} \tag{2.5.47}$$

and

$$e^{A^T t} \mathbf{q} = \begin{bmatrix} \alpha_1(t) \\ \alpha_2(t) \\ \cdot \\ \cdot \\ \alpha_n(t) \end{bmatrix} \tag{2.5.48}$$

Hence, from (2.5.46) – (2.5.48),

$$\alpha_1(t)\mathbf{r}_1 + \alpha_2(t)\mathbf{r}_2 + \ldots + \alpha_n(t)\mathbf{r}_n = 0 \tag{2.5.49}$$

Hence, columns of C^T are linearly dependent. As $rank(C) = rank(C^T)$, the matrix C is singular.

Step II: Consider that C is singular. We will show that the matrix P, Equation 2.5.28, is singular in this case.

$$e^{Av} = I + Av + \frac{A^2 v^2}{2!} + \ldots + \frac{A^{n-1} v^{n-1}}{(n-1)!} + \frac{A^n v^n}{n!} + \ldots \tag{2.5.50}$$

Recall the Cayley–Hamilton theorem:

$$A^n = -a_1 A^{n-1} - a_2 A^{n-2} - \ldots - a_n I_n \tag{2.5.51}$$

where

$$\det(sI - A) = s^n + a_1 s^{n-1} + a_2 s^{n-2} + \ldots + a_n \tag{2.5.52}$$

Using (2.5.50) and (2.5.51),

$$e^{Av} = I f_1(v) + A f_2(v) + \ldots + A^{n-1} f_n(v) \tag{2.5.53}$$

where $f_1(v), \ldots, f_n(v)$ are functions of v.

Hence,

$$e^{Av}\mathbf{b} = C\mathbf{f}(v) \tag{2.5.54}$$

where

$$\mathbf{f}(v) = \begin{bmatrix} f_1(v) \\ f_2(v) \\ . \\ . \\ f_n(v) \end{bmatrix} \tag{2.5.55}$$

Therefore, from (2.5.30),

$$P(t_f, t_0) = C[\int_0^{t_f - t_0} \mathbf{f}(v)\mathbf{f}^T(v)dv]C^T \tag{2.5.56}$$

Note that $rank(AB) \le \min\{rank(A), rank(B)\}$. Therefore, the rank of $P(t_f, t_0)$ is going to be less than n because it has been assumed that the matrix C is singular.

Controllability of the State Space Realization Obtained Using Method I

Using (2.5.42),

$$C = \begin{bmatrix} 0 & 0 & 0 & . & . & 1 \\ 0 & 0 & . & & . & 1 & 0 \\ . & . & . & . & . & . \\ . & . & . & . & . & . \\ . & 1 & -a_1 & . & . & . \\ 1 & -a_1 & -a^2 + a_1 & . & . & . \end{bmatrix} \tag{2.5.57}$$

It can be shown that

$$\det C = -1 \text{ or } +1 \tag{2.5.58}$$

Therefore, the state space realization obtained using Method I is always controllable.

Controllability of the State Space Realization Obtained Using Method II

Using (2.5.42),

$$C^T = \begin{bmatrix} b_n - b_0 a_n & . & . & . & . & b_1 - b_0 a_1 \\ -a_n(b_1 - b_0 a_1) & . & . & . & . & . \\ . & . & . & . & . & . \\ . & . & . & . & . & . \\ . & . & . & . & . & . \end{bmatrix} \qquad (2.5.59)$$

The determinant and hence the singularity of the controllability matrix will depend on system parameters. Hence, the realization may or may not be controllable.

EXAMPLE 2.6

Consider the tank system shown in Figure 2.5, where u_1 and u_2 are input mass flow rate to tank 1 and tank 2, respectively. Let p_1 and p_a be the pressure at the left end of the pipe and the atmospheric pressure, respectively. Then,

$$p_1 - p_a = \rho g h_1 \qquad (2.5.60)$$

where ρ is the fluid density.

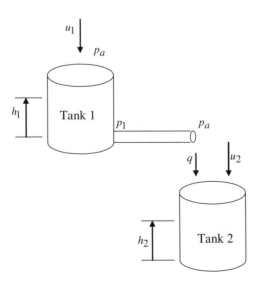

FIGURE 2.5 A tank-pipe system.

Considering a laminar pipe flow,

$$q = \frac{p_1 - p_a}{R_L} \tag{2.5.61}$$

where q and R_L are the volumetric flow rate through the pipe and the flow resistance, respectively. From the law of conservation of mass,

$$\rho A_1 \frac{dh_1}{dt} = u_1 - \rho q \tag{2.5.62}$$

$$\rho A_2 \frac{dh_2}{dt} = u_2 + \rho q \tag{2.5.63}$$

where A_1 and A_2 are cross-sectional areas of tanks 1 and 2, respectively. Using (2.5.60) and (2.5.61),

$$\begin{bmatrix} \dot{h}_1 \\ \dot{h}_2 \end{bmatrix} = \begin{bmatrix} -\alpha_1 & 0 \\ \alpha_2 & 0 \end{bmatrix} \begin{bmatrix} h_1 \\ h_2 \end{bmatrix} + \begin{bmatrix} \beta_1 \\ 0 \end{bmatrix} u_1 + \begin{bmatrix} 0 \\ \beta_2 \end{bmatrix} u_2 \tag{2.5.64}$$

where

$$\alpha_1 = \frac{\rho g}{A_1 R_L}, \quad \alpha_2 = \frac{\rho g}{A_2 R_L}, \quad \beta_1 = \frac{1}{\rho A_1}, \quad \text{and} \quad \beta_2 = \frac{1}{\rho A_2} \tag{2.5.65}$$

The controllability matrix with respect to input u_1 is

$$C = \begin{bmatrix} \beta_1 & -\alpha_1\beta_1 \\ 0 & \alpha_2\beta_1 \end{bmatrix} \tag{2.5.66}$$

Hence, the system (2.5.64) is controllable with respect to the input u_1. It is obvious that both states h_1 and h_2 are influenced by the input u_1. And, the controllability matrix with respect to the input u_2 is

$$C = \begin{bmatrix} 0 & 0 \\ \beta_2 & 0 \end{bmatrix} \tag{2.5.67}$$

Hence, the system (2.5.64) is not controllable with respect to the input u_2. It is obvious that the state h_1 cannot be influenced by the input u_2.

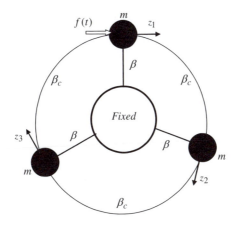

FIGURE 2.6 A periodic spring-mass system.

EXAMPLE 2.7: A PERIODIC STRUCTURE

For the system in Figure 2.6, differential equations of motion are

$$m\ddot{z}_1 + \beta z_1 + \beta_c(z_1 - z_3) + \beta_c(z_1 - z_2) = f(t) \tag{2.5.68}$$

$$m\ddot{z}_2 + \beta z_2 + \beta_c(z_2 - z_1) + \beta_c(z_2 - z_3) = 0 \tag{2.5.69}$$

$$m\ddot{z}_3 + \beta z_3 + \beta_c(z_3 - z_2) + \beta_c(z_3 - z_1) = 0 \tag{2.5.70}$$

Define

$$\mathbf{x} = [z_1 \quad z_2 \quad z_3 \quad \dot{z}_1 \quad \dot{z}_2 \quad \dot{z}_3]^T \tag{2.5.71}$$

and

$$u(t) = \frac{f(t)}{m} \tag{2.5.72}$$

Then state equations are

$$\dot{\mathbf{x}} = A\mathbf{x}(t) + \mathbf{b}u(t) \tag{2.5.73}$$

where

$$A = \begin{bmatrix} 0 & 0 & 0 & 1 & 0 & 0 \\ 0 & 0 & 0 & 0 & 1 & 0 \\ 0 & 0 & 0 & 0 & 0 & 1 \\ -p & q & q & 0 & 0 & 0 \\ q & -p & q & 0 & 0 & 0 \\ q & q & -p & 0 & 0 & 0 \end{bmatrix}, \quad b = \begin{bmatrix} 0 \\ 0 \\ 0 \\ 1 \\ 0 \\ 0 \end{bmatrix} \qquad (2.5.74a,b)$$

$$p = \frac{\beta + 2\beta_c}{m} \quad \text{and} \quad q = \frac{2\beta_c}{m} \qquad (2.5.75a,b)$$

The controllability matrix is

$$C = \begin{bmatrix} 0 & 1 & 0 & -p & 0 & p^2 + 2q^2 \\ 0 & 0 & 0 & q & 0 & q^2 - 2pq \\ 0 & 0 & 0 & p & 0 & q^2 - 2pq \\ 1 & 0 & -p & 0 & p^2 + 2q^2 & 0 \\ 0 & 0 & q & 0 & q^2 - 2pq & 0 \\ 0 & 0 & q & 0 & q^2 - 2pq & 0 \end{bmatrix} \qquad (2.5.76)$$

Therefore,

$$\det C = \det \begin{bmatrix} 1 & -p & p^2 + 2q^2 & 0 & 0 & 0 \\ 0 & q & q^2 - 2pq & 0 & 0 & 0 \\ 0 & p & q^2 - 2pq & 0 & 0 & 0 \\ 0 & 0 & 0 & 1 & -p & p^2 + 2q^2 \\ 0 & 0 & 0 & 0 & q & q^2 - 2pq \\ 0 & 0 & 0 & 0 & q & q^2 - 2pq \end{bmatrix} \qquad (2.5.77)$$

$$\det C = \det \begin{bmatrix} 1 & -p & p^2 + 2q^2 \\ 0 & q & q^2 - 2pq \\ 0 & q & q^2 - 2pq \end{bmatrix} . \det \begin{bmatrix} 1 & -p & p^2 + 2q^2 \\ 0 & q & q^2 - 2pq \\ 0 & q & q^2 - 2pq \end{bmatrix} = 0 \quad (2.5.78)$$

In other words, the state space realization is not controllable. Because of symmetry, the influence of input on states z_2 and z_3 is identical. As a result, the input cannot take the system from any initial states to any arbitrary final states.

2.6 SOME IMPORTANT SIMILARITY TRANSFORMATIONS

2.6.1 DIAGONAL FORM

Consider the similarity transformation (2.2.24). Find a nonsingular matrix T such that

$$\bar{A} = A_d = diag(\lambda_1, \lambda_2, \ldots, \lambda_n) \tag{2.6.1}$$

In other words,

$$T^{-1}AT = A_d \tag{2.6.2}$$

or

$$AT = TA_d \tag{2.6.3}$$

Define

$$T = \begin{bmatrix} \mathbf{t}_1 & \mathbf{t}_2 & \cdot & \cdot & \mathbf{t}_n \end{bmatrix} \tag{2.6.4}$$

where $\mathbf{t}_i$ is the ith column of the matrix T. Hence, from (2.6.3) and (2.6.4),

$$A\mathbf{t}_i = \lambda_i \mathbf{t}_i \tag{2.6.5}$$

Therefore, λ_i and $\mathbf{t}_i$ must be the eigenvalue and the corresponding eigenvector of the matrix A, respectively. The matrix T will be nonsingular if and only if A has n independent eigenvectors. In this context, the following two properties are stated:

If A has n distinct eigenvalues, there exists n independent eigenvectors.
If an eigenvalue of A is repeated r times, r independent eigenvectors corresponding to this eigenvalue will be found provided:

$$rank(A - \lambda_i I) = n - r \tag{2.6.6}$$

EXAMPLE 2.8

$$A = \begin{bmatrix} 5 & -1 & -1 \\ -1 & 5 & -1 \\ -1 & -1 & 5 \end{bmatrix} \tag{2.6.7}$$

Eigenvalues of the matrix A are 3, 6, and 6; i.e., the eigenvalue 6 is repeated twice.

$$A - 6I = \begin{bmatrix} -1 & -1 & -1 \\ -1 & -1 & -1 \\ -1 & -1 & -1 \end{bmatrix} \qquad (2.6.8)$$

Therefore, $rank(A - 6I) = 1$, the condition (2.6.6) is satisfied, and the matrix A can be digonalized.

2.6.2 CONTROLLABILITY CANONICAL FORM

Given a realization $\{A, \mathbf{b}, \mathbf{c}\}$, find the similarity transformation matrix T such that

$$A_c = T^{-1}AT, \qquad \mathbf{b}_c = T^{-1}\mathbf{b}, \quad \text{and} \quad \mathbf{c}_c = \mathbf{c}T \qquad (2.6.9\text{a,b,c})$$

Note that Equation 2.2.29 has been used here.
From (2.6.9a),

$$TA_c = AT \qquad (2.6.10)$$

Let

$$T = \begin{bmatrix} \mathbf{t}_1 & \mathbf{t}_2 & . & . & \mathbf{t}_n \end{bmatrix} \qquad (2.6.11)$$

Now,

$$AT = [A\mathbf{t}_1 \quad A\mathbf{t}_2 \quad A\mathbf{t}_3 \quad . \quad . \quad A\mathbf{t}_n] \qquad (2.6.12)$$

and using (2.2.9),

$$TA_c = [-a_n\mathbf{t}_n \quad \mathbf{t}_1 - a_{n-1}\mathbf{t}_n \quad \mathbf{t}_2 - a_{n-2}\mathbf{t}_n \quad . \quad . \quad \mathbf{t}_{n-1} - a_1\mathbf{t}_n] \qquad (2.6.13)$$

From (2.6.10), (2.6.12), and (2.6.13),

$$A\mathbf{t}_1 = -a_n\mathbf{t}_n$$

$$A\mathbf{t}_2 = \mathbf{t}_1 - a_{n-1}\mathbf{t}_n$$

$$A\mathbf{t}_3 = \mathbf{t}_2 - a_{n-2}\mathbf{t}_n$$

$$\vdots$$

$$At_n = t_{n-1} - a_1 t_n \tag{2.6.14}$$

As $\mathbf{b}_c = [0 \quad 0 \quad . \quad . \quad 0 \quad 1]^T$, Equation 2.6.9b yields

$$\mathbf{t}_n = \mathbf{b} \tag{2.6.15}$$

Solving equations (2.6.14) from the last equation backward,

$$\mathbf{t}_{n-1} = (A + a_1 I)\mathbf{b} = A\mathbf{b} + a_1\mathbf{b}$$

$$\mathbf{t}_{n-2} = A(A + a_1 I)\mathbf{b} + a_2\mathbf{b} = A^2\mathbf{b} + a_1 A\mathbf{b} + a_2\underline{\mathbf{b}} \tag{2.6.16}$$

$$\vdots$$

$$\mathbf{t}_2 = A^{n-2}\mathbf{b} + a_1 A^{n-3}\mathbf{b} + \ldots + a_{n-2}\mathbf{b}$$

$$\mathbf{t}_1 = A^{n-1}\mathbf{b} + a_1 A^{n-2}\mathbf{b} + \ldots + a_{n-1}\mathbf{b}$$

Representing (2.6.16) in the matrix form,

$$T = CQ \tag{2.6.17}$$

where

$$Q = \begin{bmatrix} a_{n-1} & a_{n-2} & . & . & a_1 & 1 \\ a_{n-2} & a_{n-3} & . & . & 1 & 0 \\ . & . & . & . & . & . \\ a_1 & 1 & 0 & . & . & 0 \\ 1 & 0 & 0 & . & . & 0 \end{bmatrix} \tag{2.6.18}$$

The matrix Q is always nonsingular. Therefore, the matrix T is nonsingular if and only if the controllability matrix C is nonsingular.

In summary, any given realization can be converted to the canonical form $\{A_c, \mathbf{b}_c, \mathbf{c}_c\}$, provided the controllability matrix is nonsingular.

2.7 SIMULTANEOUS CONTROLLABILITY AND OBSERVABILITY

The realization $\{A_c, \mathbf{b}_c, \mathbf{c}_c\}$ is always controllable. On the other hand, the realization $\{A_o, \mathbf{b}_o, \mathbf{c}_o\}$ is always observable. Under what conditions are $\{A_c, \mathbf{b}_c, \mathbf{c}_c\}$ and

$\{A_o, \mathbf{b}_o, \mathbf{c}_o\}$ observable and controllable, respectively? To answer this question, consider the following theorems (Kailath, 1980).

Theorem 1

A realization is both observable and controllable if and only if there are no common factors between the numerator and the denominator of the transfer function.

Theorem 2

Observability and controllability properties are preserved under similarity transformations.

Therefore, $\{A_c, \mathbf{b}_c, \mathbf{c}_c\}$ is guaranteed to be observable provided there are no common factors between the numerator and the denominator of the transfer function. Similarly, $\{A_o, \mathbf{b}_o, \mathbf{c}_o\}$ is guaranteed to be controllable provided there are no common factors between the numerator and the denominator of the transfer function. Furthermore, if one can find one realization that is both observable and controllable, all realizations are both observable and controllable. If there are no common factors between numerator and denominator, any realization can be converted to either $\{A_c, \mathbf{b}_c, \mathbf{c}_c\}$ or $\{A_o, \mathbf{b}_o, \mathbf{c}_o\}$.

2.7.1 OBSERVABILITY OF STATE SPACE REALIZATION USING METHOD I

Without any loss of generality, assume that $b_0 = 0$. Therefore,

$$\mathbf{c}_c = \begin{bmatrix} b_n & b_{n-1} & . & . & b_1 \end{bmatrix} \qquad (2.7.1)$$

Let $\mathbf{e}_i$ be the ith row of the identity matrix I_n. Then, it can be easily shown (Kailath, 1980) that

$$\mathbf{e}_i A_c = \mathbf{e}_{i+1} ; \qquad 1 \le i \le n-1 \qquad (2.7.2)$$

$$\mathbf{e}_n A_c = \begin{bmatrix} -a_n & -a_{n-1} & . & . & -a_1 \end{bmatrix} \qquad (2.7.3)$$

Consider

$$b(A_c) = b_1 A_c^{n-1} + b_2 A_c^{n-2} + \ldots + b_{n-1} A_c + b_n I_n \qquad (2.7.4)$$

Then,

$$\mathbf{e}_1 b(A_c) = b_1 \mathbf{e}_1 A_c^{n-1} + b_2 \mathbf{e}_1 A_c^{n-2} + \ldots + b_{n-1} \mathbf{e}_1 A_c + b_n \mathbf{e}_1 I_n \qquad (2.7.5)$$

Because of (2.7.2),

$$\mathbf{e}_1 A_c = \mathbf{e}_2$$

$$\mathbf{e}_1 A_c^2 = \mathbf{e}_2 A_c = \mathbf{e}_3$$

$$\vdots \qquad (2.7.6)$$

$$\mathbf{e}_1 A_c^{n-1} = \mathbf{e}_n$$

From (2.7.5) and (2.7.6),

$$\mathbf{e}_1 b(A_c) = b_1 \mathbf{e}_n + b_2 \mathbf{e}_{n-1} + \ldots + b_{n-1} \mathbf{e}_2 + b_n \mathbf{e}_1 = \mathbf{c}_c \qquad (2.7.7)$$

From (2.7.2) and (2.7.7),

$$\mathbf{e}_2 b(A_c) = \mathbf{e}_1 A_c b(A_c) = \mathbf{e}_1 b(A_c) A_c = \mathbf{c}_c A_c \qquad (2.7.8)$$

In a similar manner,

$$\mathbf{e}_3 b(A_c) = \mathbf{c}_c A_c^2, \ldots, \ \mathbf{e}_n b(A_c) = \mathbf{c}_c A_c^{n-1} \qquad (2.7.9)$$

Therefore, the observability matrix is

$$\Theta_c = \begin{bmatrix} \mathbf{c}_c \\ \mathbf{c}_c A_c \\ . \\ . \\ \mathbf{c}_c A_c^{n-1} \end{bmatrix} = \begin{bmatrix} \mathbf{e}_1 \\ \mathbf{e}_2 \\ . \\ . \\ \mathbf{e}_n \end{bmatrix} b(A_c) = I_n b(A_c) = b(A_c) \qquad (2.7.10)$$

Note that $b(A_c)$ is a polynomial in A_c. Therefore, it can easily be shown that

$$b(A_c)\mathbf{p} = b(\lambda)\mathbf{p} \qquad (2.7.11)$$

where λ and $\mathbf{p}$ are the eigenvalue and the associated eigenvector of A_c. In other words, the eigenvalue and the associated eigenvector of $b(A_c)$ are $b(\lambda)$ and $\mathbf{p}$, respectively. From (2.7.4),

$$b(\lambda) = b_1 \lambda^{n-1} + b_2 \lambda^{n-2} + \ldots + b_n \qquad (2.7.12)$$

As the determinant of a matrix equals the product of its eigenvalues, Equation 2.7.10 yields

$$\det(\Theta_c) = \det(b(A_c)) = \prod_{i=1}^{n} b(\lambda_i) \qquad (2.7.13)$$

Therefore, the determinant of the observability matrix will be zero if and only if λ_i, an eigenvalue of A_c, satisfies $b(\lambda_i) = 0$; i.e., λ_i is also a zero of the transfer function. In other words, the state space realization obtained using Method I is observable if and only if all poles and zeros of the transfer function are distinct.

EXAMPLE 2.9

The transfer function (2.2.20) can be written as

$$g(s) = 1 + h(s) \qquad (2.7.14)$$

where

$$h(s) = \frac{-2s^2 + 2s + 2}{s^3 + 2s^2 + 4s + 3} \qquad (2.7.15)$$

It can be easily seen that $\{A_c, \mathbf{b}_c, \mathbf{c}_c\}$ and $\{A_o, \mathbf{b}_o, \mathbf{c}_o\}$, Equation 2.2.21 and Equation 2.2.22, correspond to the transfer function $h(s)$. Poles of $h(s)$ are 1, and $-0.5 \pm 1.6583j$, whereas zeros are 0.618 and 1.618. In other words, there are no common factors between the numerator and denominator. Therefore, any state space realization is guaranteed to be both observable and controllable. As a result, $\{A_c, \mathbf{b}_c, \mathbf{c}_c\}$ will be observable, and $\{A_o, \mathbf{b}_o, \mathbf{c}_o\}$ will be controllable.

EXAMPLE 2.10

From Figure 2.7,

$$\frac{y(s)}{u(s)} = \frac{s+1}{s(s+4)} \qquad (2.7.16)$$

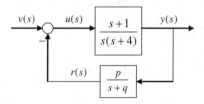

FIGURE 2.7 A feedback system.

$$\frac{r(s)}{y(s)} = \frac{p}{s+q},$$

(2.7.17)

$$u(s) = v(s) - r(s)$$

(2.7.18)

For the transfer function (2.7.16), Method I yields

$$\begin{bmatrix} \dot{x}_1 \\ \dot{x}_2 \end{bmatrix} = \begin{bmatrix} 0 & 1 \\ 0 & -4 \end{bmatrix} \begin{bmatrix} x_1 \\ x_2 \end{bmatrix} + \begin{bmatrix} 0 \\ 1 \end{bmatrix} u(t)$$

(2.7.19)

$$y(t) = \begin{bmatrix} 1 & 1 \end{bmatrix} \begin{bmatrix} x_1 \\ x_2 \end{bmatrix}$$

(2.7.20)

And, for the transfer function (2.7.17),

$$\dot{r} + qr(t) = py(t)$$

(2.7.21)

Combining (2.7.19) and (2.7.21) and using (2.7.18) and (2.7.20), the state space realization is obtained:

$$\begin{bmatrix} \dot{x}_1 \\ \dot{x}_2 \\ \dot{r} \end{bmatrix} = \begin{bmatrix} 0 & 1 & 0 \\ 0 & -4 & -1 \\ p & p & -a \end{bmatrix} \begin{bmatrix} x_1 \\ x_2 \\ r \end{bmatrix} + \begin{bmatrix} 0 \\ 1 \\ 0 \end{bmatrix} v(t)$$

(2.7.22)

$$y(t) = \begin{bmatrix} 1 & 1 & 0 \end{bmatrix} \begin{bmatrix} x_1 \\ x_2 \\ r \end{bmatrix}$$

(2.7.23)

Controllability:

$$C = \begin{bmatrix} 0 & 1 & -4 \\ 1 & -4 & 16-p \\ 0 & p & p-4p-pq \end{bmatrix}$$

(2.7.24)

$$\det C = p(1-q)$$

(2.7.25)

The realization is not controllable when $p = 0$ or $q = 1$.

Observability:

$$O = \begin{bmatrix} 1 & 1 & 0 \\ 0 & -3 & -1 \\ -p & 12-p & 3+q \end{bmatrix} \tag{2.7.26}$$

$$\det O = 3(1-q) \tag{2.7.27}$$

The realization is not observable when $q = 1$.
    Note that

$$\frac{y(s)}{v(s)} = \frac{(s+1)(s+q)}{s(s+3)(s+q)+p(s+1)} \tag{2.7.28}$$

When $p = 0$ or $q = 1$, $(s+q)$ or $(s+1)$ is a common factor between the numerator and denominator of the transfer function (2.7.28), respectively. Therefore, a state space realization cannot be both observable and controllable when $p = 0$ or $q = 1$. Here, the realization is not controllable when $p = 0$, and is neither observable nor controllable when $q = 1$.

2.8 MULTIINPUT/MULTIOUTPUT (MIMO) SYSTEMS

The transfer function matrix of a MIMO system is defined as

$$\mathbf{y}(s) = G(s)\mathbf{u}(s) \tag{2.8.1}$$

where $\mathbf{y}(s)$ and $\mathbf{u}(s)$ are $p x 1$ and $m x 1$ vectors, respectively.

$$\mathbf{y}(s) = \begin{bmatrix} y_1(s) \\ y_2(s) \\ . \\ . \\ y_p(s) \end{bmatrix} \quad \text{and} \quad \mathbf{u}(s) = \begin{bmatrix} u_1(s) \\ u_2(s) \\ . \\ . \\ u_m(s) \end{bmatrix} \tag{2.8.2a,b}$$

Accordingly, the matrix $G(s)$ is of order $p x m$. As an example, for a 2-input/2-output system,

$$\begin{bmatrix} y_1(s) \\ y_2(s) \end{bmatrix} = \begin{bmatrix} \dfrac{1}{s+1} & \dfrac{1}{s+2} \\ \dfrac{s+5}{(s+1)^2} & \dfrac{1}{s+3} \end{bmatrix} \begin{bmatrix} u_1(s) \\ u_2(s) \end{bmatrix} \tag{2.8.3}$$

Taking the least common multiple of denominators of elements of the matrix $G(s)$,

$$G(s) = \frac{1}{(s+1)^2(s+2)(s+3)} N(s) \qquad (2.8.4)$$

where

$$N(s) = \begin{bmatrix} (s+1)(s+2)(s+3) & (s+1)^2(s+3) \\ (s+2)(s+3)(s+5) & (s+1)^2(s+2) \end{bmatrix} \qquad (2.8.5)$$

The elements of the matrix $N(s)$ are polynomials in s. Hence, $N(s)$ will be called a polynomial matrix. It can also be expressed as follows:

$$N(s) = s^3 N_1 + s^2 N_2 + s N_3 + N_4 \qquad (2.8.6)$$

where

$$N_1 = \begin{bmatrix} 1 & 1 \\ 1 & 1 \end{bmatrix}; \quad N_2 = \begin{bmatrix} 6 & 5 \\ 10 & 4 \end{bmatrix}; \quad N_3 = \begin{bmatrix} 11 & 7 \\ 31 & 5 \end{bmatrix}; \quad N_4 = \begin{bmatrix} 6 & 3 \\ 30 & 2 \end{bmatrix} \qquad (2.8.7)$$

In general, a pxm transfer function matrix $G(s)$ is represented as

$$G(s) = \frac{N(s)}{d(s)} \qquad (2.8.8)$$

where

$$d(s) = s^r + d_1 s^{r-1} + \ldots + d_r \qquad (2.8.9)$$

and

$$N(s) = s^{r-1} N_1 + s^{r-2} N_2 + \ldots + s N_{r-1} + N_r \qquad (2.8.10)$$

$N_1, N_2, \ldots, N_r$ are pxm matrices.

Definitions

1. A rational transfer function matrix $G(s)$ is said to be proper if

$$\lim_{s \to \infty} G(s) < \infty \qquad (2.8.11)$$

and strictly proper if

$$\lim_{s \to \infty} G(s) = 0 \tag{2.8.12}$$

2. A vector of polynomials is called a polynomial vector. The degree of a polynomial vector equals the highest degree of all the entries of the vector. For example, the polynomial vector

$$\begin{bmatrix} s^4 + 1 \\ s^2 + s + 1 \\ s + 7 \end{bmatrix} \tag{2.8.13}$$

has degree = 4.

Lemma

If $G(s)$ is a strictly proper (proper) transfer function matrix and

$$G(s) = N(s)D^{-1}(s) \tag{2.8.14}$$

then every column (Kailath, 1980) of $N(s)$ has degree strictly less than (less than or equal to) that of the corresponding column of $D(s)$.

Definition: Column-Reduced Matrix

Let

$$k_i = \text{the degree of } i\text{th column of } m \times m \text{ matrix } D(s) \tag{2.8.15}$$

Then, it can be easily seen that

$$\text{deg } \det D(s) \leq \sum_{i=1}^{m} k_i \tag{2.8.16}$$

Inequality holds when there are cancellations of terms in the expansion of det $D(s)$. A matrix $D(s)$ for which the equality sign holds is called a *column reduced* matrix (Kailath, 1980).

EXAMPLE 2.11

$$D(s) = \begin{bmatrix} s^4 + s & s^2 + 2 \\ s^2 + s + 1 & 1 \end{bmatrix} \tag{2.8.17}$$

Here,

$$k_1 = 4 \quad \text{and} \quad k_2 = 2 \qquad (2.8.18)$$

and

$$\det D(s) = -s^3 - 3s^2 - s - 2 \qquad (2.8.19)$$

In other words,

$$\deg \ \det D(s) < \sum_{i=1}^{2} k_i \qquad (2.8.20)$$

Therefore, the matrix $D(s)$ is not column reduced.

In general, a polynomial matrix $D(s)$ can be expressed (Kailath, 1980) as

$$D(s) = D_{hc} S(s) + L(s) \qquad (2.8.21)$$

where

$$S(s) = diag[s^{k_1} \quad s^{k_2} \quad . \quad . \quad s^{k_m}] \qquad (2.8.22)$$

$$D_{hc} = \text{the highest-column-degree coefficient matrix,}$$
$$\text{or the leading (column) coefficient matrix of } D(s) \qquad (2.8.23)$$

$$L(s): \text{remaining terms} \qquad (2.8.24)$$

It can be shown that

$$\det D(s) = (\det D_{hc})s^{\sum k_i} + \text{ terms of lower degrees in } s \qquad (2.8.25)$$

When $\det D_{hc} \neq 0$,

$$\deg \ \det D(s) = \sum_{i=1}^{m} k_i \qquad (2.8.26)$$

and the matrix $D(s)$ is column reduced.

Facts

A nonsingular polynomial matrix is column reduced if and only if its leading (column) coefficient matrix is nonsingular (Kailath, 1980).

If $D(s)$ is column reduced, then $G(s) = N(s)D^{-1}(s)$ is strictly proper (proper) if and only if (Kailath, 1980) each column of $N(s)$ has degree less than (less than or equal to) the degree of the corresponding column of $D(s)$.

2.9 STATE SPACE REALIZATIONS OF A TRANSFER FUNCTION MATRIX

Two methods, similar to those for a SISO system (Section 2.2), will be presented.

METHOD I

Define

$$\xi(s) = \frac{1}{d(s)}\mathbf{u}(s) \tag{2.9.1}$$

Using Equation 2.8.9,

$$(s^r + d_1 s^{r-1} + \ldots + d_r)\xi(s) = \mathbf{u}(s) \tag{2.9.2}$$

In time-domain,

$$\frac{d^r}{dt^r}\xi(t) + d_1\frac{d^{r-1}}{dt^{r-1}}\xi(t) + \ldots + d_r\xi(t) = \mathbf{u}(t) \tag{2.9.3}$$

Define

$$\mathbf{x}_1 = \xi\ ;\quad \mathbf{x}_2 = \frac{d}{dt}\xi\ ,\ldots,\ \mathbf{x}_r = \frac{d^{r-1}}{dt^{r-1}}\xi \tag{2.9.4}$$

Equation 2.9.3 and Equation 2.9.4 can be written as

$$\dot{\mathbf{x}}_1 = \mathbf{x}_2$$

$$\dot{\mathbf{x}}_2 = \mathbf{x}_3$$

$$\vdots \tag{2.9.5}$$

$$\dot{\mathbf{x}}_{r-1} = \mathbf{x}_r$$

$$\dot{\mathbf{x}}_r = -d_1\mathbf{x}_r - d_2\mathbf{x}_{r-1} - \ldots\ldots - d_r\mathbf{x}_1 + \mathbf{u}(t)$$

In matrix form,

$$
\begin{bmatrix} \dot{\mathbf{x}}_1 \\ \dot{\mathbf{x}}_2 \\ \cdot \\ \cdot \\ \dot{\mathbf{x}}_{r-1} \\ \dot{\mathbf{x}}_r \end{bmatrix} = \begin{bmatrix} 0 & I_m & 0 & \cdot & \cdot & 0 \\ 0 & 0 & I_m & 0 & \cdot & 0 \\ \cdot & \cdot & \cdot & \cdot & \cdot & \cdot \\ \cdot & \cdot & \cdot & \cdot & \cdot & \cdot \\ 0 & 0 & \cdot & \cdot & 0 & I_m \\ -d_r I_m & -d_{r-1}I_m & \cdot & \cdot & \cdot & -d_1 I_m \end{bmatrix} \begin{bmatrix} \mathbf{x}_1 \\ \mathbf{x}_2 \\ \cdot \\ \cdot \\ \mathbf{x}_{r-1} \\ \mathbf{x}_r \end{bmatrix} + \begin{bmatrix} 0 \\ 0 \\ \cdot \\ \cdot \\ 0 \\ I_m \end{bmatrix} \mathbf{u}(t) \qquad (2.9.6)
$$

From (2.8.1), (2.8.8), and (2.9.1),

$$\mathbf{y}(s) = N(s)\xi(s) \qquad (2.9.7)$$

or

$$\mathbf{y}(s) = (s^{r-1}N_1 + s^{r-2}N_2 + \ldots + sN_{r-1} + N_r)\xi(s) \qquad (2.9.8)$$

In time-domain,

$$\mathbf{y}(t) = N_1 \frac{d^{r-1}}{dt^{r-1}}\xi(t) + N_2 \frac{d^{r-2}}{dt^{r-2}}\xi(t) + \ldots + N_r\xi(t) \qquad (2.9.9)$$

Using the definition (2.9.4),

$$\mathbf{y}(t) = \begin{bmatrix} N_r & N_{r-1} & \cdot & \cdot & N_1 \end{bmatrix} \begin{bmatrix} \mathbf{x}_1 \\ \mathbf{x}_2 \\ \cdot \\ \cdot \\ \mathbf{x}_r \end{bmatrix} \qquad (2.9.10)$$

Equation 2.9.6 and Equation 2.9.10 comprise a state space realization:

$$\dot{\mathbf{x}} = A\mathbf{x} + B\mathbf{u}(t)$$
$$\mathbf{y}(t) = C\mathbf{x}(t) \qquad (2.9.11)$$

where

$$A = \begin{bmatrix} 0 & I_m & 0 & . & . & 0 \\ 0 & 0 & I_m & 0 & . & 0 \\ . & & . & . & . & . \\ . & & . & . & . & . \\ 0 & 0 & . & . & 0 & I_m \\ -d_r I_m & -d_{r-1} I_m & . & . & . & -d_1 I_m \end{bmatrix} \qquad (2.9.12)$$

$$B = \begin{bmatrix} 0 \\ 0 \\ . \\ . \\ 0 \\ I_m \end{bmatrix}; \quad C = \begin{bmatrix} N_r & N_{r-1} & . & . & N_1 \end{bmatrix} \qquad (2.9.13\text{a,b})$$

The state vector $\mathbf{x}$ is described as

$$\mathbf{x} = \begin{bmatrix} \mathbf{x}_1 \\ \mathbf{x}_2 \\ . \\ . \\ \mathbf{x}_r \end{bmatrix} \qquad (2.9.14)$$

As the dimension of each of the block elements $\mathbf{x}_i$ is $mx1$,

$$\text{Number of states, } n = mr \qquad (2.9.15)$$

METHOD II

From (2.8.1) and (2.8.8),

$$\mathbf{y}(s) = \frac{N(s)}{d(s)} \mathbf{u}(s) \qquad (2.9.16)$$

$$(s^r + d_1 s^{r-1} + \ldots + d_r)\mathbf{y}(s) = (s^{r-1} N_1 + s^{r-2} N_2 + \ldots + s N_{r-1} + N_r)\mathbf{u}(s) \qquad (2.9.17)$$

Dividing both sides by s^r,

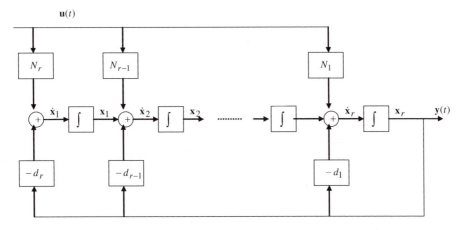

FIGURE 2.8 MIMO analog computer simulation diagram (Method II).

$$\left(1+\frac{d_1}{s}+.......+\frac{d_r}{s^r}\right)\mathbf{y}(s)=\left(\frac{N_1}{s}+\frac{N_2}{s^2}+...+\frac{N_{r-1}}{s^{r-1}}+\frac{N_r}{s^r}\right)\mathbf{u}(s) \qquad (2.9.18)$$

This equation can be written as

$$\mathbf{y}(s)=-\frac{d_1}{s}\mathbf{y}(s)-...-\frac{d_r}{s^r}\mathbf{y}(s)$$

$$+\frac{N_1}{s}\mathbf{u}(s)+\frac{N_2}{s^2}\mathbf{u}(s)+....+\frac{N_{r-1}}{s^{r-1}}\mathbf{u}(s)+\frac{N_r}{s^r}\mathbf{u}(s) \qquad (2.9.19)$$

On the basis of Equation 2.9.19, the simulation diagram is constructed as shown in Figure 2.8. Defining outputs of integrators as p-dimensional state variable vectors $\mathbf{x}_1, \mathbf{x}_2, ..., \mathbf{x}_r$, the following state space model is obtained:

$$\mathbf{y}(t) = \mathbf{x}_r(t) \qquad (2.9.20)$$

and

$$\dot{\mathbf{x}}_1 = -d_r\mathbf{y}(t)+N_r\mathbf{u}(t)=-d_r\mathbf{x}_r(t)+N_r\mathbf{u}(t)$$

$$\dot{\mathbf{x}}_2 =-d_{r-1}\mathbf{y}(t)+\mathbf{x}_1(t)+N_{r-1}\mathbf{u}(t)=-d_{r-1}\mathbf{x}_r(t)+\mathbf{x}_1(t)+N_{r-1}\mathbf{u}(t)$$

$$\vdots \qquad (2.9.21)$$

$$\dot{\mathbf{x}}_r =-d_1\mathbf{y}(t)+\mathbf{x}_{r-1}(t)+N_1\mathbf{u}(t)=-d_1\mathbf{x}_r(t)+\mathbf{x}_{r-1}(t)+N_1\mathbf{u}(t)$$

Equation 2.9.20 and Equation 2.9.21 comprise the state space model:

$$\dot{\mathbf{x}} = A\mathbf{x} + B\mathbf{u}(t)$$

$$\mathbf{y}(t) = C\mathbf{x}(t)$$

(2.9.22)

where

$$A = \begin{bmatrix} 0 & 0 & . & . & 0 & -d_r I_p \\ I_p & 0 & . & . & 0 & -d_{r-1} I_p \\ 0 & I_p & 0 & . & 0 & -d_{r-2} I_p \\ 0 & 0 & I_p & . & . & -d_{r-3} I_p \\ . & . & . & . & . & . \\ 0 & 0 & 0 & . & I_p & -d_1 I_p \end{bmatrix}$$

(2.9.23)

$$C = \begin{bmatrix} 0 \\ 0 \\ . \\ . \\ 0 \\ I_m \end{bmatrix}^T \quad ; \quad B = \begin{bmatrix} N_r & N_{r-1} & . & . & N_1 \end{bmatrix}^T$$

(2.9.24a,b)

The state vector $\mathbf{x}$ is described as

$$\mathbf{x} = \begin{bmatrix} \mathbf{x}_1 \\ \mathbf{x}_2 \\ . \\ . \\ \mathbf{x}_r \end{bmatrix}$$

(2.9.25)

As the dimension of each of the block elements $\mathbf{x}_i$ is $p \times 1$,

the number of states, $n = pr$

(2.9.26)

2.10 CONTROLLABILITY AND OBSERVABILITY OF A MIMO SYSTEM

Definition 1

A multiinput state space realization is controllable if and only if the controllability matrix

$$C = \begin{bmatrix} B & AB & A^2B & \cdots & A^{n-1}B \end{bmatrix} \qquad (2.10.1)$$

is of full rank.

Definition 2

A multioutput state space realization is observable if and only if the observability matrix

$$\Theta = \begin{bmatrix} C \\ CA \\ CA^2 \\ . \\ . \\ . \\ CA^{n-1} \end{bmatrix} \qquad (2.10.2)$$

is of full rank.

2.10.1 Controllability and Observability of Methods I and II Realizations

Method I Realization

$$C = \begin{bmatrix} B & AB & A^2B. & . & A^rB:\ldots & A^{n-1}B \end{bmatrix} \qquad (2.10.3)$$

where $n = mr$. The dimension of the matrix C is $n \times mn$. Using (2.10.3), (2.9.12), and (2.9.13a),

$$C = \begin{bmatrix} 0 & 0 & . & . & I_m: & \\ 0 & 0 & . & I_m & *: & \\ . & . & . & . & .: & \\ 0 & I_m & . & * & *: & \\ I_m & * & . & * & *: & \end{bmatrix} \qquad (2.10.4)$$

It is obvious that the $mr \times mr$ submatrix shown in (2.10.4) is nonsingular. Hence, the matrix C is of full rank, i.e.,

$$\text{Rank } (C) = n \qquad (2.10.5)$$

Hence, the realization obtained using Method I is guaranteed to be controllable. However, it need not be always observable.

Method II Realization

From (2.10.2),

$$
\Theta = \begin{bmatrix} C \\ CA \\ \cdot \\ CA^r \\ \overline{} \\ \cdot \\ \cdot \\ CA^{n-1} \end{bmatrix}
\tag{2.10.6}
$$

where $n = pr$. The dimension of the matrix Θ is $n \times pn$. Using (2.10.6), (2.9.23), and (2.9.24a),

$$
\Theta = \begin{bmatrix} 0 & 0 & \cdot & 0 & I_m \\ 0 & 0 & \cdot & I_m & * \\ \cdot & \cdot & \cdot & \cdot & \cdot \\ 0 & I_m & \cdot & * & * \\ I_m & * & \cdot & * & * \\ \cdots & \cdots & \cdots & \cdots & \cdots \\ & & & & \end{bmatrix}
\tag{2.10.7}
$$

It is obvious that the $pr \times pr$ submatrix shown in (2.10.7) is nonsingular. Hence, the matrix Θ is of full rank, i.e.,

$$
\text{Rank}(\Theta) = n
\tag{2.10.8}
$$

Hence, the realization obtained using Method II is guaranteed to be observable. However, it need not be always controllable.

2.11 MATRIX-FRACTION DESCRIPTION (MFD)

Given a transfer function matrix, two methods for obtaining state space realizations have been obtained in a manner analogous to what we did for SISO systems. The state space realization from Method I is always controllable, whereas the state space realization from Method II is always observable. Furthermore, the number of states from the two methods is different when the number of outputs does not equal number of inputs. As a result, following questions arise:

1. When will a state space realization be both controllable and observable?
2. How do we obtain a state space realization with the minimum number of states?

To answer these questions in a general way, the MFD of a transfer function is introduced. Equation 2.8.8 can also be written as

$$G(s) = N_R(s)D_R^{-1}(s) \qquad\qquad (2.11.1)$$

where

$$N_R(s) = N(s) \quad \text{and} \quad D_R(s) = d(s)I_m \qquad\qquad (2.11.2)$$

Alternatively,

$$G(s) = D_L^{-1}(s)N_L(s) \qquad\qquad (2.11.3)$$

where

$$N_L(s) = N(s) \quad \text{and} \quad D_L(s) = d(s)I_p \qquad\qquad (2.11.4)$$

Equation 2.11.1 and Equation 2.11.3 are known as right and left MFD, respectively. Any transfer function matrix can be represented by either right or left MFD. It should be noted that $D_R(s)$ and $D_L(s)$ can be viewed as denominator polynomial matrices. Similarly, $N_R(s)$ and $N_L(s)$ can be viewed as numerator polynomial matrices.

2.11.1 DEGREE OF A SQUARE POLYNOMIAL MATRIX AND GREATEST COMMON RIGHT DIVISOR (GCRD)

The *degree of a square polynomial matrix* is equal to the degree of the determinant of the polynomial matrix (Kailath, 1980).
Therefore,

$$deg\ D_R(s)\ =\ deg\ det\ D_R(s) = mr \qquad\qquad (2.11.5)$$

and

$$deg\ D_L(s)\ =\ deg\ det\ D_L(s) = pr \qquad (2.11.6)$$

It is interesting to note that the number of states in Method I realization equals $deg\ D_R(s)$, whereas the number of states in Method II realization equals $deg\ D_L(s)$. In fact, Method I and Method II are connected to the right and left MFD, respectively. Furthermore, having expressed the transfer function matrix as a right MFD, a state space realization can be developed to have the number of states equal $deg\ D_R(s)$. The same result also holds for left MFD.

For the transfer function matrix in Equation 2.8.3, a right MFD is constructed using (2.11.2):

$$G(s) = \begin{bmatrix} (s+1)(s+2)(s+3) & (s+1)^2(s+3) \\ (s+2)(s+3)(s+5) & (s+1)^2(s+2) \end{bmatrix} \times$$

$$\begin{bmatrix} (s+1)^2(s+2)(s+3) & 0 \\ 0 & (s+1)^2(s+2)(s+3) \end{bmatrix}^{-1} \qquad (2.11.7)$$

$$= N_1(s)D_1^{-1}(s)$$

where

$$deg\ D_1(s)\ =\ 8$$

Another right MFD can also be constructed:

$$G(s) = \begin{bmatrix} (s+1) & (s+3) \\ (s+5) & (s+2) \end{bmatrix} \begin{bmatrix} (s+1)^2 & 0 \\ 0 & (s+2)(s+3) \end{bmatrix}^{-1} = N_2(s)D_2^{-1}(s) \quad (2.11.8)$$

where

$$deg\ D_2(s)\ =\ 4$$

The diagonal elements of $D_2(s)$ are obtained by taking the least common multiple of each column separately. It can be easily verified that

$$N_2(s) = N_1(s)W^{-1}(s) \quad \text{and} \quad D_2(s) = D_1(s)W^{-1}(s) \qquad (2.11.9)$$

where

$$W(s) = \begin{bmatrix} (s+2)(s+3) & 0 \\ 0 & (s+1)^2 \end{bmatrix} \qquad (2.11.10)$$

Here, $W(s)$ is a nonsingular polynomial matrix, but $W^{-1}(s)$ is not a polynomial matrix. When $W^{-1}(s)$ is postmultiplied with $N_1(s)$, another polynomial matrix $N_2(s)$ is obtained. Therefore, the polynomial matrix $W(s)$ is called a **right divisor** of $N_1(s)$. Using this definition, the polynomial matrix $W(s)$ is also a **right divisor** of $D_1(s)$. In other words, $W(s)$ is a **common right divisor** (*crd*) of both $N_1(s)$ and $D_1(s)$. From (2.11.9),

$$D_2(s)W(s) = D_1(s) \qquad (2.11.11)$$

From the procedure of matrix multiplication and the definition of the determinant,

$$deg\ D_1(s) = deg\ D_2(s)W(s) = deg\ D_2(s) + deg\ W(s) \qquad (2.11.12)$$

Thus, finding a *crd* of numerator and denominator polynomial matrices, the degree of the denominator polynomial matrix (hence, the number of states) can be decreased by the degree of the *crd*. Therefore, the minimal order of state space realization will be obtained (Kailath, 1980) by finding **the greatest common right divisor** (gcrd). If the degree of a *gcrd* is zero, the number of states cannot be reduced any further. A polynomial square matrix is said to be **unimodular** if its degree equals zero. A polynomial matrix is unimodular if and only if its inverse is also a polynomial matrix.

Nonuniqueness of a gcrd

Let $W_g(s)$ be a gcrd of $N(s)$ and $D(s)$. Then,

$$\tilde{N}(s) = N(s)W_g^{-1}(s) \qquad (2.11.13)$$

$$\tilde{D}(s) = D(s)W_g^{-1}(s) \qquad (2.11.14)$$

where $\tilde{N}(s)$ and $\tilde{D}(s)$ are polynomial matrices. It can be verified that $U_a(s)W_g(s)$ is also a *gcrd* where $U_a(s)$ is any unimodular matrix.

Definitions

1. Two polynomial matrices $N(s)$ and $D(s)$ with the same number of columns will be said to be relatively right prime (or right coprime) if they only have unimodular *crd*.
2. An MFD $G(s) = N(s)D^{-1}(s)$ will be said to be irreducible if $N(s)$ and $D(s)$ are right coprime.

2.11.2 ELEMENTARY ROW AND COLUMN OPERATIONS

A general procedure for obtaining a *gcrd* is based on elementary operations on polynomial matrices defined as follows (Kailath, 1980; Rugh, 1993):

1. Interchange of any two rows or columns
2. Addition to any row (column) by a polynomial multiple of any other column (row)
3. Multiplying any row or column by any nonzero real or complex number

These elementary operations are represented by elementary matrices. Postmultiplication of a polynomial matrix by an elementary matrix results in elementary row operation, whereas premultiplication results in elementary column operation. As an example, consider the following polynomial matrix:

$$P(s) = \begin{bmatrix} s+1 & s^2+s+1 & s \\ 1 & 2s+3 & s+5 \\ s+3 & s(s+1) & 1 \end{bmatrix} \tag{2.11.15}$$

Then, various elementary operations are described as follows:

Interchange of first and second rows:

$$\begin{bmatrix} 0 & 1 & 0 \\ 1 & 0 & 0 \\ 0 & 0 & 1 \end{bmatrix} P(s) = \begin{bmatrix} 1 & 2s+3 & s+5 \\ s+1 & s^2+s+1 & s \\ s+3 & s(s+1) & 1 \end{bmatrix} \tag{2.11.16}$$

Interchange of first and second columns:

$$P(s) \begin{bmatrix} 0 & 1 & 0 \\ 1 & 0 & 0 \\ 0 & 0 & 1 \end{bmatrix} = \begin{bmatrix} s^2+s+1 & s+1 & s \\ 2s+3 & 1 & s+5 \\ s(s+1) & s+3 & 1 \end{bmatrix} \tag{2.11.17}$$

Addition of *s* times first row to the third row:

$$\begin{bmatrix} 1 & 0 & 0 \\ 0 & 1 & 0 \\ s & 0 & 1 \end{bmatrix} P(s) = \begin{bmatrix} s+1 & s^2+s+1 & s \\ 1 & 2s+3 & s+5 \\ s(s+1)+s+3 & s(s^2+s+1)+s(s+1) & s^2+1 \end{bmatrix} \tag{2.11.18}$$

Addition of s times first column to the third column:

$$P(s)\begin{bmatrix} 1 & 0 & 0 \\ 0 & 1 & 0 \\ s & 0 & 1 \end{bmatrix} = \begin{bmatrix} s+1 & s^2+s+1 & s(s+1)+s \\ 1 & 2s+3 & s+s+5 \\ s+3 & s(s+1) & s(s+3)+1 \end{bmatrix} \qquad (2.11.19)$$

Scaling of first row by 2:

$$\begin{bmatrix} 2 & 0 & 0 \\ 0 & 1 & 0 \\ 0 & 0 & 1 \end{bmatrix} P(s) = \begin{bmatrix} 2(s+1) & 2(s^2+s+1) & 2s \\ 1 & 2s+3 & s+5 \\ s+3 & s(s+1) & 1 \end{bmatrix} \qquad (2.11.20)$$

Scaling of first column by 2:

$$P(s)\begin{bmatrix} 2 & 0 & 0 \\ 0 & 1 & 0 \\ 0 & 0 & 1 \end{bmatrix} = \begin{bmatrix} (s+1) & (s^2+s+1) & 2s \\ 1 & 2s+3 & 2(s+5) \\ s+3 & s(s+1) & 2 \end{bmatrix} \qquad (2.11.21)$$

All the matrices by which $P(s)$ is pre- or postmultiplied are elementary matrices that have the following properties:

1. The inverse of an elementary matrix is also an elementary matrix.
2. Determinants of elementary matrices are nonzero constants and hence independent of s. In other words, elementary matrices are **unimodular** (see Section 2.11.1).

2.11.3 DETERMINATION OF A *gcrd*

Given $m \times m$ and $p \times m$ polynomial matrices $D(s)$ and $N(s)$, find elementary row operations or equivalently a unimodular matrix $U(s)$ such that at least the last p rows on the right-hand side of following equation are zero, i.e.,

$$U(s)\begin{bmatrix} D(s) \\ N(s) \end{bmatrix} = \begin{bmatrix} R(s) \\ 0 \end{bmatrix} \qquad (2.11.22)$$

Then the square matrix $R(s)$ is a *gcrd* (Kailath, 1980).

EXAMPLE 2.12

Find a *gcrd* of

$$N(s) = \begin{bmatrix} (s+1)(s+5) & 0 \\ 0 & (s+2)(s+5) \\ (s+2)(s+3) & (s+1)(s+4) \end{bmatrix} \qquad (2.11.23)$$

and

$$D(s) = \begin{bmatrix} (s+3)(s+5) & 0 \\ 0 & (s+4)(s+5) \end{bmatrix} \qquad (2.11.24)$$

Solution

$$\begin{bmatrix} D(s) \\ N(s) \end{bmatrix} = \begin{bmatrix} (s+3)(s+5) & 0 \\ 0 & (s+4)(s+5) \\ (s+1)(s+5) & 0 \\ 0 & (s+2)(s+5) \\ (s+2)(s+3) & (s+1)(s+4) \end{bmatrix} \qquad (2.11.25)$$

Step I: Subtract fourth row from the first row.

$$U_1(s) \begin{bmatrix} (s+3)(s+5) & 0 \\ 0 & (s+4)(s+5) \\ (s+1)(s+5) & 0 \\ 0 & (s+2)(s+5) \\ (s+2)(s+3) & (s+1)(s+4) \end{bmatrix} = \begin{bmatrix} (s+3)(s+5) & 0 \\ 0 & (s+4)(s+5) \\ (s+1)(s+5) & 0 \\ 0 & -2(s+5) \\ (s+2)(s+3) & (s+1)(s+4) \end{bmatrix} \qquad (2.11.26)$$

$$U_1(s) = \begin{bmatrix} 1 & 0 & 0 & 0 & 0 \\ 0 & 1 & 0 & 0 & 0 \\ 0 & 0 & 1 & 0 & 0 \\ 0 & -1 & 0 & 1 & 0 \\ 0 & 0 & 0 & 0 & 1 \end{bmatrix} \qquad (2.11.27)$$

Step II: Divide the fourth row by −1/2.

$$U_2(s) \begin{bmatrix} (s+3)(s+5) & 0 \\ 0 & (s+4)(s+5) \\ (s+1)(s+5) & 0 \\ 0 & -2(s+5) \\ (s+2)(s+3) & (s+1)(s+4) \end{bmatrix} = \begin{bmatrix} (s+3)(s+5) & 0 \\ 0 & (s+4)(s+5) \\ (s+1)(s+5) & 0 \\ 0 & (s+5) \\ (s+2)(s+3) & (s+1)(s+4) \end{bmatrix} \qquad (2.11.28)$$

$$U_2(s) = \begin{bmatrix} 1 & 0 & 0 & 0 & 0 \\ 0 & 1 & 0 & 0 & 0 \\ 0 & 0 & 1 & 0 & 0 \\ 0 & 0 & 0 & -1/2 & 0 \\ 0 & 0 & 0 & 0 & 1 \end{bmatrix} \qquad (2.11.29)$$

Step III: Subtract $(s + 4)$ times the fourth row from the second row.

$$U_3(s) \begin{bmatrix} (s+3)(s+5) & 0 \\ 0 & (s+4)(s+5) \\ (s+1)(s+5) & 0 \\ 0 & (s+5) \\ (s+2)(s+3) & (s+1)(s+4) \end{bmatrix} = \begin{bmatrix} (s+3)(s+5) & 0 \\ 0 & 0 \\ (s+1)(s+5) & 0 \\ 0 & (s+5) \\ (s+2)(s+3) & (s+1)(s+4) \end{bmatrix} \qquad (2.11.30)$$

$$U_3(s) = \begin{bmatrix} 1 & 0 & 0 & 0 & 0 \\ 0 & 1 & 0 & -(s+4) & 0 \\ 0 & 0 & 1 & 0 & 0 \\ 0 & 0 & 0 & 1 & 0 \\ 0 & 0 & 0 & 0 & 1 \end{bmatrix} \qquad (2.11.31)$$

Step IV: Subtract 0.5 times the first row from 0.5 times the third row.

$$U_4(s) \begin{bmatrix} (s+3)(s+5) & 0 \\ 0 & 0 \\ (s+1)(s+5) & 0 \\ 0 & (s+5) \\ (s+2)(s+3) & (s+1)(s+4) \end{bmatrix} = \begin{bmatrix} (s+3)(s+5) & 0 \\ 0 & 0 \\ (s+5) & 0 \\ 0 & (s+5) \\ (s+2)(s+3) & (s+1)(s+4) \end{bmatrix} \qquad (2.11.32)$$

$$U_4(s) = \begin{bmatrix} 1 & 0 & 0 & 0 & 0 \\ 0 & 1 & 0 & 0 & 0 \\ -1/2 & 0 & 1/2 & 0 & 0 \\ 0 & 0 & 0 & 1 & 0 \\ 0 & 0 & 0 & 0 & 1 \end{bmatrix}$$ (2.11.33)

Step V: Subtract $(s + 3)$ times the third row from the first row.

$$U_5(s) \begin{bmatrix} (s+3)(s+5) & 0 \\ 0 & 0 \\ (s+5) & 0 \\ 0 & (s+5) \\ (s+2)(s+3) & (s+1)(s+4) \end{bmatrix} = \begin{bmatrix} 0 & 0 \\ 0 & 0 \\ (s+5) & 0 \\ 0 & (s+5) \\ (s+2)(s+3) & (s+1)(s+4) \end{bmatrix}$$ (2.11.34)

$$U_5(s) = \begin{bmatrix} 1 & 0 & -(s+3) & 0 & 0 \\ 0 & 1 & 0 & 0 & 0 \\ 0 & 0 & 1 & 0 & 0 \\ 0 & 0 & 0 & 1 & 0 \\ 0 & 0 & 0 & 0 & 1 \end{bmatrix}$$ (2.11.35)

Step VI: Replace fifth row by adding $(s + 3)/3$ times the third row and $-1/3$ times the fifth row.

$$U_6(s) \begin{bmatrix} 0 & 0 \\ 0 & 0 \\ (s+5) & 0 \\ 0 & (s+5) \\ (s+2)(s+3) & (s+1)(s+4) \end{bmatrix} = \begin{bmatrix} 0 & 0 \\ 0 & 0 \\ (s+5) & 0 \\ 0 & (s+5) \\ (s+3) & -(s+1)(s+4)/3 \end{bmatrix}$$ (2.11.36)

$$U_6(s) = \begin{bmatrix} 1 & 0 & 0 & 0 & 0 \\ 0 & 1 & 0 & 0 & 0 \\ 0 & 0 & 1 & 0 & 0 \\ 0 & 0 & 0 & 1 & 0 \\ 0 & 0 & (s+3)/3 & 0 & -1/3 \end{bmatrix}$$ (2.11.37)

Step VII: Add $(s + 4)/3$ times the fourth row and to the fifth row.

$$U_7(s)\begin{bmatrix} 0 & 0 \\ 0 & 0 \\ (s+5) & 0 \\ 0 & (s+5) \\ (s+3) & -(s+1)(s+4)/3 \end{bmatrix} = \begin{bmatrix} 0 & 0 \\ 0 & 0 \\ (s+5) & 0 \\ 0 & (s+5) \\ (s+3) & 4(s+4)/3 \end{bmatrix} \qquad (2.11.38)$$

$$U_7(s) = \begin{bmatrix} 1 & 0 & 0 & 0 & 0 \\ 0 & 1 & 0 & 0 & 0 \\ 0 & 0 & 1 & 0 & 0 \\ 0 & 0 & 0 & 1 & 0 \\ 0 & 0 & 0 & (s+4)/3 & 1 \end{bmatrix} \qquad (2.11.39)$$

Step VIII: Subtract the third row from the fifth row.

$$U_8(s)\begin{bmatrix} 0 & 0 \\ 0 & 0 \\ (s+5) & 0 \\ 0 & (s+5) \\ (s+3) & 4(s+4)/3 \end{bmatrix} = \begin{bmatrix} 0 & 0 \\ 0 & 0 \\ (s+5) & 0 \\ 0 & (s+5) \\ -2 & 4(s+4)/3 \end{bmatrix} \qquad (2.11.40)$$

$$U_8(s) = \begin{bmatrix} 1 & 0 & 0 & 0 & 0 \\ 0 & 1 & 0 & 0 & 0 \\ 0 & 0 & 1 & 0 & 0 \\ 0 & 0 & 0 & 1 & 0 \\ 0 & 0 & -1 & 0 & 1 \end{bmatrix} \qquad (2.11.41)$$

Step IX: Replace fifth row by the fourth row — 3/4 times the fifth row.

$$U_9(s)\begin{bmatrix} 0 & 0 \\ 0 & 0 \\ (s+5) & 0 \\ 0 & (s+5) \\ -2 & 4(s+4)/3 \end{bmatrix} = \begin{bmatrix} 0 & 0 \\ 0 & 0 \\ (s+5) & 0 \\ 0 & (s+5) \\ 3/2 & 1 \end{bmatrix} \qquad (2.11.42)$$

$$U_9(s) = \begin{bmatrix} 1 & 0 & 0 & 0 & 0 \\ 0 & 1 & 0 & 0 & 0 \\ 0 & 0 & 1 & 0 & 0 \\ 0 & 0 & 0 & 1 & 0 \\ 0 & 0 & 0 & 1 & -3/4 \end{bmatrix} \qquad (2.11.43)$$

Step X: Subtract 2(s + 5)/3 times the fifth row from the third row.

$$U_{10}(s)\begin{bmatrix} 0 & 0 \\ 0 & 0 \\ (s+5) & 0 \\ 0 & (s+5) \\ 3/2 & 1 \end{bmatrix} = \begin{bmatrix} 0 & 0 \\ 0 & 0 \\ 0 & -2(s+5)/3 \\ 0 & (s+5) \\ 3/2 & 1 \end{bmatrix} \qquad (2.11.44)$$

$$U_{10}(s) = \begin{bmatrix} 1 & 0 & 0 & 0 & 0 \\ 0 & 1 & 0 & 0 & 0 \\ 0 & 0 & 1 & 0 & -2(s+5)/3 \\ 0 & 0 & 0 & 1 & 0 \\ 0 & 0 & 0 & 0 & 1 \end{bmatrix} \qquad (2.11.45)$$

Step XI: Add 2/3 times the fourth row to the third row.

$$U_{11}(s)\begin{bmatrix} 0 & 0 \\ 0 & 0 \\ 0 & -2(s+5)/3 \\ 0 & (s+5) \\ 3/2 & 1 \end{bmatrix} = \begin{bmatrix} 0 & 0 \\ 0 & 0 \\ 0 & 0 \\ 0 & (s+5) \\ 3/2 & 1 \end{bmatrix} \qquad (2.11.46)$$

$$U_{11}(s) = \begin{bmatrix} 1 & 0 & 0 & 0 & 0 \\ 0 & 1 & 0 & 0 & 0 \\ 0 & 0 & 1 & 2/3 & 0 \\ 0 & 0 & 0 & 1 & 0 \\ 0 & 0 & 0 & 0 & 1 \end{bmatrix} \qquad (2.11.47)$$

Step XII: Interchanging rows,

$$U_{12}(s)\begin{bmatrix} 0 & 0 \\ 0 & 0 \\ 0 & 0 \\ 0 & (s+5) \\ 3/2 & 1 \end{bmatrix} = \begin{bmatrix} 3/2 & 1 \\ 0 & (s+5) \\ 0 & 0 \\ 0 & 0 \\ 0 & 0 \end{bmatrix} \qquad (2.11.48)$$

$$U_{12}(s) = \begin{bmatrix} 0 & 0 & 0 & 0 & 1 \\ 0 & 0 & 0 & 1 & 0 \\ 0 & 0 & 1 & 0 & 0 \\ 0 & 1 & 0 & 0 & 0 \\ 1 & 0 & 0 & 0 & 0 \end{bmatrix} \qquad (2.11.49)$$

In summary, with reference to Equation 2.11.22,

$$U(s) = U_{12}(s)U_{11}(s)...U_2(s)U_1(s) \qquad (2.11.50)$$

and

$$R(s) = \begin{bmatrix} 3/2 & 1 \\ 0 & (s+5) \end{bmatrix} \qquad (2.11.51)$$

The square matrix $R(s)$ is a *gcrd* of $N(s)$ and $D(s)$.

2.12 MFD OF A TRANSFER FUNCTION MATRIX FOR THE MINIMAL ORDER OF A STATE SPACE REALIZATION

Let a transfer function be expressed as

$$G(s) = N(s)D^{-1}(s) \qquad (2.12.1)$$

First, find $W_g(s)$, a *gcrd* of $N(s)$ and $D(s)$. Then, obtain the following matrices:

$$\bar{N}(s) = N(s)W_g^{-1}(s) \qquad (2.12.2)$$

and

$$\bar{D}(s) = D(s)W_g^{-1}(s) \qquad (2.12.3)$$

The transfer function matrix can also be written as

$$G(s) = \bar{N}(s)\bar{D}^{-1}(s) \tag{2.12.4}$$

where

$$deg\ \bar{D}(s) = deg\ D(s)\ deg\ W_g(s) \tag{2.12.5}$$

Now, a state space realization based on Equation 2.12.4 will have a minimum number of states (Kailath, 1980) equal to $deg\ \bar{D}(s)$.

EXAMPLE 2.13

Let the transfer function matrix be defined as

$$G(s) = \begin{bmatrix} \dfrac{s+1}{s+3} & 0 \\[3mm] 0 & \dfrac{s+2}{s+4} \\[3mm] \dfrac{s+2}{s+5} & \dfrac{s+1}{s+5} \end{bmatrix} \tag{2.12.6}$$

With respect to (2.12.1), $N(s)$ and $D(s)$ are given by (2.11.23) and (2.11.24). From (2.11.51), a *gcrd* of $N(s)$ and $D(s)$ is

$$W_g(s) = \begin{bmatrix} 3/2 & 1 \\ 0 & (s+5) \end{bmatrix} \tag{2.12.7}$$

Therefore,

$$\bar{N}(s) = N(s)W_g^{-1}(s) = \begin{bmatrix} \dfrac{2(s+1)(s+5)}{3} & -\dfrac{2(s+1)}{3} \\[3mm] 0 & (s+2) \\[3mm] \dfrac{2(s+2)(s+3)}{3} & \dfrac{s}{3} \end{bmatrix} \tag{2.12.8}$$

and

$$\bar{D}(s) = D(s)W_g^{-1}(s) = \begin{bmatrix} \dfrac{2(s+3)(s+5)}{3} & -\dfrac{2(s+3)}{3} \\ 0 & (s+4) \end{bmatrix} \qquad (2.12.9)$$

Because $deg\ \bar{D}(s) = 3$, the minimal number of states equals 3.

2.13 CONTROLLER FORM REALIZATION FROM A RIGHT MFD

2.13.1 STATE SPACE REALIZATION

Consider a strictly proper right MFD (Kailath, 1980)

$$G(s) = N(s)D^{-1}(s) \qquad (2.13.1)$$

Rewrite (2.8.1) as

$$D(s)\xi(s) = \mathbf{u}(s) \qquad (2.13.2a)$$

$$\mathbf{y}(s) = N(s)\xi(s) \qquad (2.13.2b)$$

The matrix $D(s)$ can be written as Equation 2.8.21 where

$$L(s) = D_{lc}\Psi(s) \qquad (2.13.3)$$

Here, D_{lc} is the lower-column-degree coefficient matrix of $D(s)$ and

$$\Psi^T(s) =$$

$$\begin{bmatrix} s^{k_1-1} & . & . & s & 1 & 0 & . & . & . & 0 & . & . & 0 & . & . & . & 0 \\ 0 & . & . & . & 0 & s^{k_2-1} & . & . & s & 1 & . & . & 0 & . & . & . & 0 \\ & & . & & & & . & & & & . & & & . & & & \\ & & . & & & & . & & & & . & & & . & & & \\ 0 & . & . & . & 0 & 0 & . & . & . & 0 & . & . & s^{k_m-1} & . & . & s & 1 \end{bmatrix}$$

$$(2.13.4)$$

Note that the dimension of the matrix $\Psi^T(s)$ is $mx(k_1 + k_2 + \ldots k_m)$. Furthermore, the degree of the ith column of $N(s)$ will be at the most $k_i - 1$. Therefore, a matrix N_{lc} can be defined such that

$$\mathbf{y}(s) = N(s)\xi(s) = N_{lc}\Psi(s)\xi(s) \tag{2.13.5}$$

From (2.8.21) and (2.13.2a),

$$[D_{hc}S(s) + D_{lc}\Psi(s)]\xi(s) = \mathbf{u}(s) \tag{2.13.6}$$

Assuming that the matrix D(s) is column reduced,

$$S(s)\xi(s) = -D_{hc}^{-1}D_{lc}\mathbf{q}(s) + D_{hc}^{-1}\mathbf{u}(s) \tag{2.13.7}$$

where

$$\mathbf{q}(s) = \Psi(s)\xi(s) \tag{2.13.8}$$

Equation 2.13.8 is rewritten as

$$\mathbf{q}(s) = \Psi(s)S^{-1}(s)\mathbf{v}(s) \tag{2.13.9}$$

where

$$\mathbf{v}(s) = S(s)\xi(s) \tag{2.13.10}$$

Using (2.13.4), Equation 2.13.9 yields

$$\mathbf{q}(s) =$$

$$
\begin{bmatrix}
s^{-1} & . & . & s^{-(k_1-1)} & s^{-k_1} & 0 & . & . & . & 0 & . & . & 0 & . & . & . & 0 \\
0 & . & . & . & 0 & s^{-1} & . & . & s^{-(k_2-1)} & s^{-k_2} & . & . & 0 & . & . & . & 0 \\
 & & & & . & & & & . & & . & . & & . & & & \\
 & & & & . & & & & . & & . & . & & . & & & \\
0 & . & . & . & 0 & 0 & . & . & . & 0 & . & . & s^{-1} & . & . & s^{-(k_n-1)} & s^{-k_n}
\end{bmatrix}^T
\times
$$

$$
\begin{bmatrix}
v_1(s) \\
v_2(s) \\
. \\
. \\
v_m(s)
\end{bmatrix}
$$

$$\tag{2.13.11}$$

Let

$$s^{-1}v_1(s) = x_1,\ldots,\ s^{-(k_1-1)}v_1(s) = x_{k_1-1},\ s^{-k_1}v_1(s) = x_{k_1}$$

$$s^{-1}v_2(s) = x_{k_1+1}(s),\ldots, s^{-(k_2-1)}v_2(s) = x_{k_1+k_2-1}(s), s^{-(k_2-1)}v_2(s) = x_{k_1+k_2}(s)$$

...

... $\hspace{6cm}$ (2.13.12)

$$s^{-1}v_m(s) = x_{k_1+\ldots+k_{m-1}+1}(s),\ldots, s^{-1}v_m(s) = x_{k_1+\ldots+k_{m-1}+k_m-1}(s)$$

$$s^{-1}v_m(s) = x_{k_1+\ldots+k_{m-1}+k_m}(s)$$

In this case,

$$\dot{x}_1 = v_1, \dot{x}_2 = x_1,\ldots, \dot{x}_{k_1} = x_{k_1-1}$$

$$\dot{x}_{k_1+1} = v_2, \dot{x}_{k_1+2} = x_{k_1+1},\ldots, \dot{x}_{k_1+k_2} = x_{k_1+k_2-1}$$

... $\hspace{6cm}$ (2.13.13)

...

$$\dot{x}_{k_1+\ldots+k_{m-1}+1} = v_m, \dot{x}_{k_1+\ldots+k_{m-1}+2} = x_{k_1+\ldots+k_{m-1}+1},\ldots, \dot{x}_{k_1+\ldots+k_m} = x_{k_1+\ldots+k_m-1}$$

Define the state vector **x** as

$$\mathbf{x} = \begin{bmatrix} x_1 & \cdot & x_{k_1} & \cdot & x_{k_1+k_2} & \cdot & \cdot & \cdot & \cdot & x_{k_1+\ldots+k_{m-1}+1} & \cdot & x_{k_1+\ldots+k_m} \end{bmatrix}^T$$

$$(2.13.14)$$

Then, Equations 2.13.13 can be written as

$$\dot{\mathbf{x}} = A_c^0 \mathbf{x}(t) + B_c^0 \mathbf{v}(t) \hspace{3cm} (2.13.15)$$

where

$$A_c^0 = block\ diagonal \left\{ \begin{bmatrix} 0 & 0 & . & 0 & 0 \\ 1 & 0 & . & 0 & 0 \\ 0 & 1 & . & 0 & 0 \\ . & . & . & . & . \\ 0 & 0 & . & 1 & 0 \end{bmatrix}; k_i x k_i; i = 1, 2,\ldots, m \right\} \hspace{1cm} (2.13.16)$$

and

$$[B_c^0]^T = block\ diagonal \left\{ \begin{bmatrix} 1 & 0 & . & 0 \end{bmatrix}; 1 x k_i; i = 1, 2,\ldots, m \right\} \hspace{0.5cm} (2.13.17)$$

Utilizing (2.13.9) and (2.13.12), it can be easily seen that

$$\mathbf{q}(s) = \mathbf{x}(s) \tag{2.13.18}$$

Therefore, from (2.13.7) and (2.13.10),

$$\mathbf{v}(t) = -D_{hc}^{-1}D_{lc}\mathbf{x}(t) + D_{hc}^{-1}\mathbf{u}(t) \tag{2.13.19}$$

Substituting (2.13.19) into (2.3.15), the state equations are obtained:

$$\dot{\mathbf{x}} = A_c\mathbf{x}(t) + B_c\mathbf{u}(t) \tag{2.13.20}$$

where

$$A_c = A_c^0 - B_c^0 D_{hc}^{-1}D_{lc} \tag{2.13.21}$$

and

$$B_c = B_c^0 D_{hc}^{-1} \tag{2.13.22}$$

From (2.13.5), (2.13.14), and (2.13.18), output equations are given as

$$\mathbf{y}(t) = C_c\mathbf{x}(t) \tag{2.13.23}$$

where

$$C_c = N_{lc} \tag{2.13.24}$$

EXAMPLE 2.14

Find the minimal state space realization of transfer function matrix $G(s)$, Equation 2.12.6.

$$G(s) = \begin{bmatrix} 1 & 0 \\ 0 & 1 \\ 1 & 1 \end{bmatrix} - H(s) \tag{2.13.25}$$

where

$$H(s) = \begin{bmatrix} \dfrac{2}{s+3} & 0 \\ 0 & \dfrac{2}{s+4} \\ \dfrac{3}{s+5} & \dfrac{4}{s+5} \end{bmatrix} \qquad (2.13.26)$$

Note that $H(s)$ is a strictly proper MFD and can be expressed as

$$H(s) = N_H(s)D_H^{-1}(s) \qquad (2.13.27)$$

where

$$N_H(s) = \begin{bmatrix} 2(s+5) & 0 \\ 0 & 2(s+5) \\ 3(s+3) & 4(s+4) \end{bmatrix} \qquad (2.13.28)$$

and

$$D_H(s) = \begin{bmatrix} (s+3)(s+5) & 0 \\ 0 & (s+4)(s+5) \end{bmatrix} \qquad (2.13.29)$$

A *gcrd* of $N_H(s)$ and $D_H(s)$ is again found as

$$W_h(s) = \begin{bmatrix} 3/2 & 1 \\ 0 & s+5 \end{bmatrix} \qquad (2.13.30)$$

Then,

$$\bar{N}_H(s) = N_H(s)W_h^{-1}(s) = \begin{bmatrix} \dfrac{4(s+5)}{3} & -\dfrac{4}{3} \\ 0 & 2 \\ 2(s+3) & 2 \end{bmatrix} \qquad (2.13.31)$$

and

$$\bar{D}_H(s) = D_H(s)W_h^{-1}(s) = \begin{bmatrix} \dfrac{2(s+3)(s+5)}{3} & -\dfrac{2(s+3)}{3} \\ 0 & (s+4) \end{bmatrix} \qquad (2.13.32)$$

For $\bar{D}_H(s)$ in Equation 2.13.32, $k_1 = 2$ and $k_2 = 1$. And,

$$D_{hc} = \begin{bmatrix} 2/3 & -2/3 \\ 0 & 1 \end{bmatrix} \quad \text{and} \quad D_{lc} = \begin{bmatrix} 16/3 & 10 & -2 \\ 0 & 0 & 4 \end{bmatrix} \qquad (2.13.32)$$

$$A_c^0 = \begin{bmatrix} 0 & 0 & 0 \\ 1 & 0 & 0 \\ 0 & 0 & 0 \end{bmatrix}, \ B_c^0 = \begin{bmatrix} 1 & 0 \\ 0 & 0 \\ 0 & 1 \end{bmatrix} \qquad (2.13.33)$$

For $\bar{N}_H(s)$ in Equation 2.13.31,

$$N_{lc} = \begin{bmatrix} 4/3 & 20/3 & -4/3 \\ 0 & 0 & 2 \\ 2 & 6 & 2 \end{bmatrix} \qquad (2.13.34)$$

From (2.13.21), (2.13.22), and (2.13.24),

$$A_c = \begin{bmatrix} -8 & -15 & -1 \\ 1 & 0 & 0 \\ 0 & 0 & -4 \end{bmatrix} \ B_c = \begin{bmatrix} 1.5 & 1 \\ 0 & 0 \\ 0 & 1 \end{bmatrix} \quad \text{and} \quad C_c = N_{lc} \qquad (2.13.35)$$

2.13.2 SIMILARITY TRANSFORMATION TO CONVERT ANY STATE SPACE REALIZATION {A,B,C} TO THE CONTROLLER FORM REALIZATION

Given *any* state space realization $\{A,B,C\}$, the objective is to find the nonsingular matrix T such that

$$T^{-1}AT = \hat{A}_c \quad \text{and} \quad T^{-1}B = \hat{B}_c \qquad (2.13.36)$$

where the structures of $\hat{A}_c$ and $\hat{B}_c$ match those of A_c and B_c given by Equation 2.13.21 and Equation 2.3.22, respectively.

TABLE 2.1
Search for Independent Vectors

$\mathbf{b}_1$ X	$\mathbf{b}_2$ X	$\mathbf{b}_3$ X
$A\mathbf{b}_1$ X	$A\mathbf{b}_2$ X	$A\mathbf{b}_3$ X
$A^2\mathbf{b}_1$ 0	$A^2\mathbf{b}_2$ 0	$A^2\mathbf{b}_3$ 0
$A^3\mathbf{b}_1$	$A^3\mathbf{b}_2$	$A^3\mathbf{b}_3$
$A^4\mathbf{b}_1$	$A^4\mathbf{b}_2$	$A^4\mathbf{b}_3$
$A^5\mathbf{b}_1$	$A^5\mathbf{b}_2$	$A^5\mathbf{b}_3$

First, consider the controllability matrix

$$C = [B \quad AB \quad A^2B \quad . \quad . \quad A^{n-1}B] \tag{2.13.37}$$

The order of the matrix is $n \times nm$. Out of these nm column vectors of the matrix, only n column vectors are independent. To find these n independent column vectors, Table 2.1 is created for the state equations with three inputs and six states as an example. The search for independent columns is conducted (Kailath, 1980) row-wise from left to right, and a sign "X" is introduced to indicate that it is independent of all the previous vectors that have been found to be independent. The sign "0" is introduced to indicate that the associated vector is not independent of previously found independent columns. There is no need to consider vectors in the table column below the vector with the "0" sign because they are guaranteed to be dependent on previously found independent vectors due to the Cayley–Hamilton theorem. Next, each vector with the sign "0" is expanded in terms of independent vectors, i.e.,

$$A^2\mathbf{b}_i = \sum_{j=1}^{3} \alpha_{ji}\mathbf{b}_j + \sum_{j=1}^{3} \beta_{ji}A\mathbf{b}_j ; \quad i = 1, 2, \text{ and } 3 \tag{2.13.38}$$

where α_j and β_j are coefficients to be determined. Using the matrix notation, Equation 2.13.38 can be written as

$$C_I[\alpha_{1i} \quad \alpha_{2i} \quad \alpha_{3i} \quad \beta_{1i} \quad \beta_{2i} \quad \beta_{3i}]^T = A^2\mathbf{b}_i \tag{2.13.39}$$

where

$$C_I = \begin{bmatrix} \mathbf{b}_1 & \mathbf{b}_2 & \mathbf{b}_3 & A\mathbf{b}_1 & A\mathbf{b}_2 & A\mathbf{b}_3 \end{bmatrix} \tag{2.13.40}$$

Solving (2.13.39),

$$[\alpha_{1i} \quad \alpha_{2i} \quad \alpha_{3i} \quad \beta_{1i} \quad \beta_{2i} \quad \beta_{3i}]^T = C_I^{-1} A^2 \mathbf{b}_i \qquad (2.13.41)$$

Then, for each vector with the sign "0", terms having A are collected on the left side and the matrix A is factored out as follows:

$$A\mathbf{t}_{i2} = \sum_{j=1}^{3} \alpha_{ji} \mathbf{b}_j \qquad (2.13.42)$$

where

$$\mathbf{t}_{i2} = [A\mathbf{b}_i - \sum_{j=1}^{3} \beta_{ji} \mathbf{b}_j] \qquad (2.13.43)$$

Again, for $\mathbf{t}_{i2}$, terms having A are collected on the left side and the matrix A is factored out as follows:

$$A\mathbf{t}_{i1} = \mathbf{t}_{i2} + \sum_{j=1}^{3} \beta_{ji} \mathbf{b}_j \qquad (2.13.44)$$

where

$$\mathbf{t}_{i1} = \mathbf{b}_i \qquad (2.13.45)$$

Lastly, the transformation matrix T is defined as

$$T = [\mathbf{t}_{11} \quad \mathbf{t}_{12} \quad \mathbf{t}_{21} \quad \mathbf{t}_{22} \quad \mathbf{t}_{31} \quad \mathbf{t}_{32}] \qquad (2.13.46)$$

2.14 POLES AND ZEROS OF A MIMO TRANSFER FUNCTION MATRIX

2.14.1 SMITH FORM

For any *pxm* polynomial matrix $P(s)$, unimodular matrices $U(s)$ and $V(s)$ can be found (Kailath, 1980) such that

$$U(s)P(s)V(s) = \Lambda(s) \qquad (2.14.1)$$

where

$$\Lambda(s) = \begin{bmatrix} \lambda_1(s) & 0 & . & 0 & 0 & . & 0 & 0 \\ 0 & \lambda_2(s) & . & 0 & 0 & . & 0 & 0 \\ . & . & . & 0 & 0 & . & 0 & 0 \\ 0 & . & . & \lambda_r(s) & 0 & . & 0 & 0 \\ 0 & 0 & . & 0 & 0 & . & 0 & 0 \\ 0 & 0 & . & 0 & 0 & . & 0 & 0 \\ . & . & . & . & . & . & . & . \\ 0 & 0 & 0 & 0 & 0 & . & 0 & 0 \end{bmatrix} \qquad (2.14.2)$$

r is the normal rank of $P(s)$, and $\lambda_i(s)$; $i = 1, 2, \ldots, r$ are unique monic polynomials obeying the division property

$$\lambda_i(s) \big| \lambda_{i+1}(s); \quad i = 1, 2, \ldots, r-1 \qquad (2.14.3)$$

The pxm polynomial matrix $\Lambda(s)$ is known as the Smith form of $P(s)$. The notation (2.14.3) indicates that the polynomial $\lambda_i(s)$ is divisible by the polynomial $\lambda_{i+1}(s)$.

2.14.2 SMITH–MCMILLAN FORM

A transfer function matrix $G(s)$ can be written (Kailath, 1980) as

$$G(s) = \frac{1}{d(s)} N(s) \qquad (2.14.4)$$

where $d(s)$ is the monic least common multiple of the denominators of elements of $G(s)$. From (2.14.4),

$$d(s)G(s) = N(s) \qquad (2.14.5)$$

Expressing $N(s)$ in the Smith form (2.14.1),

$$N(s) = U(s)\Lambda(s)V(s) \qquad (2.14.6)$$

where $U(s)$ and $V(s)$ are unimodular matrices. From (2.14.5) and (2.14.6),

$$U^{-1}(s)G(s)V^{-1}(s) = \frac{1}{d(s)}\Lambda(s) \qquad (2.14.7)$$

As the inverse of a unimodular matrix is also a unimodular matrix, any transfer function matrix can be converted to the form on the right-hand side of (2.14.7) via elementary row and column operations. Now, from (2.14.2), nonzero diagonal elements of the matrix on the right-hand side of (2.14.7) will be

$$\frac{\lambda_i(s)}{d(s)} \; ; \quad i = 1, 2, \ldots, r \tag{2.14.8}$$

where r is the normal rank of $G(s)$.

Eliminating common factors between $\lambda_i(s)$ and $d(s)$,

$$\frac{\lambda_i(s)}{d(s)} = \frac{\varepsilon_i(s)}{\psi_i(s)} \tag{2.14.9}$$

There are no common factors between $\varepsilon_i(s)$ and $\psi_i(s)$. Define

$$M(s) =$$

$$\begin{bmatrix} \varepsilon_1(s)/\psi_1(s) & 0 & . & 0 & 0 & . & 0 & 0 \\ 0 & \varepsilon_2(s)/\psi_2(s) & . & 0 & 0 & . & 0 & 0 \\ . & . & . & 0 & 0 & . & 0 & 0 \\ 0 & . & . & \varepsilon_r(s)/\psi_r(s) & 0 & . & 0 & 0 \\ 0 & 0 & . & 0 & 0 & . & 0 & 0 \\ 0 & 0 & . & 0 & 0 & . & 0 & 0 \\ . & . & . & . & . & . & . & . \\ 0 & 0 & 0 & 0 & 0 & . & 0 & 0 \end{bmatrix} \tag{2.14.10}$$

Then, from (2.14.7) and (2.14.10),

$$U_1^{-1}(s)G(s)U_2^{-1}(s) = M(s) \tag{2.14.11}$$

In other words, a transfer function matrix $G(s)$ can be converted to have the structure of the matrix $M(s)$ via elementary row and column operations. Polynomials in $M(s)$ satisfy the following properties:

$$\psi_{i+1}(s)\big|\psi_i(s) \; ; \quad i = 1, 2, \ldots, r - 1 \tag{2.14.12}$$

$$\varepsilon_i(s)\big|\varepsilon_{i+1}(s) \; ; i = 1, 2, \ldots, r - 1 \tag{2.14.13}$$

$$d(s) = \psi_1(s) \tag{2.14.14}$$

Notations (2.14.12) and (2.14.13) indicate that polynomials $\psi_{i+1}(s)$ and $\varepsilon_i(s)$ are divisible by polynomials $\psi_i(s)$ and $\varepsilon_{i+1}(s)$, respectively.

2.14.3 Poles and Zeros via Smith–McMillan Form

Poles of a multivariable transfer function $G(s)$ are roots of denominator polynomials $\psi_i(s)$ in the Smith–McMillan form $M(s)$, Equation 2.4.10. Similarly, zeros of a multivariable transfer function $G(s)$ are roots of numerator polynomials $\varepsilon_i(s)$ in the Smith–McMillan form $M(s)$.

Example 2.15

Find poles and zeros of the following transfer function matrix:

$$G(s) = \frac{1}{s^2 + 4s + 5} P(s) \qquad (2.14.15)$$

where

$$P(s) = \begin{bmatrix} 2 & (s+1)^2 \\ (s+1)^2 & -2 \end{bmatrix} \qquad (2.14.16)$$

Using elementary operations, $P(s)$ will be first converted to Smith form.

Step I: Divide the first row of $P(s)$ by 1/2:

$$U_1(s)P(s) = \begin{bmatrix} 1 & 0.5(s+1)^2 \\ (s+1)^2 & -2 \end{bmatrix} \qquad (2.14.17)$$

where

$$U_1(s) = \begin{bmatrix} 0.5 & 0 \\ 0 & 1 \end{bmatrix} \qquad (2.14.18)$$

Step II: Subtract $(s+1)^2$ times the first row from the second row:

$$U_2(s)U_1(s)P(s) = \begin{bmatrix} 1 & 0.5(s+1)^2 \\ 0 & -2 - 0.5(s+1)^4 \end{bmatrix} \qquad (2.14.19)$$

where

$$U_2(s) = \begin{bmatrix} 1 & 0 \\ -(s+1)^2 & 1 \end{bmatrix} \tag{2.14.20}$$

Step III: Subtract $0.5(s+1)^2$ times the first column from the second column:

$$U_2(s)U_1(s)P(s)V_3(s) = \begin{bmatrix} 1 & 0 \\ 0 & -2-0.5(s+1)^4 \end{bmatrix} \tag{2.14.21}$$

where

$$V_3(s) = \begin{bmatrix} 1 & -0.5(s+1)^2 \\ 0 & 1 \end{bmatrix} \tag{2.14.22}$$

Step IV: Multiplying the second column by -2,

$$U_2(s)U_1(s)P(s)V_3(s)V_4(s) = \begin{bmatrix} 1 & 0 \\ 0 & 4+(s+1)^4 \end{bmatrix} \tag{2.14.23}$$

where

$$V_4(s) = \begin{bmatrix} 1 & 0 \\ 0 & -2 \end{bmatrix} \tag{2.14.24}$$

Now, it can be verified that Equation 2.14.23 has the Smith form (2.14.1), i.e.,

$$U(s)P(s)V(s) = \begin{bmatrix} 1 & 0 \\ 0 & 4+(s+1)^4 \end{bmatrix} \tag{2.14.25}$$

$$U(s) = U_2(s)U_1(s) = \begin{bmatrix} 0.5 & 0 \\ -0.5(s+1)^2 & 1 \end{bmatrix} \tag{2.14.26}$$

and

$$V(s) = V_3(s)V_4(s) = \begin{bmatrix} 1 & (s+1)^2 \\ 0 & -2 \end{bmatrix} \tag{2.14.27}$$

Therefore, from (2.14.15) and (2.14.25),

$$U^{-1}(s)G(s)V^{-1}(s) = \frac{1}{s^2 + 4s + 5}\begin{bmatrix} 1 & 0 \\ 0 & 4 + (s+1)^4 \end{bmatrix} \qquad (2.14.28)$$

As $(s+1)^4 + 4 = (s^2 + 4s + 5)(s^2 + 1)$,

$$U^{-1}(s)G(s)V^{-1}(s) = \begin{bmatrix} \dfrac{1}{s^2 + 4s + 5} & 0 \\ 0 & \dfrac{s^2 + 1}{1} \end{bmatrix} \qquad (2.14.29)$$

Equation 2.14.29 has the Smith–McMillan form. Therefore, poles are $-2 \pm j$, which are the roots of $s^2 + 4s + 5 = 0$. Further, zeros are $\pm j$, which are the roots of $s^2 + 1 = 0$.

2.14.4 POLES AND ZEROS VIA AN IRREDUCIBLE MFD

Consider the following irreducible MFD of the multivariable transfer function $G(s)$:

$$G(s) = N(s)D^{-1}(s) \qquad (2.14.30)$$

Poles of $G(s)$ are roots of $\det D(s) = 0$. Zeros of $G(s)$ are values of s for which $N(s)$ is not of full rank.

EXAMPLE 2.16

For the transfer function matrix (2.12.6), the irreducible MFD is represented by

$$G(s) = \bar{N}(s)\bar{D}^{-1}(s) \qquad (2.14.31)$$

where $\bar{N}(s)$ and $\bar{D}(s)$ are represented by (2.12.8) and (2.12.9), respectively. Now,

$$\det \bar{D}(s) = \frac{2}{3}(s + 3)(s + 4)(s + 5) \qquad (2.14.32)$$

Therefore, poles are located at -3, -4, and -5. There is no s for which the rank of $\bar{N}(s)$ is less than 2. Therefore, there is no zero.

2.15 STABILITY ANALYSIS

The stability of a linear and time-invariant system is not influenced by external inputs $\mathbf{u}(t)$. Therefore, $\mathbf{u}(t)$ will be set to zero, and the following initial-value problem is considered:

$$\dot{\mathbf{x}} = A\mathbf{x}(t) \; ; \quad \mathbf{x}(0) \neq 0 \qquad (2.15.1)$$

The system (2.15.1) is described to be stable when

$$\mathbf{x}(t) \to 0 \quad \text{as} \quad t \to \infty \qquad (2.15.2)$$

This property is also referred to as asymptotic stability. The condition (2.15.2) is satisfied when all the eigenvalues of A have negative (nonzero) real parts, i.e., they are located in the left half (excluding the imaginary axis) of the complex plane.

Define a positive definite (Appendix B) function $V(t)$ as

$$V(t) = \mathbf{x}^T(t)P\mathbf{x}(t) \quad \text{where} \quad P = P^T > 0 \qquad (2.15.3)$$

Differentiating (2.15.3) with respect to time,

$$\dot{V}(t) = \dot{\mathbf{x}}^T(t)P\mathbf{x}(t) + \mathbf{x}^T(t)P\dot{\mathbf{x}}(t) \qquad (2.15.4)$$

Substituting (2.15.1) into (2.15.4),

$$\dot{V}(t) = \mathbf{x}^T(t)[A^T P + PA]\mathbf{x}(t) \qquad (2.15.5)$$

For a stable system, $\dot{V}(t)$ is a negative definite function. Therefore,

$$A^T P + PA = -Q \qquad (2.15.6)$$

where $Q = Q^T > 0$.

Equation 2.15.6 is known as the Lyapunov equation, often written as

$$A^T P + PA + Q = 0 \qquad (2.15.7)$$

An Important Property of the Lyapunov Equation

When $Q = Q^T > 0$, and all the eigenvalues of A have negative (nonzero) real parts, the Lyapunov equation 2.15.7 has a unique solution for P, satisfying $P = P^T > 0$ as follows (Kailath, 1980):

$$P = \int_0^\infty e^{A^T t} Q e^{At} dt \qquad (2.15.8)$$

EXERCISE PROBLEMS

P2.1   Consider the following transfer function:

$$G(s) = \frac{4s^3 + 25s^2 + 45s + 34}{s^3 + 6s^2 + 10s + 8}$$

Develop state space realizations using Methods I and II. Write state space equations and draw the block diagram for each method.

P2.2   Using the properties of the determinant, find the characteristic polynomial of the following matrix:

$$\begin{bmatrix} 0 & 1 & 0 & . & . & 0 \\ 0 & 0 & 1 & . & . & 0 \\ . & . & . & . & . & . \\ 0 & 0 & 0 & . & . & 1 \\ -a_n & -a_{n-1} & & . & . & -a_1 \end{bmatrix}$$

P2.3

  a. If $\bar{A} = T^{-1}AT$, show that $e^{\bar{A}t} = Te^{\bar{A}t}T^{-1}$.

  b. If there are n independent eigenvectors for the $n \times n$ matrix A, develop an algorithm to compute e^{At} using the result in part (a).

P2.4   Matlab Exercise

  Consider the system shown in the Figure P.2.4 where

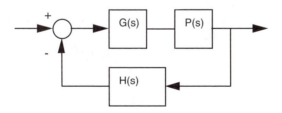

FIGURE P.2.4 A Feedback System.

$$G(s) = \frac{s+2}{s^2 + 6s + 8} \qquad P(s) = \frac{s^3 + 2s^2 + 9s + 10}{s^3 + s^2 + 10s + 20}$$

$$H(s) = \frac{s+1}{s+5}$$

  a. Develop state space models for both open-loop and closed-loop systems.

  b. Examine controllability and observability of these state space models.

  c. Determine the stability of open- and closed-loop systems.

P2.5 Consider a single input/single output system with the following transfer function:

$$\frac{s+\alpha}{(s+1)(s+2)(s+3)}$$

  a. Construct a state space realization $\{A, \mathbf{b}, \mathbf{c}\}$ that is controllable for all values of α. You should also show the corresponding simulation diagram.

  b. Find the values of α for which the state space realization developed in part (a) is not observable.

P2.6 Consider a linear system described by the following state space equation:

$$\frac{dx_1}{dt} = x_2$$

$$\frac{dx_2}{dt} = -x_1 + u$$

$$y(t) = x_1(t) + x_2(t)$$

Given $y(0) = 0$, $\dfrac{dy}{dt}(0) = 1$, and $u(0) = 0$.

  a. Determine $x_1(0)$ and $x_2(0)$.

  b. Determine the response $x_1(t)$ and $x_2(t)$ via the matrix exponential when $u(t) = 1$ for $t > 0$.

P2.7 Consider a single input/single output system with the following transfer function:

$$\frac{s+2}{(s+1)(s+2)(s+3)}$$

  a. Construct a state space realization $\{A, \mathbf{b}, \mathbf{c}\}$ that is controllable. You should also show the corresponding simulation diagram.

  b. Answer the following questions:

   i. Will it be possible to construct a realization that is both controllable and observable?

   ii. Will it be possible to construct a realization that is neither controllable nor observable?

P2.8   Consider the following system:

$$\frac{d\mathbf{x}}{dt} = A\mathbf{x}(t) + \mathbf{b}u(t)$$

where

$$A = \begin{bmatrix} 0 & 1 \\ 0 & 0 \end{bmatrix}, \quad \mathbf{b} = \begin{bmatrix} 0 \\ 1 \end{bmatrix}, \quad \text{and} \quad \mathbf{x}(0) = \begin{bmatrix} 0 \\ 0 \end{bmatrix}$$

Find $u(t)$, $0 \le t \le 1$, such that $\mathbf{x}(1) = [1 \quad 0]^T$.

P2.9   Consider the state space realization using Method II. Show that this realization is controllable, provided there is no common factor between numerator and denominator of the transfer function.

P2.10   Consider the following state space realization:

$$\frac{d\mathbf{x}}{dt} = \begin{bmatrix} -2 & 1 \\ 1 & -2 \end{bmatrix} \mathbf{x}(t) + \begin{bmatrix} 1 \\ 0 \end{bmatrix} u(t)$$

$$y(t) = \begin{bmatrix} 1 & 2 \end{bmatrix} \mathbf{x}(t)$$

i.   Find the transfer function of the system.
ii. Determine the observability and controllability of the state space realization.
iii. Let $y(0) = 1$, $\dfrac{dy}{dt}(0) = 0$, and $u(0) = 0$. Find $\mathbf{x}(0)$.
iv. Find e^{At}.
v.   Solve the state equations when $\mathbf{x}(0) = \mathbf{0}$ and $u(t) = 1$ for $t \ge 0$.

P2.11   Consider the following transfer function matrix:

$$H(s) = \begin{bmatrix} \dfrac{s+1}{s^2 + s + 1} & \dfrac{s+2}{s^3 + 2s^2 + 1} \end{bmatrix}$$

a. Develop the state space realization using Method I. Determine its controllability and observability.
b. Develop the state space realization using Method II. Determine its controllability and observability.

P2.12 Consider the problem of the control of a tire-tread extrusion line:

$$G(s) = \begin{bmatrix} \dfrac{0.071}{s+0.19} & \dfrac{0.007}{(s+0.23)^2} \\ \dfrac{0.24}{(s+0.23)^2} & \dfrac{0.027}{(s+0.3)^2} \end{bmatrix}$$

    a. Find the controller-form state space realization using a suitable MFD.
    b. Determine controllability and observability of your state space realiza-
        tion and discuss your results.
    c. Find poles and zeros of the system.

P2.13 Consider a system with the following transfer function matrix:

$$G(s) = \begin{bmatrix} -\dfrac{s}{(s+1)} & \dfrac{1}{(s+1)} \\ \dfrac{(2s+1)}{s(s+1)} & \dfrac{1}{s+1} \end{bmatrix}$$

    i. Find an irreducible right MFD.
    ii. What is the order of a minimal realization?
    iii. Find the poles and transmission zeros.

P2.14 Consider the linearized dynamics of a spark ignition engine (Abate et al.,
      1994):

$$\begin{bmatrix} y_1(s) \\ y_2(s) \end{bmatrix} = \frac{1}{d(s)} \begin{bmatrix} \alpha_1\alpha_4 & \alpha_6(s+\alpha_2) \\ \alpha_1(Js+\alpha_7-\alpha_5) & -\alpha_3\alpha_6 \end{bmatrix} \begin{bmatrix} u_1(s) \\ u_2(s) \end{bmatrix}$$

where $y_1(t)$ = engine speed, $y_2(t)$ = relative air pressure of mani-
fold, $u_1(t)$ = duty cycle of the throttle valve, and $u_2(t)$ = spark advance
position. The polynomial $d(s)$ is defined as follows:

$$d(s) = Js^2 + (J\alpha_2 + \alpha_7 - \alpha_5)s + (\alpha_3\alpha_4 + \alpha_2\alpha_7 - \alpha_2\alpha_5)$$

Parameters J (mass-moment of inertia) and $\alpha_1 - \alpha_7$ are provided in Table
2.P.1 for three different operating conditions I, II, and III.

TABLE 2.P1
Parameters of a Spark Engine

	J	α_1	α_2	α_3	α_4	α_5	α_6	α_7
I	1	2.1608	0.1027	0.0357	0.5607	2.0183	4.4962	2.0283
II	1	3.4329	0.1627	0.1139	0.2539	1.7993	2.0247	1.8201
III	10	2.1608	0.1027	0.0357	0.5607	1.7993	4.4962	1.8201

a. Develop a minimum-order state space realization.
b. Find poles and zeros of the transfer function matrix for each operating condition.

3 State Feedback Control and Optimization

First, the design of a full state feedback control system is presented for a single input/single output (SISO) system along with its impact on poles and zeros of the closed-loop system. Next, the full state feedback control system is presented for a multiinput/multioutput (MIMO) system. The necessary conditions for the optimal control are then derived and used to develop the linear quadratic (LQ) control theory and the minimum time control.

3.1 STATE VARIABLE FEEDBACK FOR A SINGLE INPUT SYSTEM

Consider the state space realization $\{A, \mathbf{b}, \mathbf{c}\}$; i.e.,

$$\frac{d\mathbf{x}}{dt} = A\mathbf{x}(t) + \mathbf{b}u(t) \qquad (3.1.1)$$

$$y(t) = \mathbf{c}\mathbf{x}(t) \qquad (3.1.2)$$

The open-loop transfer function is given as

$$\frac{y(s)}{u(s)} = \mathbf{c}(sI - A)^{-1}\mathbf{b} = \frac{\mathbf{c}Adj(sI - A)\mathbf{b}}{\det(sI - A)} \qquad (3.1.3)$$

The state variable feedback system is shown in Figure 3.1, in which

$$u(t) = v(t) - \mathbf{k}\mathbf{x}(t) \qquad (3.1.4)$$

where $v(t)$ is the external input and $\mathbf{k}$ is the state feedback vector. Using (3.1.1) and (3.1.4),

$$\frac{d\mathbf{x}}{dt} = (A - \mathbf{b}\mathbf{k})\mathbf{x}(t) + \mathbf{b}v(t) \qquad (3.1.5)$$

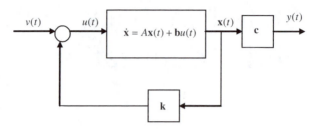

FIGURE 3.1 A state feedback control system.

Equation 3.1.5 represents the state dynamics of the closed-loop system. The transfer function of the closed-loop system is

$$\frac{y(s)}{v(s)} = \mathbf{c}(sI - A + \mathbf{bk})^{-1}\mathbf{b} = \frac{\mathbf{c}Adj(sI - A + \mathbf{bk})\mathbf{b}}{\det(sI - A + \mathbf{bk})} \qquad (3.1.6)$$

It is desirable to study the effects of state feedback on poles and zeros of the closed-loop transfer function.

3.1.1 EFFECTS OF STATE FEEDBACK ON POLES OF THE CLOSED-LOOP TRANSFER FUNCTION: COMPUTATION OF STATE FEEDBACK GAIN VECTOR

It will be shown here that poles of the closed-loop transfer function can be located anywhere in the s-plane by the full state feedback, provided the state space realization is controllable. In other words, it is always possible to find the state feedback vector $\mathbf{k}$ corresponding to any desired locations of closed-loop poles, provided the state space realization is controllable.

Let

$$\frac{y(s)}{u(s)} = \frac{b_1 s^{n-1} + \ldots + b_{n-1}s + b_n}{s^n + a_1 s^{n-1} + a_2 s^{n-2} + \ldots + a_{n-1}s + a_n} \qquad (3.1.7)$$

Hence, the characteristic polynomial of the open-loop system is

$$a(s) = \det(sI - A) = s^n + a_1 s^{n-1} + a_2 s^{n-2} + \ldots + a_{n-1}s + a_n \qquad (3.1.8)$$

Let the desired characteristic polynomial of the closed-loop system be

$$\alpha(s) = \det(sI - A + \mathbf{bk}) = s^n + \alpha_1 s^{n-1} + \alpha_2 s^{n-2} + \ldots + \alpha_{n-1}s + \alpha_n \qquad (3.1.9)$$

where $\alpha_1, \alpha_2, \ldots, \alpha_n$ are arbitrary real numbers.

Now,

$$\alpha(s) = \det(sI - A + \mathbf{bk}) = \det\{(sI - A)(I + (sI - A)^{-1}\mathbf{bk})\}$$
$$= a(s).\det(I + (sI - A)^{-1}\mathbf{bk}) \tag{3.1.10}$$

Using the identity $\det(I_n - PQ) = \det(I_m - QP)$, where P and Q are $n \times m$ and $m \times n$ matrices, respectively, Equation 3.1.10 can be written (Kailath, 1980) as

$$\alpha(s) = a(s)[1 + \mathbf{k}(sI - A)^{-1}\mathbf{b}] \tag{3.1.11}$$

or

$$\alpha(s) - a(s) = \mathbf{k}\,adj(sI - A)\mathbf{b} \tag{3.1.12}$$

Using Equation 2.3.4 and equating coefficients of s^{n-1}, s^{n-2}, ..., s^0 on both sides, we obtain

$$\alpha_1 - a_1 = \mathbf{kb}$$
$$\alpha_2 - a_2 = \mathbf{kAb} + a_1\mathbf{kb}$$
$$\alpha_3 - a_3 = \mathbf{kA^2b} + a_1\mathbf{kAb} + a_2\mathbf{kb} \tag{3.1.13}$$
$$\vdots$$
$$\alpha_n - a_n = \mathbf{kA^{n-1}b} + a_1\mathbf{kA^{n-2}b} + a_2\mathbf{kA^{n-3}b} + \ldots + a_{n-1}\mathbf{kb}$$

Representing the system of equations in matrix form,

$$\boldsymbol{\alpha} - \mathbf{a} = \mathbf{k}CU_t \tag{3.1.14}$$

where C is the controllability matrix,

$$\boldsymbol{\alpha} = \begin{bmatrix} \alpha_1 & \alpha_2 & \alpha_3 & . & . & \alpha_n \end{bmatrix} \tag{3.1.15}$$

$$\mathbf{a} = \begin{bmatrix} a_1 & a_2 & a_3 & . & . & a_n \end{bmatrix} \tag{3.1.16}$$

and U_t is the upper triangular Toeplitz matrix defined as

$$U_t = \begin{bmatrix} 1 & a_1 & a_2 & . & . & a_{n-1} \\ 0 & 1 & a_1 & a_2 & . & a_{n-2} \\ . & . & . & . & . & . \\ 0 & 0 & . & . & 0 & 1 \end{bmatrix} \tag{3.1.17}$$

It can be seen that U_t is always nonsingular. Hence, for any choice of $\boldsymbol{\alpha}$, $\mathbf{k}$ can be found, provided C is nonsingular (or the system is controllable). The result is

$$\mathbf{k} = (\boldsymbol{\alpha} - \mathbf{a})U_t^{-1}C^{-1} \tag{3.1.18}$$

This is also known as the Bass–Gura formula for computing $\mathbf{k}$.

The fact that the eigenvalues can be arbitrarily relocated by state feedback is known as *modal controllability*. This can be understood by realizing that states of a system possess all the information about system dynamics. Hence, feeding back all the states is equivalent to using all the information in deciding the control input.

Another important point to note is that the order of the closed-loop system is the same as that of the open-loop system, which is not always true for the output feedback via a compensator.

EXAMPLE 3.1: BALANCING A POINTER

Consider the balancing of a pointer (Kailath, 1980) on your fingertip (Figure 3.2). If you try to do it yourself, you will find that the pointer will fall down when the fingertip does not move. Furthermore, the fingertip must move to balance the pointer. In fact, the acceleration of the fingertip serves as the control input. To develop a mathematical model, the following assumptions are made (Kailath, 1980):

1. The mass of the pointer is concentrated at the top end.
2. The angle ϕ is small.
3. The force F from the fingertip is applied only along the direction of the pointer.

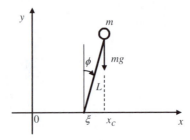

FIGURE 3.2 A pointer on the fingertip.

Applying Newton's second law of motion,

$$m\ddot{x}_c = F \sin \phi \approx F\phi \qquad (3.1.19)$$

where x_c is the x-coordinate of the center of mass. For a small ϕ, the acceleration of the center of mass along the y-direction can be neglected. Therefore,

$$mg = F \cos \phi \approx F \qquad (3.1.20)$$

From (3.1.19) and (3.1.20),

$$\ddot{x}_c = g\phi \qquad (3.1.21)$$

Now,

$$x_c(t) = \xi(t) + L\phi(t) \qquad (3.1.22)$$

where $\xi(t)$ is the x-coordinate of the fingertip. Substituting (3.1.22) into (3.1.21),

$$\ddot{\phi} = \frac{g}{L}\phi - u(t) \qquad (3.1.23)$$

where

$$u(t) = \frac{\ddot{\xi}}{L} \qquad (3.1.24)$$

The variable $u(t)$, which is proportional to the fingertip acceleration $\ddot{\xi}$, will be treated as the control input. The state variable equations are described as

$$\begin{bmatrix} \dot{q}_1 \\ \dot{q}_2 \end{bmatrix} = \begin{bmatrix} 0 & 1 \\ \dfrac{g}{L} & 0 \end{bmatrix} \begin{bmatrix} q_1 \\ q_2 \end{bmatrix} + \begin{bmatrix} 0 \\ -1 \end{bmatrix} u \qquad (3.1.25)$$

where

$$q_1 = \phi \quad \text{and} \quad q_2 = \dot{\phi} \qquad (3.1.26)$$

Furthermore,

$$A = \begin{bmatrix} 0 & 1 \\ \dfrac{g}{L} & 0 \end{bmatrix} \quad \text{and} \quad \mathbf{b} = \begin{bmatrix} 0 \\ -1 \end{bmatrix} \tag{3.1.27}$$

Open-Loop Stability

$$\det(sI - A) = s^2 - \frac{g}{L} \tag{3.1.28}$$

Open-loop eigenvalues are $\sqrt{\dfrac{g}{L}}$ and $-\sqrt{\dfrac{g}{L}}$. Hence, the open-loop system is unstable which is exhibited by the fact that the pointer falls down if the fingertip does not have any acceleration, i.e., $u(t) = 0$. Furthermore, if L is smaller, the magnitude of the unstable eigenvalue $\sqrt{\dfrac{g}{L}}$ is larger. This matches with the fact that it is harder to balance a small pointer. Try it.

Controllability

The controllability matrix is

$$C = \begin{bmatrix} \mathbf{b} & A\mathbf{b} \end{bmatrix} = \begin{bmatrix} 0 & -1 \\ -1 & 0 \end{bmatrix} \tag{3.1.29}$$

The matrix C is nonsingular. Hence, the state space realization is controllable.

State Feedback Control

$$u(t) = -k_1 q_1 - k_2 q_2 = -\begin{bmatrix} k_1 & k_2 \end{bmatrix}\begin{bmatrix} q_1 \\ q_2 \end{bmatrix} \tag{3.1.30}$$

$$A - b\mathbf{k} = \begin{bmatrix} 0 & 1 \\ \dfrac{g}{L} & 0 \end{bmatrix} - \begin{bmatrix} 0 \\ -1 \end{bmatrix}\begin{bmatrix} k_1 & k_2 \end{bmatrix} = \begin{bmatrix} 0 & 1 \\ \dfrac{g}{L} + k_1 & k_2 \end{bmatrix} \tag{3.1.31}$$

$$\det(sI - A + b\mathbf{k}) = \det\begin{bmatrix} s & -1 \\ -\left(\dfrac{g}{L} + k_1\right) & s - k_2 \end{bmatrix} = s^2 - k_2 s - \left(\dfrac{g}{L} + k_1\right) \tag{3.1.32}$$

Let the desired closed-loop poles be -1 and -2. In this case, the desired closed-loop characteristic equation will be

$$\det(sI - A + b\mathbf{k}) = (s + 1)(s + 2) = s^2 + 3s + 2 \tag{3.1.33}$$

Matching the coefficients of polynomials,

$$-k_2 = 3 \Rightarrow k_2 = -3 \tag{3.1.34}$$

$$-\left(\frac{g}{L} + k_1\right) = 2 \Rightarrow k_1 = -\frac{g}{L} - 2 \tag{3.1.35}$$

Comments

1. To determine state feedback vector, the Bass–Gura formula (3.1.18) can also be used. But for a low-order problem, it is more straightforward to directly match the coefficients of polynomials. The Bass–Gura formula is extremely useful to solve higher order problems via a computer software such as MATLAB®.
2. The control input is $u(t) = -k_1\phi - k_2\dot{\phi}$. Its implementation requires measurements of ϕ and $\dot{\phi}$, and real-time computation of $-k_1\phi - k_2\dot{\phi}$. When a human being tries to balance a pointer, one can say that the eyes are estimating the values of ϕ and $\dot{\phi}$, and the brain is deciding on a suitable control input and instructing the motor to move the fingertip with proper acceleration.

3.1.2 Effects of State Feedback on Zeros of the Closed-Loop Transfer Function

If the state space realization is controllable, it can be converted to the controller canonical form $\{A_c, \mathbf{b}_c, \mathbf{c}_c\}$ described in Chapter 2, Section 2.2. Under the state feedback, the system matrix of the closed-loop system is given as

$$A_c - \mathbf{b}_c\mathbf{k} = \begin{bmatrix} 0 & 1 & 0 & . & . & 0 \\ 0 & 0 & 1 & 0 & . & 0 \\ & & & & & \\ 0 & 0 & 0 & . & . & 1 \\ -a_n - k_1 & -a_{n-1} - k_2 & . & . & . & -a_1 - k_n \end{bmatrix} \tag{3.1.36}$$

Hence, the realization $\{A_c - \mathbf{b}_c\mathbf{k}, \mathbf{b}_c, \mathbf{c}_c\}$ remains in the controller canonical form. Therefore, the transfer function of the closed-loop system is given as

$$\mathbf{c}_c(sI - A_c + \mathbf{b}_c\mathbf{k})^{-1}\mathbf{b}_c = \frac{b_1 s^{n-1} + \ldots + b_{n-1}s + b_n}{s^n + (a_1 + k_n)s^{n-1} + \ldots + (a_n + k_1)} \tag{3.1.37}$$

This expression clearly indicates that the numerator polynomial of the transfer function remains unchanged. In other words, the state feedback has no influence on the zeros of the transfer function (Kailath, 1980).

EXAMPLE 3.2: BICYCLE DYNAMICS

Consider a simple model of a bicycle (Lowell and McKell, 1982; Astrom et al., 2005), in which the rider, wheels, front-fork assembly, and the rear frame are treated as a single rigid body as shown by the plane in Figure 3.3. The total mass is m and the location of the center of gravity (cg) is shown in the figure. Assume that the forward velocity v of the bicycle is a constant. The small angle θ is the perturbation from the bicycle's upright position. As the weight mg will further try to increase this angle, the rider turns the handlebar by a small angle α, so that the bicycle begins to travel in a circle with the instant radius r and the instant center of rotation O. This circular travel is represented by the angle ϕ about the vertical axis passing through the rear-wheel contact point. Therefore, the acceleration of the cg in the direction normal to the plane is

$$h\ddot{\theta} + b\ddot{\phi} + \frac{v^2}{r} \tag{3.1.38}$$

where v^2/r is the centripetal acceleration, and b is the distance of cg from the vertical axis passing through the rear-wheel contact point. Applying Newton's second law,

$$m\left(h\ddot{\theta} + b\ddot{\phi} + \frac{v^2}{r}\right) = mg\theta \tag{3.1.39}$$

where $mg\theta$ is the component of the weight along the direction normal to the frame for small θ.

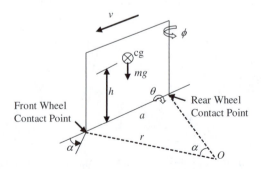

FIGURE 3.3 Fundamentals of bicycle dynamics.

From the definition of the instantaneous center for rotation:

$$r\dot{\phi} = v \quad \text{and} \quad r\alpha = a \qquad (3.1.40a,b)$$

Substituting (3.1.40) into (3.1.39), the input/output equation is obtained:

$$\ddot{\theta} - \frac{g}{h}\theta = -\frac{bv}{ha}\dot{\alpha} - \frac{v^2}{ah}\alpha \qquad (3.1.41)$$

where angles $\theta(t)$ and $\alpha(t)$ are the output and the input, respectively. Taking the Laplace transform of (3.1.41) with zero initial conditions:

$$\frac{\theta(s)}{\alpha(s)} = \frac{b_1 s + b_2}{s^2 + a_1 s + a_2} \qquad (3.1.42)$$

where

$$b_1 = -\frac{bv}{ha} \; ; \quad b_2 = -\frac{v^2}{ha} \; ; \quad a_1 = 0 \; ; \quad a_2 = -\frac{g}{h} \qquad (3.1.43)$$

Now, a state space model can be constructed via either Method I or Method II in Chapter 2, Section 2.2.

There are two poles of the system:

$$-\sqrt{\frac{g}{h}} \quad \text{and} \quad -\sqrt{\frac{g}{h}} \qquad (3.1.44)$$

And there is one zero:

$$-\frac{v}{b} \qquad (3.1.45)$$

The open-loop system is unstable, and has a zero in the left half plane. The system can be stabilized by a simple feedback law:

$$\alpha = k_p \theta \qquad (3.1.46)$$

Substituting (3.1.46) into (3.1.39), the closed-loop system is

$$\ddot{\theta} + \frac{bv}{ha} k_p \dot{\theta} + \left(\frac{v^2}{ah} k_p - \frac{g}{h} \right) \theta = 0 \qquad (3.1.47)$$

For stability,

$$k_p > \frac{ag}{v^2} \qquad (3.1.48)$$

Note that the control law (3.1.46) has to be implemented by the rider using his or her eye, brain, or hand. Equation 3.1.48 indicates that the lower bound of k_p is smaller at a higher velocity. In fact, the damping is also higher at a higher velocity, Equation 3.1.47. Therefore, the rider finds it easier to stabilize the bicycle at a higher velocity.

State Feedback Control

Using Method I (Chapter 2, Section 2.2):

$$\dot{x} = Ax(t) + bu(t) \qquad (3.1.49)$$

$$y(t) = cx(t) \qquad (3.1.50)$$

where

$$A = \begin{bmatrix} 0 & 1 \\ g/h & 0 \end{bmatrix}; \quad b = \begin{bmatrix} 0 \\ 1 \end{bmatrix} \quad \text{and} \quad c = \begin{bmatrix} b_1 & b_2 \end{bmatrix} \qquad (3.1.51)$$

and

$$\det(sI - A + bk) = \det \begin{bmatrix} s & -1 \\ -g/h + k_1 & s + k_2 \end{bmatrix} = s^2 + k_2 s + (k_1 - g/h) \qquad (3.1.52)$$

Let the desired characteristic polynomial be

$$\det(sI - A + bk) = s^2 + \alpha_1 s + \alpha_2 \qquad (3.1.53)$$

Matching coefficients of polynomials in (3.1.52) and (3.1.53),

$$k_1 = \alpha_2 + g/h \qquad (3.1.54)$$

and

$$k_2 = \alpha_1 \qquad (3.1.55)$$

And, the closed-loop transfer function is

$$\mathbf{c}(sI - A + \mathbf{bk})^{-1}\mathbf{b} = \frac{b_1 s + b_2}{s^2 + \alpha_1 s + \alpha_2} \qquad (3.1.56)$$

Note that the zero of the transfer function remains unchanged, whereas the poles have been located arbitrarily via state feedback control:

$$u(t) = u_r(t) - \mathbf{kx}(t) \qquad (3.1.57)$$

where $u_r(t)$ is the reference input, which will be zero in this case. Compared to the state feedback law (3.1.57), the proportional controller (3.1.46) is simpler and hence easier to be implemented by a rider. This may explain why a person can ride a bicycle for hours, whereas he or she can balance a pointer on a finger only for a few minutes as full state feedback (Equation 3.1.30) is necessary for stability.

3.1.3 STATE FEEDBACK CONTROL FOR A NONZERO AND CONSTANT OUTPUT

Let the external input $v(t)$ be a constant v_0. From Equation 3.1.5, the steady state value of the state vector $\mathbf{x}_s$ is given by

$$0 = (A - \mathbf{bk})\mathbf{x}_s + \mathbf{b}v_0 \qquad (3.1.58)$$

or

$$\mathbf{x}_s = -(A - \mathbf{bk})^{-1}\mathbf{b}v_0 \qquad (3.1.59)$$

Hence, the steady state output is given as

$$y_s = \mathbf{cx}_s = -\mathbf{c}(A - \mathbf{bk})^{-1}\mathbf{b}v_0 \qquad (3.1.60)$$

If the desired value of the output is y_d, the corresponding command input v_0 is obtained by setting $y_s = y_d$ and solving (3.1.60). The result is

$$v_0 = -\frac{y_d}{\mathbf{c}(A - \mathbf{bk})^{-1}\mathbf{b}} \quad \text{provided} \quad \mathbf{c}(A - \mathbf{bk})^{-1}\mathbf{b} \neq 0 \qquad (3.1.61)$$

Now, let us examine the conditions under which $c(A - bk)^{-1}b \neq 0$. Substituting $s = 0$ in Equation 3.1.6, it is clear that $c(A - bk)^{-1}b \neq 0$ if there is no zero of the closed-loop transfer function at $s = 0$. As locations of zeros remain unchanged under state feedback, the command input for the nonzero set point can be found, provided the open-loop transfer function does not have any zero located at $s = 0$ (Kailath, 1980).

EXAMPLE 3.3: SPRING-MASS SYSTEM

For the spring-mass-damper system (Chapter 2, Figure 2.4), state equations are defined by (2.5.16) and (2.5.17).

Case I: Position Output

If the position of the mass, x, is measured by a sensor,

$$y(t) = cx(t) \tag{3.1.62}$$

where

$$c = [1 \quad 0] \tag{3.1.63}$$

Let the state feedback control law be

$$u(t) = v_0 - kx(t) \tag{3.1.64}$$

It can be shown that

$$c(A - bk)^{-1}b = -\left(k_1 + \frac{\beta}{m}\right)^{-1} \tag{3.1.65}$$

From Equation 3.1.61, to achieve the desired set point y_d for the output,

$$v_0 = -\frac{y_d}{c(A - bk)^{-1}b} = y_d\left(k_1 + \frac{\beta}{m}\right) \tag{3.1.66}$$

If f_0 is the force corresponding to v_0,

$$v_0 = \frac{f_0}{m} \tag{3.1.67}$$

From (3.1.66) and (3.1.67),

$$f_0 = y_d(mk_1 + \beta) \tag{3.1.68}$$

From (3.1.64),

$$f(t) = f_0 - mk_1 x - mk_2 \dot{x} \tag{3.1.69}$$

Substituting (3.1.69) into the differential Equation 2.5.14,

$$m\ddot{x} + (\alpha + mk_2)\dot{x} + (\beta + mk_1)x = f_0 \tag{3.1.70}$$

Therefore, position and velocity feedback coefficients k_1 and k_2 add to the stiffness and damping coefficient, respectively. Furthermore, conditions (3.1.66) and (3.1.68) represent the static equilibrium condition.

Case II: Velocity Output

If the velocity of the mass, $\dot{x}$, is measured by a sensor,

$$y(t) = \mathbf{c}\mathbf{x}(t) \tag{3.1.71}$$

where

$$\mathbf{c} = [0 \quad 1] \tag{3.1.72}$$

In this case,

$$\mathbf{c}(A - \mathbf{bk})^{-1}\mathbf{b} = 0 \tag{3.1.73}$$

And a nonzero set point y_d cannot be achieved. Mathematically, this is due to the presence of a zero at $s = 0$ as the transfer function is

$$\frac{y(s)}{u(s)} = \frac{ms}{ms^2 + \alpha s + \beta} \tag{3.1.74}$$

Physically, nonzero constant velocity is not possible when the applied force is a constant.

3.1.4 State Feedback Control under Constant Input Disturbances: Integral Action

Consider the following system:

$$\frac{dx}{dt} = Ax + bu(t) + w \tag{3.1.75}$$

$$y(t) = cx(t) \tag{3.1.76}$$

where $\mathbf{w}$ is a constant disturbance vector of unknown magnitudes. The objective is to develop a state feedback control algorithm such that $\lim_{t \to \infty} y(t) = 0$. The reader can verify that the algorithm (3.1.4) will not be able to achieve this goal. The required algorithm is developed by defining a new variable $q(t)$ as follows:

$$\frac{dq}{dt} = y(t) = cx(t); \qquad q(0) = 0 \tag{3.1.77}$$

Defining a new state vector,

$$\mathbf{p}(t) = \begin{bmatrix} \mathbf{x}(t) \\ q(t) \end{bmatrix} \tag{3.1.78}$$

Using (3.1.75) to (3.1.77),

$$\frac{d\mathbf{p}}{dt} = A_a \mathbf{p}(t) + \mathbf{b}_a u(t) + \begin{bmatrix} \mathbf{w} \\ 0 \end{bmatrix} \tag{3.1.79}$$

where

$$A_a = \begin{bmatrix} A & 0 \\ \mathbf{c} & 0 \end{bmatrix}; \quad \mathbf{b}_a = \begin{bmatrix} \mathbf{b} \\ 0 \end{bmatrix} \tag{3.1.80}$$

Lemma

The augmented system $\{A_a, \mathbf{b}_a\}$ is controllable provided that the original system $\{A, \mathbf{b}\}$ is controllable and

$$rank \begin{bmatrix} A & \mathbf{b} \\ \mathbf{c} & 0 \end{bmatrix} = n + 1 \tag{3.1.81}$$

Proof

The controllability matrix (Gopal, 1984) for the system $\{A_a, \mathbf{b}_a\}$ is

$$C_a = \begin{bmatrix} \mathbf{b} & A\mathbf{b} & A^2\mathbf{b} & . & . & A^n\mathbf{b} \\ 0 & \mathbf{cb} & \mathbf{c}A\mathbf{b} & . & . & \mathbf{c}A^{n-1}\mathbf{b} \end{bmatrix} \tag{3.1.82}$$

$$= \begin{bmatrix} A & \mathbf{b} \\ \mathbf{c} & 0 \end{bmatrix} \begin{bmatrix} 0 & C \\ 1 & 0 \end{bmatrix} \tag{3.1.83}$$

The matrix

$$\begin{bmatrix} 0 & C \\ 1 & 0 \end{bmatrix} \tag{3.1.84}$$

is nonsingular if the original system $\{A, \mathbf{b}\}$ is controllable; i.e., the matrix C is nonsingular. Therefore, C_a is nonsingular if the condition (3.1.81) is satisfied. Now, the state feedback control for the augmented system is

$$u(t) = -\mathbf{kx}(t) - k_q q(t) \tag{3.1.85}$$

Substituting (3.1.85) into (3.1.79),

$$\begin{bmatrix} \dfrac{d\mathbf{x}}{dt} \\[2mm] \dfrac{dq}{dt} \end{bmatrix} = \begin{bmatrix} A - \mathbf{bk} & -\mathbf{b}k_q \\ \mathbf{c} & 0 \end{bmatrix} \begin{bmatrix} \mathbf{x}(t) \\ q(t) \end{bmatrix} + \begin{bmatrix} \mathbf{w} \\ 0 \end{bmatrix} \tag{3.1.86}$$

Assuming that C_a is nonsingular, it is always possible to find $\mathbf{k}$ and k_q such that eigenvalues of system (3.1.86) are in the left half of the s-plane. Because $\mathbf{w}$ is a constant vector, steady state values of $\mathbf{x}(t)$ and q(t) will be constants for a stable closed-loop system. Hence, in steady state, $\dfrac{dq}{dt} = 0$. From Equation 3.1.77, $\lim_{t\to\infty} y(t) = 0$. Finally, the control law (3.1.85) can be represented as

$$u(t) = -\mathbf{kx}(t) - k_q \int_0^t y(t)dt \tag{3.1.87}$$

Hence, by feeding back the integral of the output in addition to states, we can have the output go to zero in the presence of a constant disturbance vector with unknown magnitudes (Kailath, 1980).

EXAMPLE 3.4: INTEGRAL OUTPUT FEEDBACK

For the spring-mass-damper system (Chapter 2, Figure 2.4), state equations are defined by (2.5.16) and (2.5.17). In addition, consider a constant but unknown disturbance force acting on the mass.

Case I: Position Output

If the position of the mass, x, is measured by a sensor,

$$y(t) = \mathbf{c}\mathbf{x}(t) \tag{3.1.88}$$

where

$$\mathbf{c} = [1 \quad 0] \tag{3.1.89}$$

In this case,

$$rank \begin{bmatrix} A_a & \mathbf{b} \\ \mathbf{c} & 0 \end{bmatrix} = rank \begin{bmatrix} 0 & 1 & 0 \\ -\alpha/m & -\beta/m & 1 \\ 1 & 0 & 0 \end{bmatrix} = 3 \tag{3.1.90}$$

Therefore, the condition (3.1.81) is satisfied and the augmented system is controllable. For the control law (3.1.52), define

$$\mathbf{k}_a = [\mathbf{k} \quad k_q] = [k_1 \quad k_2 \quad k_q] \tag{3.1.91}$$

Then, the closed-loop characteristic polynomial is

$$\det(sI - (A_a - \mathbf{b}_a\mathbf{k}_a)) = s^3 + \left(\frac{\alpha}{m} + k_2\right)s^2 + \left(\frac{\beta}{m} + k_1\right)s + k_q \tag{3.1.92}$$

Let the desired characteristic polynomial be

$$\det(sI - (A_a - \mathbf{b}_a\mathbf{k}_a)) = s^3 + \alpha_1 s^2 + \alpha_2 s + \alpha_3 \tag{3.1.93}$$

Matching the coefficients of polynomials,

$$k_1 = \alpha_2 - \frac{\beta}{m}; \quad k_2 = \alpha_1 - \frac{\alpha}{m}; \quad \text{and} \quad k_q = \alpha_3 \tag{3.1.94}$$

Case II: Velocity Output

If the velocity of the mass, $\dot{x}$, is measured by a sensor,

$$y(t) = \mathbf{c}\mathbf{x}(t) \tag{3.1.95}$$

where

$$\mathbf{c} = [0 \quad 1] \tag{3.1.96}$$

In this case,

$$rank \begin{bmatrix} A_a & \mathbf{b} \\ \mathbf{c} & 0 \end{bmatrix} = rank \begin{bmatrix} 0 & 1 & 0 \\ -\alpha/m & -\beta/m & 1 \\ 0 & 1 & 0 \end{bmatrix} = 2 \tag{3.1.97}$$

Therefore, the condition (3.1.48) is not satisfied, and the augmented system is not controllable.

3.2 COMPUTATION OF STATE FEEDBACK GAIN MATRIX FOR A MULTIINPUT SYSTEM

The state feedback control law is

$$\mathbf{u}(s) = \mathbf{r}(s) - K\mathbf{x}(s) \tag{3.2.1}$$

where $\mathbf{r}(s)$ and K are reference input vector and state feedback gain matrix, respectively. Using (2.13.8) and (2.13.18),

$$\mathbf{u}(s) = \mathbf{r}(s) - K\Psi(s)\xi(s) \tag{3.2.2}$$

Substituting (3.2.2) into (2.13.6),

$$\left[D_{hc}S(s) + (D_{lc} + K)\Psi(s) \right]\xi(s) = \mathbf{r}(s) \tag{3.2.3}$$

Equation 3.2.3 yields

$$\xi(s) = D_k^{-1}(s)\mathbf{r}(s) \tag{3.2.4}$$

where

$$D_k(s) = D_{hc}S(s) + (D_{lc} + K)\Psi(s) = D(s) + K\Psi(s) \tag{3.2.5}$$

Substituting (3.2.4) into (2.13.5),

$$\mathbf{y}(s) = G_K(s)\mathbf{r}(s) \tag{3.2.6}$$

where

$$G_K(s) = N(s)D_K^{-1}(s) \tag{3.2.7}$$

$G_K(s)$ is the closed-loop transfer function with full state feedback. Examining the structures of (3.2.5) and (3.2.7), the following points (Kailath, 1980) should be noted:

- State feedback does not alter the numerator polynomial $N(s)$.
- State feedback does not alter D_{hc}. Therefore, the state feedback control cannot change the column degrees of $D(s)$.

Let the characteristic polynomial of the closed-loop system be described by the following monic polynomial:

$$\alpha(s) = s^n + \alpha_1 s^{n-1} + \ldots + \alpha_{n-1}s + \alpha_n \tag{3.2.8}$$

Assume that the column degrees (Chapter 2, Section 2.8) of $D(s)$ are arranged as

$$k_1 \leq k_2 \leq \ldots \leq k_m; \quad \sum_{i=1}^{m} k_i = n \tag{3.2.9}$$

Rewrite the characteristic polynomial (3.2.8) as follows:

$$\alpha(s) = s^n + \alpha_1(s)s^{n-k_1} + \alpha_2(s)s^{n-k_1-k_2} + \ldots + \alpha_m(s) \tag{3.2.10}$$

where $\alpha_i(s)$ is a polynomial of degree less than k_i. Then, it can be verified that

$$\det \begin{bmatrix} s^{k_1} + \alpha_1(s) & \alpha_2(s) & . & 0 & \alpha_m(s) \\ -1 & s^{k_2} & . & 0 & 0 \\ . & . & . & . & . \\ 0 & 0 & . & s^{k_{m-1}} & 0 \\ 0 & 0 & . & -1 & s^{k_m} \end{bmatrix} = \alpha(s) \tag{3.2.11}$$

From (3.2.7), the closed-loop characteristic polynomial is given by $\det(D_K(s))$. Because $\alpha(s)$ is a monic polynomial,

$$\det(sI - A_c + B_c K) = \det(D_{hc}^{-1} D_K) = \alpha(s) \tag{3.2.12}$$

From (3.2.5),

$$\det(S(s) + D_{hc}^{-1}(D_{lc} + K)\Psi(s)) = \alpha(s) \tag{3.2.13}$$

Comparing (3.2.11) and (3.2.13),

$$\begin{bmatrix} s^{k_1} + \alpha_1(s) & \alpha_2(s) & . & 0 & \alpha_m(s) \\ -1 & s^{k_2} & . & 0 & 0 \\ . & . & . & . & . \\ 0 & 0 & . & s^{k_{m-1}} & 0 \\ 0 & 0 & . & -1 & s^{k_m} \end{bmatrix} = S(s) + D_{hc}^{-1}(D_{lc} + K)\Psi(s) \tag{3.2.14}$$

Using the definition of $S(s)$, Equation 3.2.14 yields

$$K\Psi(s) = D_{hc} \begin{bmatrix} \alpha_1(s) & \alpha_2(s) & . & 0 & \alpha_m(s) \\ -1 & 0 & . & 0 & 0 \\ . & . & . & . & . \\ 0 & 0 & . & 0 & 0 \\ 0 & 0 & . & -1 & 0 \end{bmatrix} - D_{lc}\Psi(s) \tag{3.2.15}$$

Factoring out $\Psi(s)$ on the right-hand side yields the state feedback gain matrix K. Equation 3.2.11 is not a unique way to express the desired characteristic polynomial $\alpha(s)$. As shown in Example 3.5, there are other matrices with their determinants equal to $\alpha(s)$. Therefore, the state feedback gain matrix K is not unique to obtain the desired closed-loop characteristic polynomial $\alpha(s)$ for a multiinput system (Kailath, 1980).

EXAMPLE 3.5

For $\bar{D}_H(s)$ in Equation 2.13.32, $k_1 = 2$ and $k_2 = 1$. And,

$$D_{hc} = \begin{bmatrix} 2/3 & -2/3 \\ 0 & 1 \end{bmatrix} \quad \text{and} \quad D_{lc} = \begin{bmatrix} 16/3 & 10 & -2 \\ 0 & 0 & 4 \end{bmatrix} \tag{3.2.16}$$

Let the desired characteristic polynomial be

$$\alpha(s) = s^3 + 30s^2 + 300s + 1000 \tag{3.2.17}$$

Because $k_1 = 2$ and $k_2 = 3$, this polynomial can be rearranged to be

$$\alpha(s) = s^3 + \alpha_1(s)s + \alpha_2(s) \tag{3.2.18}$$

where

$$\alpha_1(s) = 30s + 300 \quad \text{and} \quad \alpha_2(s) = 1000 \tag{3.2.19}$$

From (3.2.15),

$$(K_c + D_{lc})\Psi(s) = D_{hc}\begin{bmatrix} \alpha_1(s) & \alpha_2(s) \\ -1 & 0 \end{bmatrix} \tag{3.2.20}$$

where

$$\Psi(s) = \begin{bmatrix} s & 0 \\ 1 & 0 \\ 0 & 1 \end{bmatrix} \tag{3.2.21}$$

From (3.2.20),

$$(K_c + D_{lc})\Psi(s) = D_{hc}\begin{bmatrix} 30 & 300 & 1000 \\ 0 & -1 & 0 \end{bmatrix}\Psi(s) \tag{3.2.22}$$

Equating coefficients of $\Psi(s)$ on both sides,

$$K_c + D_{lc} = D_{hc}\begin{bmatrix} 30 & 300 & 1000 \\ 0 & -1 & 0 \end{bmatrix} \tag{3.2.23}$$

Therefore,

$$K_c = \begin{bmatrix} 44/3 & 572/3 & 2006/3 \\ 0 & -1 & -4 \end{bmatrix} \tag{3.2.24}$$

Alternatively,

$$(K_c + D_{lc})\Psi(s) = D_{hc}\begin{bmatrix} \alpha_1(s) & \alpha_2(s)/10 \\ -10 & 0 \end{bmatrix}\quad(3.2.25)$$

Using (3.2.19),

$$(K_c + D_{lc})\Psi(s) = D_{hc}\begin{bmatrix} 30 & 300 & 100 \\ 0 & -10 & 0 \end{bmatrix}\Psi(s)\quad(3.2.26)$$

Equating coefficients of $\Psi(s)$ on both sides,

$$K_c + D_{lc} = D_{hc}\begin{bmatrix} 30 & 300 & 100 \\ 0 & -10 & 0 \end{bmatrix}\quad(3.2.27)$$

Therefore,

$$K_c = \begin{bmatrix} 44/3 & 50/3 & 206/3 \\ 0 & -10 & -4 \end{bmatrix}\quad(3.2.28)$$

There exists another K_c matrix to achieve the desired characteristic polynomial (3.2.18).

3.3 STATE FEEDBACK GAIN MATRIX FOR A MULTIINPUT SYSTEM FOR DESIRED EIGENVALUES AND EIGENVECTORS

The closed-loop system is described by

$$G_K(s) = N(s)D_K^{-1}(s)\quad(3.3.1)$$

where

$$D_k(s) = D_{hc}S(s) + (D_{lc} + K)\Psi(s) = D(s) + K\Psi(s)\quad(3.3.2)$$

and

$$D(s) = D_{hc}S(s) + D_{lc}\Psi(s)\quad(3.3.3)$$

Let the desired eigenvalues of the closed-loop system be μ_i; $i = 1, 2, ..., n$. In this case,

$$\det D_k(\mu_i) = 0; \; i = 1, 2, ..., n \tag{3.3.4}$$

It can be shown that eigenvectors $\mathbf{f}_i$ associated with the system matrix of the controller form realization (Chapter 2, Section 2.13) of the transfer matrix $G_K(s)$ can be expressed as

$$\mathbf{f}_i = \Psi(\mu_i)\mathbf{p}_i \tag{3.3.5}$$

where the vector $\mathbf{p}_i$ satisfies

$$D_K(\mu_i)\mathbf{p}_i = 0 \tag{3.3.6}$$

Using (3.3.2) and (3.3.6),

$$D(\mu_i)\mathbf{p}_i + K\Psi(\mu_i)\mathbf{p}_i = 0 \tag{3.3.7}$$

Substituting (3.3.5) into (3.3.7),

$$K\mathbf{f}_i = -\mathbf{g}_i; \quad i = 1, 2, ..., n \tag{3.3.8}$$

where

$$\mathbf{g}_i = D(\mu_i)\mathbf{p}_i \tag{3.3.9}$$

Equation 3.3.8 can be put in the following form (Kailath, 1980):

$$K\begin{bmatrix} \mathbf{f}_1 & \mathbf{f}_2 & . & . & \mathbf{f}_n \end{bmatrix} = -\begin{bmatrix} \mathbf{g}_1 & \mathbf{g}_2 & . & . & \mathbf{g}_n \end{bmatrix} \tag{3.3.10}$$

Assuming that eigenvectors $\{\mathbf{f}_i\}$ are linearly independent,

$$K = -\begin{bmatrix} \mathbf{g}_1 & \mathbf{g}_2 & . & . & \mathbf{g}_n \end{bmatrix}\begin{bmatrix} \mathbf{f}_1 & \mathbf{f}_2 & . & . & \mathbf{f}_n \end{bmatrix}^{-1} \tag{3.3.11}$$

Linear independence of eigenvectors $\mathbf{f}_i$ is guaranteed when eigenvalues μ_i are distinct. Although it is possible to get a unique solution for K, even when some of the eigenvalues are repeated, it will be assumed that eigenvalues μ_i are distinct for the following discussion.

To use Equation 3.3.11, $\mathbf{g}_i$ and $\mathbf{f}_i$ are obtained (Kailath, 1980) as follows:

1. Choose unrepeated closed-loop eigenvalues μ_i.
2. Choose the desired closed-loop eigenvectors $\mathbf{f}_i$ to satisfy the following constraints:
   a. With reference to Equation 3.3.5, $\mathbf{f}_i$ must belong to the range space of $\Psi(\mu_i)$.
   b. All the eigenvectors $\mathbf{f}_i$ are linearly independent.
   c. If $\mathbf{f}_i$ corresponds to a complex eigenvalue μ_i, the complex conjugate of $\mathbf{f}_i$ must correspond to the eigenvalue which is the complex conjugate of μ_i.
3. Use Equation 3.3.5 to solve for $\mathbf{p}_i$. Premultiply both sides of Equation 3.3.5 by $\Psi^T(\mu_i)$:

$$\Psi^T(\mu_i)\Psi(\mu_i)\mathbf{p}_i = \Psi^T(\mu_i)\mathbf{f}_i \tag{3.3.12}$$

Because the matrix $\Psi(\mu_i)$ is of full rank, the square matrix $\Psi^T(\mu_i)\Psi(\mu_i)$ is nonsingular. Therefore,

$$\mathbf{p}_i = (\Psi^T(\mu_i)\Psi(\mu_i))^{-1}\Psi^T(\mu_i)\mathbf{f}_i \tag{3.3.13}$$

4. From Equation 3.3.9, $\mathbf{g}_i$ is computed.

EXAMPLE 3.6: ELECTRONICS NAVIGATION OR GYRO BOX (SCHULTZ AND INMAN, 1994)

An electronics navigation or gyro box is mounted on passive spring-damper isolators located at the bottom four corners (Figure 3.4). There are isolators along x and z directions at each bottom four corner. There are also three actuators providing active control forces u_1, u_2, and u_3 as shown in Figure 3.4. Free body diagram of the system is shown in Figure 3.5.

Applying Newton's law, the system of differential equations of motion is written as

$$M\ddot{\mathbf{q}} + E\dot{\mathbf{q}} + K_s\mathbf{q}(t) = \mathbf{v}(t) \tag{3.3.14}$$

$$\mathbf{v}(t) = B_f\mathbf{u}(t) \tag{3.3.15}$$

where the mass matrix M, the damping matrix E, the stiffness matrix K_s, and the input force matrix B_f are expressed as:

$$M = \begin{bmatrix} m & 0 & 0 \\ 0 & m & 0 \\ 0 & 0 & I_{yy} \end{bmatrix} \quad E = \begin{bmatrix} 4c_z & 0 & 0 \\ 0 & 4c_x & -4\ell_z c_x \\ 0 & -4\ell_z c_x & 4(c_z\ell_x^2 + c_x\ell_z^2) \end{bmatrix} \tag{3.3.16a,b}$$

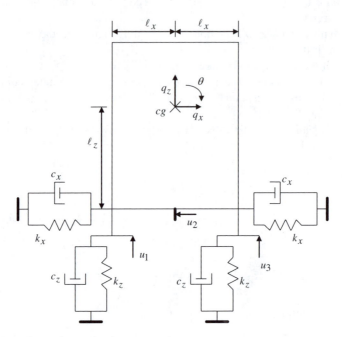

FIGURE 3.4 Electronics navigation or gyro box.

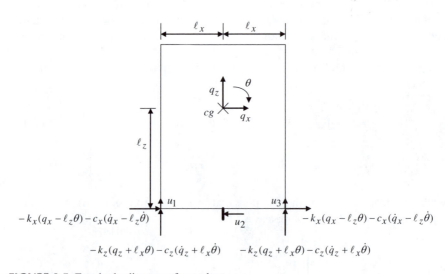

FIGURE 3.5 Free body diagram of gyro box.

$$K_s = \begin{bmatrix} 4k_z & 0 & 0 \\ 0 & 4k_x & -4\ell_z k_x \\ 0 & -4\ell_z k_x & 4(k_z\ell_x^2 + k_x\ell_z^2) \end{bmatrix} \quad B_f = \begin{bmatrix} 1 & 0 & 1 \\ 0 & -1 & 0 \\ \ell_x & \ell_z & -\ell_x \end{bmatrix} \quad (3.3.17a,b)$$

For a mechanical system, it is usual to define states as

$$\mathbf{p}_1 = \mathbf{q} \quad \text{and} \quad \mathbf{p}_2 = \dot{\mathbf{q}} \qquad (3.3.18a,b)$$

Then,

$$\dot{\mathbf{p}}_1 = \mathbf{p}_2 \qquad (3.3.19)$$

and from (3.3.14),

$$\dot{\mathbf{p}}_2 = -M^{-1}K_s\mathbf{p}_1 - M^{-1}E\mathbf{p}_2 + M^{-1}\mathbf{v}(t) \qquad (3.3.20)$$

Putting (3.3.19) and (3.3.20) in the matrix form,

$$\dot{\mathbf{p}} = A_p\mathbf{p}(t) + B_p\mathbf{v}(t) \qquad (3.3.21)$$

where

$$\mathbf{p} = \begin{bmatrix} \mathbf{p}_1 \\ \mathbf{p}_2 \end{bmatrix}; \quad A_p = \begin{bmatrix} 0 & I \\ -M^{-1}K_s & -M^{-1}E \end{bmatrix}; \quad B_p = \begin{bmatrix} 0 \\ M^{-1} \end{bmatrix} \qquad (3.3.22a,b,c)$$

Expressing (3.3.14) in the matrix fraction description (MFD) form,

$$\mathbf{q}(s) = N(s)D^{-1}(s)\mathbf{v}(s) \qquad (3.3.23)$$

where

$$N(s) = I \quad \text{and} \quad D(s) = (Ms^2 + Es + K_s) \qquad (3.3.24a,b)$$

Here,

$$D(s) = D_{hc}\begin{bmatrix} s^2 & 0 & 0 \\ 0 & s^2 & 0 \\ 0 & 0 & s^2 \end{bmatrix} + D_{\ell c}\Psi(s) \qquad (3.3.25)$$

where

$$D_{hc} = M \tag{3.3.26}$$

$$D_{\ell c} =$$

$$\begin{bmatrix} 4c_z & 4k_z & 0 & 0 & 0 & 0 \\ 0 & 0 & 4c_x & 4k_x & -4\ell_z c_x & -4\ell_z k_x \\ 0 & 0 & -4\ell_z c_x & -4\ell_z k_x & 4(c_z\ell_x^2 + c_x\ell_z^2) & 4(k_z\ell_x^2 + k_x\ell_z^2) \end{bmatrix} \tag{3.3.27}$$

and

$$\Psi(s) = \begin{bmatrix} s & 0 & 0 \\ 1 & 0 & 0 \\ 0 & s & 0 \\ 0 & 1 & 0 \\ 0 & 0 & s \\ 0 & 0 & 1 \end{bmatrix} \tag{3.3.28}$$

Let the desired characteristic polynomial for the closed-loop system be

$$\alpha(s) = s^6 + \beta_1 s^5 + \beta_2 s^4 + \beta_3 s^3 + \beta_4 s^2 + \beta_5 s + \beta_6 \tag{3.3.29}$$

Here, $k_1 = k_2 = k_3 = 2$. Therefore, Equation 3.3.29 should be expressed as

$$\alpha(s) = s^6 + s^4 \alpha_1(s) + s^2 \alpha_2(s) + \alpha_3(s) \tag{3.3.30}$$

where

$$\alpha_1(s) = \beta_1 s + \beta_2, \quad \alpha_2(s) = \beta_3 s + \beta_4, \quad \text{and} \quad \alpha_3(s) = \beta_5 s + \beta_6 \tag{3.3.31}$$

It can be easily verified that

$$\alpha(s) = \det \begin{bmatrix} s^2 + \alpha_1(s) & \dfrac{\alpha_2(s)}{\gamma} & \dfrac{\alpha_3(s)}{\gamma\lambda} \\ -\gamma & s^2 & 0 \\ 0 & -\lambda & s^2 \end{bmatrix} \tag{3.3.32}$$

where γ and λ are any arbitrary real numbers. Factoring out $\Psi(s)$ on both sides of Equation 3.2.15,

$$K =$$

$$D_{hc} \begin{bmatrix} \beta_1 & \beta_2 & \beta_3/\gamma & \beta_4/\gamma & \beta_5/(\gamma\lambda) & \beta_6/(\gamma\lambda) \\ 0 & -\gamma & 0 & 0 & 0 & 0 \\ 0 & 0 & 0 & -\lambda & 0 & 0 \end{bmatrix} - D_{\ell c} \qquad (3.3.33)$$

This result clearly indicates the gain matrix K is not unique because γ and λ can be chosen arbitrarily. Furthermore, this gain matrix is associated with the controller form state space realization (Chapter 2, Section 2.13), which is developed below.

Following definitions (2.13.16) and (2.13.17),

$$A_c^0 = \begin{bmatrix} 0 & 0 & 0 & 0 & 0 & 0 \\ 1 & 0 & 0 & 0 & 0 & 0 \\ 0 & 0 & 0 & 0 & 0 & 0 \\ 0 & 0 & 1 & 0 & 0 & 0 \\ 0 & 0 & 0 & 0 & 0 & 0 \\ 0 & 0 & 0 & 0 & 1 & 0 \end{bmatrix}, \quad B_c^0 = \begin{bmatrix} 1 & 0 & 0 \\ 0 & 0 & 0 \\ 0 & 1 & 0 \\ 0 & 0 & 0 \\ 0 & 0 & 1 \\ 0 & 0 & 0 \end{bmatrix} \qquad (3.3.34)$$

Therefore,

$$A_c = A_c^0 - B_c^0 D_{hc}^{-1} D_{lc} =$$

$$\begin{bmatrix} \dfrac{4c_z}{m} & \dfrac{4k_z}{m} & 0 & 0 & 0 & 0 \\ 1 & 0 & 0 & 0 & 0 & 0 \\ 0 & 0 & \dfrac{4c_x}{m} & \dfrac{4k_x}{m} & \dfrac{-4\ell_z c_x}{m} & \dfrac{-4\ell_z k_x}{m} \\ 0 & 0 & 1 & 0 & 0 & 0 \\ 0 & 0 & \dfrac{-4\ell_z c_x}{I_{yy}} & \dfrac{-4\ell_z k_x}{I_{yy}} & \dfrac{4(c_z\ell_x^2 + c_x\ell_z^2)}{I_{yy}} & \dfrac{4(k_z\ell_x^2 + k_x\ell_z^2)}{I_{yy}} \\ 0 & 0 & 0 & 0 & 1 & 0 \end{bmatrix} \qquad (3.3.35)$$

$$B_c = B_c^0 D_{hc}^{-1} = \begin{bmatrix} 1/m & 0 & 0 \\ 0 & 0 & 0 \\ 0 & 1/m & 0 \\ 0 & 0 & 0 \\ 0 & 0 & 1/I_{yy} \\ 0 & 0 & 0 \end{bmatrix} \qquad (3.3.36)$$

The state equations are

$$\dot{\mathbf{x}} = A_c \mathbf{x}(t) + B_c \mathbf{u}(t) \qquad (3.3.37)$$

with the control law

$$\mathbf{u}(t) = -K\mathbf{x}(t) \qquad (3.3.38)$$

where K is given by Equation 3.3.33. The control law is not directly implementable because sensors provide the states $\mathbf{p}(t)$, not $\mathbf{x}(t)$. Therefore, to implement this controller, it is necessary to find the similarity transformation that will convert the state space realization (3.3.21) to the state space realization (3.3.37). This is where the similarity transformation (2.13.46) can be used for parameter values as follows.

$$\ell_z = 0.11 \text{ m}, \quad k_z = 17500 \text{ N/m}, \quad c_z = 0.002 k_z$$

$$\ell_x = 0.08 \text{ m}, \quad k_x = 8750 \text{ N/m}, \quad c_x = 0.002 k_x \qquad (3.3.39)$$

$$m = 10 \text{ kg}, \quad I_{yy} = 0.0487 \text{ kg-m-m}$$

Then

$$A_c = \begin{bmatrix} -14 & 700 & 0 & 0 & 0 & 0 \\ 1 & 0 & 0 & 0 & 0 & 0 \\ 0 & 0 & -7 & -350 & 0.77 & 385 \\ 0 & 0 & 1 & 0 & 0 & 0 \\ 0 & 0 & 158.11 & 79055.44 & -35.79 & -17895.277 \\ 0 & 0 & 0 & 0 & 1 & 0 \end{bmatrix}$$

Define

$$\mathbf{p}(t) = T\mathbf{z}(t) \tag{3.3.40}$$

where the similarity transformation matrix T is defined by Equation 2.13.46. Substituting (3.3.40) into (3.3.21),

$$\dot{\mathbf{z}} = A_{cc}\mathbf{z}(t) + B_{cc}\mathbf{v}(t) \tag{3.3.41}$$

where

$$A_{cc} = T^{-1}A_pT =$$

$$
\begin{bmatrix}
-14 & 700 & 0 & 0 & 0 & 0 \\
1 & 0 & 0 & 0 & 0 & 0 \\
0 & 0 & -7 & -350 & 158.11 & 79055.44 \\
0 & 0 & 1 & 0 & 0 & 0 \\
0 & 0 & 0.77 & 385 & -35.79 & -17895.277 \\
0 & 0 & 0 & 0 & 1 & 0
\end{bmatrix} \tag{3.3.42}
$$

and

$$B_{cc} = T^{-1}B_p = B_c^0 \tag{3.3.43}$$

It is seen that structures of A_{cc} and B_{cc} are similar to those of A_c and B_c, respectively. Using (2.13.21) and (2.3.22), new $D_{hc} = D_{hcc}$ and new $D_{\ell c} = D_{\ell cc}$ are obtained as follows:

$$D_{hcc} = I_6 \tag{3.3.44}$$

and

$$
D_{\ell cc} =
\begin{bmatrix}
14 & -700 & 0 & 0 & 0 & 0 \\
0 & 0 & 7 & 350 & -158.11 & -79055.44 \\
0 & 0 & -0.77 & -385 & 35.79 & 17895.277
\end{bmatrix} \tag{3.3.45}
$$

Now, formulae (3.2.15) and (3.3.11) should be used with these new $D_{hc} = D_{hcc}$ and new $D_{\ell c} = D_{\ell cc}$. A Matlab code 3.1 is attached to do the following:

1. Locate the eigenvalues of the closed-loop system, μ_i, such that the damping factor of each vibratory mode is five times the corresponding value of the open-loop system, and undamped natural frequencies remain unchanged.
2. Locate the eigenvalues of the closed-loop system as described in part (1), and eigenvectors as:

$$
\mathbf{f}_1 = \begin{bmatrix} \mu_1 \\ 1 \\ 0 \\ 0 \\ 0 \\ 0 \end{bmatrix} ; \quad
\mathbf{f}_2 = \begin{bmatrix} \mu_2 \\ 1 \\ 0 \\ 0 \\ 0 \\ 0 \end{bmatrix} ; \quad
\mathbf{f}_3 = \begin{bmatrix} 0 \\ 0 \\ \mu_3 \\ 1 \\ 0 \\ 0 \end{bmatrix} ;
$$

$$
\mathbf{f}_4 = \begin{bmatrix} 0 \\ 0 \\ \mu_4 \\ 1 \\ 0 \\ 0 \end{bmatrix} ; \quad
\mathbf{f}_5 = \begin{bmatrix} 0 \\ 0 \\ 0 \\ 0 \\ \mu_5 \\ 1 \end{bmatrix} ; \quad
\mathbf{f}_6 = \begin{bmatrix} 0 \\ 0 \\ 0 \\ 0 \\ \mu_6 \\ 1 \end{bmatrix} \qquad (3.3.46)
$$

It should be noted the choice of eigenvectors (3.3.46) satisfies the required constraints.

MATLAB PROGRAM 3.1: FULL STATE FEEDBACK CONTROL OF ELECTRONICS NAVIGATION OR GYRO BOX

```
%
clear all
close all
%
lz=0.11;
m=10;
kz=17500;
Iyy=0.0487;
lx=0.08;
kx=8750;
%
```

```
M=diag([m m Iyy]);
Ks=[4*kz 0 0;0 4*kx -4*lz*kx;0, -4*lz*kx
4*((kx*lz*lz)+(kz*lx*lx))];
E=0.002*Ks;
Bf=[1 0 1;0 -1 0;lx lz -lz];
%
Ac0=zeros(6,6);
Ac0(2,1)=1.;
Ac0(4,3)=1;
Ac0(6,5)=1;
%
Bc0=zeros(6,3);
Bc0(1,1)=1;
Bc0(3,2)=1;
Bc0(5,3)=1;
%
Dhc=M;
Dlc=[E(1,1) Ks(1,1) 0 0 0 0;0 0 E(2,2) Ks(2,2) E(2,3)
Ks(2,3);0 0 E(3,2) Ks(3,2) E(3,3) Ks(3,3)];
Ac=Ac0-Bc0*inv(Dhc)*Dlc;
Bc=Bc0*inv(Dhc);
%
Ap=[0*eye(3) eye(3);-inv(M)*Ks -inv(M)*E];
eigop=eig(Ap);
Bp=[0*eye(3);inv(M)];
CI=[Bp Ap*Bp];
%Soloution of Eq. (2.13.39)
Coeff=inv(CI)*Ap*Ap*Bp;
Beta=Coeff(4:6,:);
%
%Similarity Transformation Matrix T, Eq. (2.13.46)
```

```
for i=1:3
  T(:,2*i-1)=Bp(:,i);
  T(:,2*i)=Ap*Bp(:,i)-Bp*Beta(:,i);
end
%
Acc=inv(T)*Ap*T;
Bcc=inv(T)*Bp;
%
%New Dhc=Dhcc and New Dlc=Dlcc
%
Accd=Acc-Ac0;
Dhcc=eye(3);
Dlcc=-[Accd(1,:);Accd(3,:);Accd(5,:)];
%
%Find Closed-Loop Eigenvalues
%
im=sqrt(-1);
for i=1:6
  rp=real(eigop(i));
  if (imag(eigop(i))<0.)imm=-im;
  end
  if (imag(eigop(i))>0.)imm=im;
  end
%Open Loop Frequencies and Damping Factors
  ogg(i)=abs(eigop(i));
  zeta(i)=-rp/ogg(i);
%Closed Loop Damping Factor=5* Open Loop Damping
Factor
  zetacl(i)=5*zeta(i);
```

```
%Desired Closed Loop Eigenvalues

  mu(i)=-zetacl(i)*ogg(i)+imm*ogg(i)*sqrt(1.-
zetacl(i)^2);

end

%

% Desired Closed-Loop Characteristic Polynomial

%

chcl=poly(mu);

%

%Non-unique State Feedback Gain Matrix KU for
%Specified Eigenvalues

%gamma and lambda can be chosen arbitrarily.

%

gamma=10;

lambda=1000;

%

ve1=[chcl(2)  chcl(3)];

ve2=[chcl(4)  chcl(5)]/gamma;

ve3=[chcl(6)  chcl(7)]/(gamma*lambda);

%

mave=[ve1 ve2 ve3;0 -gamma 0 0 0 0;0 0 0 -lambda 0 0];

%

KU=Dhcc*mave-Dlcc;

%

% Unique State Feedback Gain Matrix KK for Specified
%Eigenvalues and Eigenvectors

%Choose eigenvectors appropriately

ff(:,1)=[mu(1)  1  0  0  0  0].';

ff(:,2)=[mu(2)  1  0  0  0  0].';

ff(:,3)=[0  0  mu(3)  1  0  0].';
```

```
ff(:,4)=[0 0 mu(4) 1 0 0].';
ff(:,5)=[0 0 0 0 mu(5) 1].';
ff(:,6)=[0 0 0 0 mu(6) 1].';
%
for i=1:6
 psi=[mu(i) 1 0 0 0 0;0 0 mu(i) 1 0 0;0 0 0 0 mu(i)
 1].';
 DD=Dhcc*(mu(i)^2)+Dlcc*psi;
 pp=inv(psi.'*psi)*psi.'*ff(:,i);
%Equation (3.3.9)
 gg(:,i)=DD*pp;
end
%
%Solution of Eq. (3.3.11)
%
KK=-gg*inv(ff);
```

3.4 FUNDAMENTALS OF OPTIMAL CONTROL THEORY

Consider a general dynamic system of order n,

$$\frac{d\mathbf{x}}{dt} = \mathbf{f}(\mathbf{x}(t), \mathbf{u}(t)); \qquad \mathbf{x}(0) = \mathbf{x}_0 \tag{3.4.1}$$

where $\mathbf{x}(t)$ and $\mathbf{u}(t)$ are n-dimensional state and m-dimensional input vectors, respectively. The objective is to determine $\mathbf{u}(t)$; $0 \le t \le t_f$, such that the following objective function (Ray, 1981) is minimized or maximized:

$$I(\mathbf{u}(t)) = G(\mathbf{x}(t_f)) + \int_0^{t_f} F(\mathbf{x}, \mathbf{u}) dt \tag{3.4.2}$$

Note that state equations serve as constraints for the optimization of I. In addition, constraints on the input of the following types will be considered:

$$u_{i*} \leq u_i \leq u_i^* \tag{3.4.3}$$

3.4.1 NECESSARY CONDITIONS FOR OPTIMALITY

Let $\mathbf{u}_o(t)$ be a candidate for the optimal input vector, and let the corresponding state vector be $\mathbf{x}_o(t)$, i.e.,

$$\frac{d\mathbf{x}_0(t)}{dt} = \mathbf{f}(\mathbf{x}_o(t), \mathbf{u}_o(t)) \tag{3.4.4}$$

In order to see whether $\mathbf{u}_o(t)$ is indeed an optimal solution, this candidate optimal input is perturbed (Ray, 1981) by a small amount $\delta\mathbf{u}(t)$; i.e.,

$$\mathbf{u}(t) = \mathbf{u}_o(t) + \delta\mathbf{u}(t) \tag{3.4.5}$$

The change in the value of the objective function can be written as

$$\delta I = I(\mathbf{u}_o(t) + \delta\mathbf{u}(t)) - I(\mathbf{u}_o(t))$$

$$= \left(\frac{\partial G}{\partial \mathbf{x}}\right)\delta\mathbf{x}(t_f) + \int_0^{t_f}\left[\left(\frac{\partial F}{\partial \mathbf{x}}\right)\delta\mathbf{x} + \left(\frac{\partial F}{\partial \mathbf{u}}\right)\delta\mathbf{u}\right]dt + \left[F(t_f) + \left(\frac{\partial G}{\partial \mathbf{x}}\right)\mathbf{f}(t_f)\right]\delta t_f \tag{3.4.6}$$

If the solution of (3.4.1) with $\mathbf{u}(t)$ given by (3.4.5) is $\mathbf{x}_o(t) + \delta\mathbf{x}(t)$,

$$\frac{d(\mathbf{x}_o(t) + \delta\mathbf{x}(t))}{dt} = \mathbf{f}(\mathbf{x}_o(t) + \delta\mathbf{x}(t), \mathbf{u}_o(t) + \delta\mathbf{u}(t)) \tag{3.4.7}$$

Linearizing (3.4.7),

$$\frac{d(\delta\mathbf{x})}{dt} = (\frac{\partial \mathbf{f}}{\partial \mathbf{x}})\delta\mathbf{x}(t) + (\frac{\partial \mathbf{f}}{\partial \mathbf{u}})\delta\mathbf{u}(t) \tag{3.4.8}$$

Multiplying Equation 3.4.8 by $\boldsymbol{\lambda}^T(t)$ and integrating from 0 to t_f,

$$\int_0^{t_f} \boldsymbol{\lambda}^T(t)\frac{d(\delta\mathbf{x})}{dt}dt - \int_0^{t_f} \boldsymbol{\lambda}^T(t)\left[\left(\frac{\partial \mathbf{f}}{\partial \mathbf{x}}\right)\delta\mathbf{x}(t) + \left(\frac{\partial \mathbf{f}}{\partial \mathbf{u}}\right)\delta\mathbf{u}(t)\right]dt = 0 \tag{3.4.9}$$

where $\boldsymbol{\lambda}(t)$ is an n-dimensional vector. Adding (3.4.9) to (3.4.6) and evaluating the first integral in (3.4.9) by parts (Ray, 1981),

$$\delta I = \left[F(t_f) + \left(\frac{\partial G}{\partial \mathbf{x}} \right) \mathbf{f}(t_f) \right] \delta t_f + \boldsymbol{\lambda}^T(0) \delta \mathbf{x}(0) + \left[\left(\frac{\partial G}{\partial \mathbf{x}} \right) - \boldsymbol{\lambda}^T(t_f) \right] \delta \mathbf{x}(t_f)$$

$$+ \int_0^{t_f} \left[\left(\frac{\partial H}{\partial \mathbf{x}} \right) \delta \mathbf{x} + \left(\frac{\partial H}{\partial \mathbf{u}} \right) \delta \mathbf{u}(t) + \frac{d\boldsymbol{\lambda}^T}{dt} \delta \mathbf{x} \right] dt$$
(3.4.10)

where the function H is known as *Hamiltonian*, defined as follows:

$$H = F(\mathbf{x}, \mathbf{u}) + \boldsymbol{\lambda}^T(t) \mathbf{f}(\mathbf{x}, \mathbf{u})$$
(3.4.11)

Because $\boldsymbol{\lambda}(t)$ is arbitrary, it is chosen to satisfy

$$\frac{d\boldsymbol{\lambda}^T}{dt} = -\left(\frac{\partial H}{\partial \mathbf{x}} \right)$$
(3.4.12)

Terms outside the integral in (3.4.10) are known as boundary conditions terms, which are removed for the specified problem. For example, if t_f and $\mathbf{x}(0)$ are specified, $\delta t_f = 0$ and $\delta \mathbf{x}(0) = 0$, and the third outside term in (3.4.10) vanishes under the following condition:

$$\boldsymbol{\lambda}^T(t_f) = \left(\frac{\partial G}{\partial \mathbf{x}} \right)_{t=t_f}$$
(3.4.13)

Hence, the Equation 3.4.10 can be written as

$$\delta I = \int_0^{t_f} \left(\frac{\partial H}{\partial \mathbf{u}} \right) \delta \mathbf{u} \, dt$$
(3.4.14)

Minimization of *I*

If $\mathbf{u}_o(t)$ is an optimal solution,

$$\delta I \geq 0 \quad \text{for any perturbation } \delta \mathbf{u}(t)$$
(3.4.15)

Case I: No Constraint on Input
For condition (3.4.15) to be true,

$$\left(\frac{\partial H}{\partial \mathbf{u}} \right) = 0 \quad \text{for every } t$$
(3.4.16)

This is the necessary condition for optimality.

Case II: With Constraints (3.4.3) on Inputs
In view of the condition (3.4.15),

$$\text{If } u_{io} = u_i^*, \quad \left(\frac{\partial H}{\partial u_i}\right) \leq 0 \tag{3.4.17}$$

$$\text{If } u_{io} = u_{i*}, \quad \left(\frac{\partial H}{\partial u_i}\right) \geq 0 \tag{3.4.18}$$

Maximization of *I*

If $\mathbf{u}_o(t)$ is an optimal solution,

$$\delta I \leq 0 \quad \text{for any perturbation } \delta\mathbf{u}(t) \tag{3.4.19}$$

Case I: No Constraint on Input
For condition (3.4.19) to be true,

$$\left(\frac{\partial H}{\partial \mathbf{u}}\right) = 0 \text{ for every } t \tag{3.4.20}$$

This is the necessary condition for optimality and is identical to Equation 3.4.16.

Case II: With Constraints (3.4.3) on Inputs
In view of the condition (3.4.19),

$$\text{If } u_{io} = u_i^*, \quad \left(\frac{\partial H}{\partial u_i}\right) \geq 0 \tag{3.4.21}$$

$$\text{If } u_{io} = u_{i*}, \quad \left(\frac{\partial H}{\partial u_i}\right) \leq 0 \tag{3.4.22}$$

3.4.2 PROPERTIES OF HAMILTONIAN FOR AN AUTONOMOUS SYSTEM

For an autonomous system, the function $\mathbf{f}$ is not an explicit function of time. Therefore, from Equation 3.4.11,

$$\frac{dH}{dt} = \frac{\partial H}{\partial \mathbf{x}}\frac{d\mathbf{x}}{dt} + \frac{\partial H}{\partial \mathbf{u}}\frac{d\mathbf{u}}{dt} + \frac{\partial H}{\partial \boldsymbol{\lambda}}\frac{d\boldsymbol{\lambda}}{dt} \tag{3.4.23}$$

But

$$\frac{\partial H}{\partial \boldsymbol{\lambda}} = \mathbf{f}^T \qquad (3.4.24)$$

Substituting (3.4.1) and (3.4.12) into (3.4.23),

$$\frac{dH}{dt} = \frac{\partial H}{\partial \mathbf{u}} \frac{d\mathbf{u}}{dt} \qquad (3.4.25)$$

Without any constraints on inputs, $\frac{\partial H}{\partial \mathbf{u}} = 0$. When there are constraints, inputs can be either maximum or minimum constant values according to Equation 3.14.17 and Equation 3.4.18 or Equation 3.14.21 and Equation 3.4.22, respectively. In this case, $\frac{d\mathbf{u}}{dt} = 0$. Therefore, for optimal inputs,

$$\frac{dH}{dt} = 0 \qquad (3.4.26)$$

In other words, Hamiltonian H is a constant along an optimal trajectory for an autonomous system.

Special Case

Final time t_f not specified; i.e., $\delta t_f \neq 0$. In this case, to remove the corresponding boundary condition term in Equation 3.4.10,

$$F(t_f) + \left(\frac{\partial G}{\partial \mathbf{x}} \right) \mathbf{f}(t_f) = 0 \qquad (3.4.27)$$

When final conditions on states are not specified, the condition (3.4.13) holds, and Equation 3.4.27 reduces to

$$H(t_f) = F(t_f) + \boldsymbol{\lambda}^T(t_f)\mathbf{f}(t_f) = 0 \qquad (3.4.28)$$

The condition (3.4.28) is valid even when some or all states are specified at the final time. For example, consider that all states are specified at the final time. In this case, with the first-order term in the Taylor series expansion (Ray, 1981),

$$\mathbf{x}_o(t_f) = \mathbf{x}(t_f + \delta t_f) = \mathbf{x}(t_f) + \dot{\mathbf{x}}(t_f)\delta t_f \qquad (3.4.29)$$

or

$$\delta \mathbf{x}(t_f) = \mathbf{x}(t_f) - \mathbf{x}_o(t_f) = -\mathbf{f}(t_f)\delta t_f \qquad (3.4.30)$$

It is interesting to note that $\delta \mathbf{x}(t_f) \neq 0$. Substituting Equation 3.4.30 into (3.4.10) and using (3.4.12),

$$\delta I = [F(t_f) + \boldsymbol{\lambda}^T(t_f)\mathbf{f}(t_f)]\delta t_f + \boldsymbol{\lambda}^T(0)\delta \mathbf{x}(0) + + \int_0^{t_f}\left(\frac{\partial H}{\partial \mathbf{u}}\right)\delta \mathbf{u}(t)dt \qquad (3.4.31)$$

Therefore, the condition (3.4.28) is again needed to remove boundary condition terms.

In summary, $H(t_f) = 0$ when t_f is not specified. Because H is a constant, it is concluded that $H(t) = 0$ along an optimal trajectory for an autonomous system when t_f is not specified.

EXAMPLE 3.7: TRAVEL OVER MAXIMUM DISTANCE IN SPECIFIED TIME WITHOUT CONSTRAINT ON FINAL VELOCITY

Consider the simple mass m, which is subjected to force $f(t)$. The differential equation of motion is

$$\ddot{x} = u \qquad (3.4.32)$$

where

$$u(t) = \frac{f(t)}{m} \qquad (3.4.33)$$

Find the optimal control input $u(t)$, such that the vehicle covers the maximum distance (Biegler, 1982) in a fixed time t_f, subject to the following constraints:

$$a \leq u(t) \leq b \qquad (3.4.34)$$

Without any loss of generality take $x(0) = 0$. Assume that the vehicle starts from rest; i.e.,

$$\dot{x}(0) = 0 \qquad (3.4.35)$$

Solution

Define the following state variables:

$$x_1 = x \quad \text{and} \quad x_2 = \dot{x} \qquad (3.4.36)$$

Hence, state equations are

$$\begin{bmatrix} \dot{x}_1 \\ \dot{x}_2 \end{bmatrix} = \begin{bmatrix} x_2 \\ u \end{bmatrix} = \mathbf{f} \tag{3.4.37}$$

The objective in this problem is to maximize

$$I = x_1(t_f) \tag{3.4.38}$$

Therefore,

$$G = x_1(t) \quad \text{and} \quad F = 0 \tag{3.4.39}$$

The Hamiltonian H for this problem is

$$H = F + \boldsymbol{\lambda}^T \mathbf{f} = \begin{bmatrix} \lambda_1 & \lambda_2 \end{bmatrix} \begin{bmatrix} x_2 \\ u \end{bmatrix} = \lambda_1 x_2 + \lambda_2 u \tag{3.4.40}$$

Adjoint equations are

$$\frac{d\lambda_1}{dt} = -\frac{\partial H}{\partial x_1} = 0 \tag{3.4.41}$$

$$\frac{d\lambda_2}{dt} = -\frac{\partial H}{\partial x_2} = -\lambda_1 \tag{3.4.42}$$

Then, conditions on state variables are as follows:

$$x_1(0) = 0 \Rightarrow \delta x_1(0) = 0 \tag{3.4.43a}$$

$$x_2(0) = 0 \Rightarrow \delta x_2(0) = 0 \tag{3.4.43b}$$

$$x_1(t_f) \text{unspecified} \Rightarrow \delta x_1(t_f) \neq 0 \tag{3.4.43c}$$

$$x_2(t_f) \text{unspecified} \Rightarrow \delta x_2(t_f) \neq 0 \tag{3.4.44d}$$

Also, because the final time t_f is fixed,

$$\delta t_f = 0 \qquad (3.4.45)$$

For the boundary condition terms to be zero,

$$\boldsymbol{\lambda}^T(t_f) = \left(\frac{\partial G}{\partial \mathbf{x}}\right)_{t=t_f} \qquad (3.4.46)$$

Therefore,

$$\lambda_1(t_f) = \left(\frac{\partial G}{\partial x_1}\right)_{t=t_f} = 1 \qquad (3.4.47a)$$

$$\lambda_2(t_f) = \left(\frac{\partial G}{\partial x_2}\right)_{t=t_f} = 0 \qquad (3.4.47b)$$

Solutions of adjoint equations are

$$\lambda_1(t) = c \quad \text{and} \quad \lambda_2(t) = -ct + d \qquad (3.4.48a,b)$$

where c and d are constants. To satisfy the final conditions,

$$c = 1 \quad \text{and} \quad d = t_f \qquad (3.4.49a,b)$$

In other words,

$$\lambda_1(t) = 1 \quad \text{and} \quad \lambda_2(t) = -t + t_f \qquad (3.4.50a,b)$$

Optimality condition:

$$\frac{\partial H}{\partial u} = \lambda_2 \qquad (3.4.51)$$

$\frac{\partial H}{\partial u} = \lambda_2$ is plotted in Figure 3.6. $\frac{\partial H}{\partial u}$ is not equal to zero for a finite time interval. This implies that the optimal u cannot take any intermediate value between a and b. Furthermore, $\frac{\partial H}{\partial u} > 0$ for $0 \le t < t_f$. Equation 3.4.21 implies that the optimal control input is

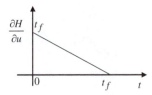

FIGURE 3.6 $\dfrac{\partial H}{\partial u} = \lambda_2$ vs. time t.

$$u = b \quad \text{for } 0 \le t \le t_f \tag{3.4.52}$$

The optimal strategy is the maximum acceleration, which is employed by a rider out of common sense, when the objective is to cover the maximum distance in a fixed time, without any constraint on the vehicle velocity at the final time. Now, the optimal state trajectory can be easily obtained by solving the state equations with given initial conditions:

$$x_2(t) = bt \quad \text{and} \quad x_1(t) = \frac{1}{2}bt^2 \tag{3.4.53}$$

Therefore, the maximum value of the objective function I is $\dfrac{1}{2}bt_f^2$. Along the optimal trajectory,

$$H = bt + (t_f - t)b = bt_f \tag{3.4.54}$$

As expected, the Hamiltonian H is a constant along the optimal trajectory.

EXAMPLE 3.8: TRAVEL OVER MAXIMUM DISTANCE IN SPECIFIED TIME WITH CONSTRAINT THAT FINAL VELOCITY IS EQUAL TO ZERO.

Consider the system, which is same as that in Example 3.7. Find the optimal control input $u(t)$ such that the vehicle covers the maximum distance in a fixed time t_f and ends at rest (Biegler, 1982) subject to the following constraints:

$$a \le u(t) \le b \tag{3.4.55}$$

Here, it is also given that $a < 0$ and $b > 0$.

Solution

Define the following state variables:

$$x_1 = x \quad \text{and} \quad x_2 = \dot{x} \tag{3.4.56}$$

Hence, state equations are

$$\begin{bmatrix} \dot{x}_1 \\ \dot{x}_2 \end{bmatrix} = \begin{bmatrix} x_2 \\ u \end{bmatrix} = \mathbf{f} \tag{3.4.57}$$

The objective in this problem is to maximize

$$I = x_1(t_f) \tag{3.4.58}$$

Therefore,

$$G = x_1(t) \quad \text{and} \quad F = 0 \tag{3.4.59a,b}$$

The Hamiltonian H for this problem is

$$H = F + \boldsymbol{\lambda}^T \mathbf{f} = \begin{bmatrix} \lambda_1 & \lambda_2 \end{bmatrix} \begin{bmatrix} x_2 \\ u \end{bmatrix} = \lambda_1 x_2 + \lambda_2 u \tag{3.4.60}$$

Adjoint equations are

$$\frac{d\lambda_1}{dt} = -\frac{\partial H}{\partial x_1} = 0 \tag{3.4.61}$$

$$\frac{d\lambda_2}{dt} = -\frac{\partial H}{\partial x_2} = -\lambda_1 \tag{3.4.62}$$

Then, conditions on state variables are as follows:

$$x_1(0) = 0 \Rightarrow \delta x_1(0) = 0 \tag{3.4.63a}$$

$$x_2(0) = 0 \Rightarrow \delta x_2(0) = 0 \tag{3.4.63b}$$

$$x_1(t_f) \textit{unspecified} \Rightarrow \delta x_1(t_f) \neq 0 \tag{3.4.63c}$$

$$x_2(t_f) = 0 \Rightarrow \delta x_2(t_f) = 0 \tag{3.4.63d}$$

Also, because the final time t_f is fixed,

$$\delta t_f = 0 \tag{3.4.64}$$

Now, consider the following boundary condition term:

$$\left[\left(\frac{\partial G}{\partial \mathbf{x}}\right) - \boldsymbol{\lambda}^T(t_f)\right]\delta \mathbf{x}(t_f) =$$

$$\left[\left(\frac{\partial G}{\partial x_1}\right) - \lambda_1(t_f)\right]\delta x_1(t_f) + \left[\left(\frac{\partial G}{\partial x_2}\right) - \lambda_2(t_f)\right]\delta x_2(t_f) \tag{3.4.65}$$

Because $\delta x_2(t_f) = 0$,

$$\left[\left(\frac{\partial G}{\partial \mathbf{x}}\right) - \boldsymbol{\lambda}^T(t_f)\right]\delta \mathbf{x}(t_f) = \left[\left(\frac{\partial G}{\partial x_1}\right) - \lambda_1(t_f)\right]\delta x_1(t_f) \tag{3.4.66}$$

Hence, for all boundary condition terms to be zero,

$$\lambda_1(t_f) = \left(\frac{\partial G}{\partial x_1}\right)_{t=t_f} = 1 \tag{3.4.67}$$

Solutions of adjoint equations are

$$\lambda_1(t) = c \quad \text{and} \quad \lambda_2(t) = -ct + d \tag{3.4.68}$$

where c and d are constants. To satisfy the final condition,

$$c = 1 \tag{3.4.69}$$

In other words,

$$\lambda_1(t) = 1 \quad \text{and} \quad \lambda_2(t) = -t + d \tag{3.4.70a,b}$$

Optimality condition:

$$\frac{\partial H}{\partial u} = \lambda_2 = -t + d \tag{3.4.71}$$

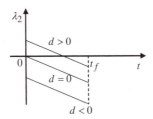

FIGURE 3.7 $\dfrac{\partial H}{\partial u} = \lambda_2$ vs. time t.

The value of d is not known. Any of these three possibilities exists: $d < 0$, $d = 0$, or $d > 0$. $\dfrac{\partial H}{\partial u} = \lambda_2$ is plotted in Figure 3.7 for all these possibilities. The first thing to note is that $\dfrac{\partial H}{\partial u}$ is not equal to zero for a finite time interval. This implies that the optimal $u(t)$ cannot take any intermediate value between a and b. Secondly, $d < 0$ and $d = 0$ are ruled out because they will lead to $u(t) = a < 0$ (Equation 3.4.22). Hence, $d > 0$ and the optimal control input is

$$u(t) = \begin{cases} b & \text{for} \quad t \le t_s \\ a & \text{for} \quad t > t_s \end{cases} \tag{3.4.72}$$

where t_s is the switching instant, which is, when $\lambda_2(t)$ changes sign in Figure 3.7, $d > 0$.

Solving the second state equation,

$$x_2(t_s) = bt_s \quad \text{and} \quad x_2(t_f) = bt_s + a(t_f - t_s) \tag{3.4.73}$$

For $x_2(t_f) = 0$,

$$t_s = \frac{-a}{b - a} t_f \tag{3.4.74}$$

3.5 LINEAR QUADRATIC REGULATOR (LQR) PROBLEM

Consider the linear system

$$\frac{d\mathbf{x}}{dt} = A\mathbf{x}(t) + B\mathbf{u}(t); \quad \mathbf{x}(0) = \mathbf{x}_0 \tag{3.5.1}$$

The objective is to drive the state vector $\mathbf{x}(t)$ to the origin of the state space (zero state vector) from any nonzero initial values of states. If a state feedback control law is used, $\mathbf{x}(t)$ will quickly die out provided closed-loop poles are located far inside the left half of the s-plane. However, elements of the feedback gain vector can be large in magnitudes and the control cost can be high. On the other hand, if closed-loop poles are located close to the open-loop poles, there will not be much increase in the rate of decay of $\mathbf{x}(t)$ and a relatively small amount of control action will be required. Hence, the location of closed-loop poles is a trade-off between the rate of decay of $\mathbf{x}(t)$ and the magnitude of control input. To make this trade-off, the following objective function is chosen:

$$I = \frac{1}{2}\mathbf{x}^T(t_f)S_f\mathbf{x}(t_f) + \frac{1}{2}\int_0^{t_f}[\mathbf{x}^T(t)Q\mathbf{x}(t) + \mathbf{u}^T(t)R\mathbf{u}(t)]dt \qquad (3.5.2)$$

where the final time t_f is fixed. Without any loss of generality, matrices S_f, Q, and R are chosen to be symmetric (Appendix B). In addition, S_f and Q are chosen to be positive semidefinite, and R to be positive definite. Symbolically, these are expressed as

$$S_f = S_f^T \geq 0, \quad Q = Q^T \geq 0, \quad \text{and} \quad R = R^T > 0$$

Problem

Find $\mathbf{u}(t)$; $0 \leq t \leq t_f$, such that the objective function (3.5.2) is minimized.

3.5.1 SOLUTION (OPEN-LOOP OPTIMAL CONTROL)

For this optimal control problem, the Hamiltonian (3.4.11) is

$$H = \frac{1}{2}(\mathbf{x}^T(t)Q\mathbf{x}(t) + \mathbf{u}^T(t)R\mathbf{u}(t)) + \boldsymbol{\lambda}^T(t)(A\mathbf{x}(t) + B\mathbf{u}(t)) \qquad (3.5.3)$$

The necessary condition (3.4.16) for optimality yields

$$\frac{\partial H}{\partial \mathbf{u}} = R\mathbf{u}(t) + B^T\boldsymbol{\lambda}(t) = 0$$

or

$$\mathbf{u}(t) = -R^{-1}B^T\boldsymbol{\lambda}(t) \qquad (3.5.4)$$

The dynamics of $\lambda(t)$ is given by Equation 3.4.12 with final conditions (3.4.13). Hence,

$$\frac{d\lambda}{dt} = -Q\mathbf{x}(t) - A^T\lambda(t); \qquad \lambda(t_f) = S_f\mathbf{x}(t_f) \qquad (3.5.5)$$

Equation 3.5.1 and Equation 3.5.5 represent a two-point boundary value problem (TPBVP) which can be solved to find $\lambda(t)$ and $\mathbf{x}(t)$. Putting (3.5.1) and (3.5.5) in the matrix form,

$$\begin{bmatrix} \dot{\mathbf{x}} \\ \dot{\lambda} \end{bmatrix} = M \begin{bmatrix} \mathbf{x} \\ \lambda \end{bmatrix} \qquad (3.5.6)$$

where

$$M = \begin{bmatrix} A & -BR^{-1}B^T \\ -Q & -A^T \end{bmatrix} \qquad (3.5.7)$$

Solving (3.5.6),

$$\begin{bmatrix} \mathbf{x}(t) \\ \lambda(t) \end{bmatrix} = e^{Mt} \begin{bmatrix} \mathbf{x}(0) \\ \lambda(0) \end{bmatrix} \qquad (3.5.8)$$

To determine $\lambda(0)$, the matrix e^{Mt} is partitioned as follows:

$$e^{Mt} = \begin{bmatrix} E_{11}(t) & E_{11}(t) \\ E_{21}(t) & E_{22}(t) \end{bmatrix} \qquad (3.5.9)$$

From (3.5.8) and (3.5.9),

$$\mathbf{x}(t) = E_{11}(t)\mathbf{x}(0) + E_{12}(t)\lambda(0) \qquad (3.5.10)$$

$$\lambda(t) = E_{21}(t)\mathbf{x}(0) + E_{22}(t)\lambda(0) \qquad (3.5.11)$$

Imposing the condition $\lambda(t_f) = S_f\mathbf{x}(t_f)$,

$$E_{21}(t_f)\mathbf{x}(0) + E_{22}(t_f)\lambda(0) = S_f[E_{11}(t_f)\mathbf{x}(0) + E_{12}(t_f)\lambda(0)] \qquad (3.5.12)$$

From (3.5.12),

$$\lambda(0) = [E_{22}(t_f) - S_f E_{12}(t_f)]^{-1}[S_f E_{11}(t_f) - E_{21}(t_f)]\mathbf{x}(0) \qquad (3.5.13)$$

Substituting this $\lambda(0)$ into (3.5.8), $\lambda(t)$ is obtained. Then, the optimal $\mathbf{u}(t)$ is found from (3.5.4). However, this TPBVP must be solved again if initial conditions change. Furthermore, the control inputs are implementable in an open-loop fashion only, as they are not in the forms of functions of states.

3.5.2 SOLUTION (CLOSED-LOOP OPTIMAL CONTROL)

Use the following transformation (Ray, 1981):

$$\lambda(t) = S(t)\mathbf{x}(t) \qquad (3.5.14)$$

where $S(t)$ is a symmetric $n \times n$ matrix. Substituting (3.5.14) into (3.5.5),

$$\frac{dS}{dt}\mathbf{x}(t) + S(t)\frac{d\mathbf{x}}{dt} = -Q\mathbf{x}(t) - A^T S(t)\mathbf{x}(t) \qquad (3.5.15)$$

Using (3.5.1),

$$\left(\frac{dS}{dt} + SA - SBR^{-1}B^T S + Q + A^T S\right)\mathbf{x}(t) = 0 \qquad (3.5.16)$$

Because Equation 3.5.16 is true for all $\mathbf{x}(t)$,

$$\frac{dS}{dt} + SA - SBR^{-1}B^T S + Q + A^T S = 0 \qquad (3.5.17)$$

Equation 3.4.13 and Equation 3.5.14 yield

$$S(t_f) = S_f \qquad (3.5.17b)$$

Equation 3.5.17 is known as the *Riccati equation*. This nonlinear differential equation can be numerically solved backward in time to determine $S(t)$. From (3.5.4) and (3.5.14),

$$\mathbf{u}(t) = -K(t)\mathbf{x}(t) \qquad (3.5.18)$$

where

$$K(t) = R^{-1}B^T S(t) \tag{3.5.19}$$

The structure of (3.5.18) indicates that the $K(t)$ is the optimal state feedback gain matrix. Because the solution of $S(t)$ does not depend on system states, this gain is optimal for all initial conditions on states.

3.5.3 CROSS TERM IN THE OBJECTIVE FUNCTION

Consider a more general form of the quadratic objective function (Anderson and Moore, 1990):

$$I = \frac{1}{2}\mathbf{x}^T(t_f)S_f\mathbf{x}(t_f) + \frac{1}{2}\int_0^{t_f}[\mathbf{x}^T(t)Q\mathbf{x}(t) + \mathbf{u}^T(t)R\mathbf{u}(t) + 2\mathbf{x}^T(t)N\mathbf{u}(t)]dt \tag{3.5.20}$$

It can be seen that

$$\mathbf{x}^T(t)Q\mathbf{x}(t) + 2\mathbf{x}^T(t)N\mathbf{u}(t) + \mathbf{u}^T(t)R\mathbf{u}(t) = \mathbf{x}^T(t)Q_m\mathbf{x}(t) + \mathbf{v}^T(t)R\mathbf{v}(t) \tag{3.5.21}$$

where

$$Q_m = Q - NR^{-1}N^T \tag{3.5.22}$$

and

$$\mathbf{v}(t) = \mathbf{u}(t) + R^{-1}N^T\mathbf{x}(t) \tag{3.5.23}$$

Equation 3.5.21 can be proved by simply multiplying out the terms on the right-hand side. Hence, Equation 3.5.20 can be rewritten as

$$I = \frac{1}{2}\mathbf{x}^T(t_f)S_f\mathbf{x}(t_f) + \frac{1}{2}\int_0^{t_f}[\mathbf{x}^T(t)Q_m\mathbf{x}(t) + \mathbf{v}^T(t)R\mathbf{v}(t)]dt \tag{3.5.24}$$

and the plant equation (3.5.1) is modified with Equation 3.5.23:

$$\dot{\mathbf{x}} = A_m\mathbf{x}(t) + B\mathbf{v}(t) \tag{3.5.25}$$

where

$$A_m = A - BR^{-1}N^T \tag{3.5.26}$$

Assuming that $Q_m \geq 0$, Equation 3.5.24 and Equation 3.5.25 constitute a standard LQ problem for which the optimal state feedback control law is

$$\mathbf{v}(t) = -R^{-1}B^T S_m(t)\mathbf{x}(t) \tag{3.5.27}$$

where

$$\dot{S}_m = -S_m A_m - A_m^T S_m + S_m BR^{-1}B^T S_m - Q_m; \quad S_m(t_f) = S_f \tag{3.5.28}$$

From (3.5.23) and (3.5.27),

$$\mathbf{u}(t) = -K_m(t)\mathbf{x}(t) \tag{3.5.29}$$

where the optimal state feedback gain matrix is given by

$$K_m(t) = R^{-1}(B^T S_m(t) + N^T) \tag{3.5.30}$$

EXAMPLE 3.9: MINIMUM ENERGY CONTROL OF A DC MOTOR

Consider the position controller shown in Figure 3.8. The actuator is an armature-controlled DC motor (Kuo, 1995). The torque produced by the motor $T_m(t)$ is proportional to the armature current $i_a(t)$; i.e.,

$$T_m(t) = k_i i_a(t) \tag{3.5.31}$$

Let the back emf developed across the armature be $e_b(t)$, i.e.,

$$e_b(t) = k_b \dot{\theta}_m(t) \tag{3.5.32}$$

where k_b is the back emf constant and $\theta_m(t)$ is the angular position of the rotor. Applying Kirchoff's law to the armature,

$$u(t) = R_a i_a(t) + e_b(t) \tag{3.5.33}$$

where $u(t)$ is the input voltage and $R_a(t)$ is the armature resistance. The inductance of the armature windings has been neglected.

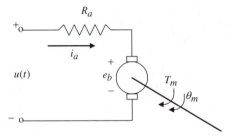

FIGURE 3.8 An armature-controlled DC motor.

Applying Newton's second law,

$$J_m \ddot{\theta}_m + B_m \dot{\theta}_m = T_m(t) \tag{3.5.34}$$

where J_m and B_m are mass moment of inertia and equivalent viscous damping, respectively. From (3.5.31) to (3.5.33),

$$T_m(t) = \frac{k_i}{R_a} u(t) - \frac{k_i k_b}{R_a} \dot{\theta}_m \tag{3.5.35}$$

From (3.5.34) and (3.5.35),

$$\ddot{\theta}_m + \alpha \dot{\theta}_m = \beta u(t) \tag{3.5.36}$$

where

$$\alpha = \frac{B_m}{J_m} + \frac{k_i k_b}{R_a J_m} \tag{3.5.37}$$

and

$$\beta = \frac{k_i}{R_a J_m} \tag{3.5.38}$$

From (3.5.36), state equations are

$$\dot{\mathbf{x}} = A\mathbf{x}(t) + bu(t) \tag{3.5.39}$$

where

$$\mathbf{x}(t) = \begin{bmatrix} \theta_m(t) \\ \dot{\theta}_m(t) \end{bmatrix}; \quad A = \begin{bmatrix} 0 & 1 \\ 0 & -\alpha \end{bmatrix}; \quad \mathbf{b} = \begin{bmatrix} 0 \\ \beta \end{bmatrix} \tag{3.5.40}$$

Let us define an optimal control problem: Find the input $u(t)$, $0 \le t \le t_f$, such that the energy consumed by the DC motor is minimized over a fixed final time t_f, with the following initial and final conditions on the states:

$$\mathbf{x}(0) = \begin{bmatrix} 0 \\ 0 \end{bmatrix} \quad \text{and} \quad \mathbf{x}(t_f) = \begin{bmatrix} \theta_f \\ 0 \end{bmatrix} \tag{3.5.41}$$

The energy consumed by the DC motor over a fixed final time t_f is given by the following integral:

$$I_d = \int_{t=0}^{t_f} T_m(t)d\theta_m = \int_0^{t_f} T_m \dot{\theta}_m dt \tag{3.5.42}$$

Using (3.5.35), (3.5.40), and (3.5.42)

$$I_d = \int_0^{t_f} \frac{k_i}{R_a}(ux_2 - k_b x_2^2)dt \tag{3.5.43}$$

To satisfy the requirements for the existence of solution to the LQ control, the objective function (3.5.43) is modified to be

$$I = \int_0^{t_f} \frac{k_i}{R_a}(ux_2 - k_b x_2^2) + \gamma x_2^2 + \rho u^2 dt ; \quad \gamma > 0 \quad \text{and} \quad \rho > 0 \tag{3.5.44}$$

Hence, with respect to the definition (3.5.20),

$$I = \frac{1}{2}\int_0^{t_f} \mathbf{x}^T Q\mathbf{x} + 2\mathbf{x}^T N\mathbf{u} + \mathbf{u}^T R\mathbf{u} \, dt \tag{3.5.45}$$

$$Q = \begin{bmatrix} 0 & 0 \\ 0 & 2(\rho - \dfrac{k_i k_b}{R_a}) \end{bmatrix}; \quad N = \begin{bmatrix} 0 \\ \dfrac{k_i}{R_a} \end{bmatrix}; \quad R = 2\rho \tag{3.5.46}$$

and

$$Q_m = \begin{bmatrix} 0 & 0 \\ 0 & (2\gamma - \dfrac{2k_i k_b}{R_a} - \dfrac{k_i^2}{2R_a^2 \rho}) \end{bmatrix} \tag{3.5.47}$$

For $Q_m \geq 0$, ρ and γ must be chosen to satisfy

$$2\gamma - \frac{2k_i k_b}{R_a} - \frac{k_i^2}{2R_a^2 \rho} \geq 0 \tag{3.5.48}$$

In order to have final conditions on states to be zero, the following transformation of the state vector is introduced:

$$\tilde{\mathbf{x}}(t) = \mathbf{x}(t) - \mathbf{x}(t_f) \tag{3.5.49}$$

Then,

$$\tilde{\mathbf{x}}(0) = -\mathbf{x}(t_f) \quad \text{and} \quad \tilde{\mathbf{x}}(t_f) = 0 \tag{3.5.50a,b}$$

Substituting (3.5.49) into (3.5.39), and noting that $A\mathbf{x}(t_f) = 0$,

$$\dot{\tilde{\mathbf{x}}} = A\tilde{\mathbf{x}}(t) + \mathbf{b}u(t) \tag{3.5.51}$$

Because the objective function does not contain x_1 and $x_2 = \tilde{x}_2$, Equation 3.5.45 can be rewritten as

$$I = \frac{1}{2} \int_0^{t_f} \tilde{\mathbf{x}}^T Q\tilde{\mathbf{x}} + 2\tilde{\mathbf{x}}^T N\mathbf{u} + \mathbf{u}^T R\mathbf{u}\, dt \tag{3.5.52}$$

In the LQ control, there is no constraint on the final conditions. Therefore, the objective function is modified to contain the final state vector:

$$I = \frac{1}{2}\tilde{\mathbf{x}}^T(t_f)S_f\tilde{\mathbf{x}}(t_f) + \frac{1}{2} \int_0^{t_f} \tilde{\mathbf{x}}^T Q\tilde{\mathbf{x}} + 2\tilde{\mathbf{x}}^T N\mathbf{u} + \mathbf{u}^T R\mathbf{u}\, dt \tag{3.5.53}$$

where

$$S_f = \begin{bmatrix} a & 0 \\ 0 & b \end{bmatrix}; \quad a > 0 \text{ and } b > 0 \tag{3.5.54}$$

Relaxing constraints on final states, the LQ problem can be solved to minimize (3.5.53) for the linear system (3.5.51) with nonzero $\tilde{\mathbf{x}}(0)$. By choosing a and b to be large numbers, $\tilde{\mathbf{x}}(t_f)$ can be forced to be close to zero.

3.5.4 IMPORTANT CASE: INFINITE FINAL TIME

When the final time $t_f \to \infty$, the optimal gain $K_m(t)$ or $S_m(t)$ turns out to be a constant (Kwakernaak and Sivan, 1972). The system of differential equations (3.5.28) reduces to a system of algebraic equations:

$$S_m A_m - S_m B R^{-1} B^T S_m + Q_m + A_m^T S_m = 0 \tag{3.5.55}$$

This is known as the algebraic Riccati equation (ARE) and is probably one of the most commonly used equations in modern control theory. Since the matrix S_m is a constant, the feedback gain matrix also turns out to be a constant as follows:

$$K_o = R^{-1}(B^T S_m + N^T) \tag{3.5.56}$$

Hence from (3.5.1) and (3.5.56), the closed-loop system dynamics is represented by

$$\frac{d\mathbf{x}}{dt} = (A - BK_o)\mathbf{x}(t) \tag{3.5.57}$$

The eigenvalues of the matrix $A - BK_o$ are the optimal closed-loop poles. In the next section, a method is described to determine these optimal closed-loop poles first via root locus plots. Then, the optimal feedback gain can be obtained via pole placement techniques.

Because $Q_m \geq 0$, there exists a matrix H known as the square root of Q_m, such that

$$Q_m = HH^T \tag{3.5.58}$$

If (A, H) is observable, S_m is positive definite and is the only solution of ARE, Equation 3.5.55, with this property. Furthermore, the optimal closed-loop system (3.5.57) is asymptotically stable (Anderson and Moore, 1990; Kailath, 1980).

EXAMPLE 3.10: A FIRST-ORDER SYSTEM

Consider a first-order system

$$\dot{x} = x + u \tag{3.5.59}$$

and the objective function

$$I = \frac{5}{2} x^2(t_f) + \frac{1}{2} \int_0^{t_f} 2x^2 + 3u^2 \, dt \tag{3.5.60}$$

Here,

$$A = 1, \ \mathbf{b} = 1, \ Q = 2, \ R = 3, \ S_f = 5 \tag{3.5.61}$$

Therefore, the Riccati equation becomes

$$\dot{S} = -2S + \frac{S^2}{3} - 2 \ ; \quad S(t_f) = 5 \tag{3.5.62}$$

Nonlinear differential equation (3.5.62) has to be numerically solved backward in time to determine $S(t)$ for $0 \le t \le t_f$. Matlab routine ODE23 or ODE45 can be used for this purpose. Then, the optimal control law will be

$$u(t) = -\frac{S(t)}{3} x(t) \ ; \quad 0 \le t \le t_f \tag{3.5.63}$$

If $t_f \to \infty$, ARE becomes

$$0 = -2S + \frac{S^2}{3} - 2 \tag{3.5.64}$$

Solving (3.5.64),

$$S = 3 \pm \sqrt{15} \tag{3.5.65}$$

In order to have a stable closed-loop system, a "+" sign must be chosen. The optimal state feedback law is

$$u(t) = -Kx(t) \tag{3.5.66}$$

where

$$K = 1 + \frac{\sqrt{15}}{3} \tag{3.5.67}$$

Substituting (3.5.66) into (3.5.59), the closed-loop system dynamics is

$$\dot{x} = -\frac{\sqrt{15}}{3} x \tag{3.5.68}$$

which is stable.

EXAMPLE 3.11: ACTIVE SUSPENSION WITH OPTIMAL LINEAR STATE FEEDBACK (THOMSON, 1976)

Differential equations of motion for the linear vehicle model shown in Figure 3.9 are

$$m_2 \ddot{x}_2 = u \tag{3.5.69}$$

$$m_1 \ddot{x}_1 + \lambda_1 (x_1 - x_0) = -u \tag{3.5.70}$$

Defining $x_3 = \dot{x}_1$ and $x_4 = \dot{x}_2$,

$$\dot{x}_1 = x_3 \tag{3.5.71}$$

$$\dot{x}_2 = x_4 \tag{3.5.72}$$

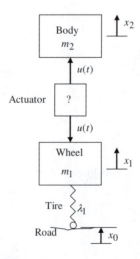

FIGURE 3.9 A quarter car model.

$$\dot{x}_3 = -\frac{\lambda_1}{m_1}(x_1 - x_0) - \frac{u(t)}{m_1} \qquad (3.5.73)$$

$$\dot{x}_4 = \frac{u(t)}{m_2} \qquad (3.5.74)$$

Assume that the road profile is a step function. In this case, x_0 is a constant, and a new set of state variables is defined as

$$\hat{x}_1 = x_1 - x_0 \qquad (3.5.75)$$

$$\hat{x}_2 = x_2 - x_0 \qquad (3.5.76)$$

$$\hat{x}_3 = \dot{\hat{x}}_1 = \dot{x}_1 \qquad (3.5.77)$$

$$\hat{x}_4 = \dot{\hat{x}}_2 = \dot{x}_2 \qquad (3.5.75)$$

and the new set of state equations can be written as

$$
\begin{bmatrix} \dot{\hat{x}}_1 \\ \dot{\hat{x}}_2 \\ \dot{\hat{x}}_3 \\ \dot{\hat{x}}_4 \end{bmatrix} = \begin{bmatrix} 0 & 0 & 1 & 0 \\ 0 & 0 & 0 & 1 \\ -\frac{\lambda_1}{m_1} & 0 & 0 & 0 \\ 0 & 0 & 0 & 0 \end{bmatrix} \begin{bmatrix} \hat{x}_1 \\ \hat{x}_2 \\ \hat{x}_3 \\ \hat{x}_4 \end{bmatrix} + \begin{bmatrix} 0 \\ 0 \\ -\frac{1}{m_1} \\ \frac{1}{m_2} \end{bmatrix} u(t) \qquad (3.5.76)
$$

The performance index is chosen as

$$I = \frac{1}{2} \int_0^\infty [\rho u^2 + q_1(x_0 - x_1)^2 + q_2(x_1 - x_2)^2] dt \qquad (3.5.77)$$

Note the following:

$$(x_0 - x_1): \text{ tire dynamic deflection} \qquad (3.5.78)$$

$$(x_1 - x_2): \text{ relative wheel travel} \qquad (3.5.79)$$

Moreover, the actuator force $u(t)$, which is proportional to the vertical acceleration of the body, is also a measure of the ride discomfort. Hence, the minimization of the objective function I will result in a trade-off between the minimization of the actuator input or ride discomfort and minimization of the weighted sum of tire dynamic deflection and relative wheel travel. Weighting parameters are ρ, q_1, and q_2. Because

$$q_1(x_0 - x_1)^2 + q_2(x_1 - x_2)^2 = q_1\hat{x}_1^2 + q_2(\hat{x}_1 - \hat{x}_2)^2$$
$$= (q_1 + q_2)\hat{x}_1^2 + q_2\hat{x}_2^2 - 2q_2\hat{x}_1\hat{x}_2 \tag{3.5.80}$$

the objective function I can be expressed as

$$I = \frac{1}{2}\int_0^\infty [\rho u^2 + \hat{\mathbf{x}}^T Q\hat{\mathbf{x}}]dt \tag{3.5.81}$$

where

$$Q = \begin{bmatrix} q_1 + q_2 & -q_2 & 0 & 0 \\ -q_2 & q_2 & 0 & 0 \\ 0 & 0 & 0 & 0 \\ 0 & 0 & 0 & 0 \end{bmatrix} \quad \text{and} \quad \hat{\mathbf{x}} = \begin{bmatrix} \hat{x}_1 \\ \hat{x}_2 \\ \hat{x}_3 \\ \hat{x}_4 \end{bmatrix} \tag{3.5.82a,b}$$

The optimal control law is

$$u(t) = -\mathbf{k}\hat{\mathbf{x}} = -\begin{bmatrix} k_1 & k_2 & k_3 & k_4 \end{bmatrix}\hat{\mathbf{x}} = -k_1\hat{x}_1 - k_2\hat{x}_2 - k_3\hat{x}_3 - k_4\hat{x}_4 \tag{3.5.83}$$

where

$$\mathbf{k} = \rho^{-1}\mathbf{b}^T S \tag{3.5.84}$$

and

$$SA + A^T S - Sb\rho^{-1}\mathbf{b}^T S + Q = 0 \tag{3.5.85}$$

Physical realization of the control law:

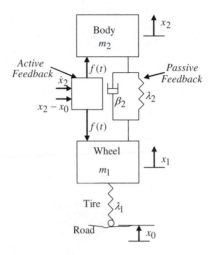

FIGURE 3.10 Active or passive vehicle suspension system.

$$u(t) = -k_1(\hat{x}_1 - \hat{x}_2) - k_3(\hat{x}_3 - \hat{x}_4) - (k_1 + k_2)\hat{x}_2 - (k_3 + k_4)\hat{x}_4$$

$$= u_p + f(t) \tag{3.5.86}$$

where

$$u_p(t) = \lambda_2(x_1 - x_2) + \beta_2(\dot{x}_1 - \dot{x}_2) \tag{3.5.87}$$

and

$$f(t) = -(k_1 + k_2)(x_2 - x_0) - (k_3 + k_4)\dot{x}_2 \tag{3.5.88}$$

The part of the control input $u_p(t)$ has been applied by a spring and a damper that provide passive feedback of $(x_1 - x_2)$ and $(\dot{x}_1 - \dot{x}_2)$ (Figure 3.10). The remaining part $f(t)$ is applied by active feedback via sensors to measure $(x_2 - x_0)$ and $\dot{x}_2$, and an actuator to apply the force $f(t)$. In the study by Thompson (1976), the relative distance $(x_2 - x_0)$ is measured by an ultrasonic transmitter/receiver, and the velocity $\dot{x}_2$ is obtained by integrating the signal from an accelerometer. The actuator is shown to be an electrohydraulic system.

3.6 SOLUTION OF LQR PROBLEM VIA ROOT LOCUS PLOT: SISO CASE

Let the quadratic objective function be

$$I = \frac{1}{2}\int_0^\infty y^2(t) + ru^2(t)dt \tag{3.6.1}$$

where $u(t)$ and $y(t)$ are input and output of a SISO linear system. Note that

$$y^2(t) = y^T(t)y(t) = \mathbf{x}^T(t)\mathbf{c}^T\mathbf{c}\mathbf{x}(t) \tag{3.6.2}$$

Referring to (3.5.2),

$$Q = \mathbf{c}^T\mathbf{c} \quad \text{and} \quad R = r \tag{3.6.3}$$

Hence, from Equation 3.5.1, Equation 3.5.4, and Equation 3.5.5,

$$\begin{bmatrix} \dfrac{d\mathbf{x}(t)}{dt} \\ \dfrac{d\boldsymbol{\lambda}(t)}{dt} \end{bmatrix} = M\begin{bmatrix} \mathbf{x}(t) \\ \boldsymbol{\lambda}(t) \end{bmatrix} \tag{3.6.4}$$

where

$$M = \begin{bmatrix} A & -r^{-1}\mathbf{b}\mathbf{b}^T \\ -\mathbf{c}^T\mathbf{c} & -A^T \end{bmatrix} \tag{3.6.5}$$

The matrix M is known as the Hamiltonian matrix. It can be shown (Kwakernaak and Sivan, 1972) that

$$(-1)^n\Delta(s) = a_1(s)a_1(-s) + r^{-1}a_2(s)a_2(-s) \tag{3.6.6}$$

where

$$\Delta(s) = \det(sI - M) \tag{3.6.7}$$

and $a_1(s)$ and $a_2(s)$ are denominator and numerator polynomials of the open-loop transfer function, i.e.,

$$\mathbf{c}(sI - A)^{-1}\mathbf{b} = \frac{a_2(s)}{a_1(s)} \tag{3.6.8}$$

It will be assumed that there are no common factors between $a_2(s)$ and $a_1(s)$. Equation 3.6.6 establishes the fact that if s^* is a root of $\Delta(s) = 0$, $-s^*$ would also be the root. Hence, eigenvalues of M are symmetric with respect to the imaginary axis of the s-plane. Recall that eigenvalues of any real matrix are always symmetric with respect to the real axis. Hence, eigenvalues of M are symmetric with respect to both real and imaginary axes. Furthermore,

$$(-1)^n \Delta(j\omega) = a_1(j\omega)a_1(-j\omega) + r^{-1}a_2(j\omega)a_2(-j\omega)$$
$$= \left|a_1(j\omega)\right|^2 + r^{-1}\left|a_2(j\omega)\right|^2 \tag{3.6.9}$$

Note that $\left|a_i(j\omega)\right|$, $i = 1$ and 2, will be zero only when $s = 0$ is a root of $a_i(s) = 0$. Because it has been assumed that there are no common factors between $a_1(s)$ and $a_2(s)$, $\left|a_1(j\omega)\right|$ and $\left|a_2(j\omega)\right|$ cannot simultaneously be zero. As a result, $\left|\Delta(j\omega)\right| > 0$; i.e., none of the eigenvalues is on the imaginary axis of the s-plane.

Fact

The optimal poles are the stable eigenvalues of M (Kailath, 1980).

THE SYMMETRIC ROOT LOCUS

From (3.6.6),

$$a_1(s)a_1(-s) + r^{-1}a_2(s)a_2(-s) = 0 \tag{3.6.10}$$

Hence,

$$r^{-1}\frac{a_2(s)a_2(-s)}{a_1(s)a_1(-s)} = -1 \tag{3.6.11}$$

Let

$$a_2(s) = \prod_{i=1}^{m}(s - z_i) \tag{3.6.12}$$

$$a_1(s) = \prod_{i=1}^{n}(s - p_i) \tag{3.6.13}$$

$$r^{-1}(-1)^{m-n}\frac{\displaystyle\prod_{i=1}^{m}(s-z_i)(s+z_i)}{\displaystyle\prod_{i=1}^{n}(s-p_i)(s+p_i)}=-1 \tag{3.6.14}$$

Equation 3.6.14 is in the standard root locus form (Kuo, 1995). When $n-m$ is even, the 180° root locus plot will be constructed. If $n-m$ is odd, the 0° root locus will be constructed.

Case I: High Cost of Control ($r \to \infty$)

When the cost of control action is high, it is desired to use a small value of input $u(t)$. This can be achieved by selecting a very large value of r. When $r \to \infty$, Equation 3.6.6 yields

$$\Delta(s) = a_1(s)a_1(-s) = 0 \tag{3.6.15}$$

Because the optimal poles are the stable roots of $\Delta(s) = 0$, we have the following two situations:

1. Stable open-loop system: If all the eigenvalues of the matrix A are in the left half of the s-plane, optimal closed-loop poles are the same as open-loop poles.
2. Unstable open-loop system: Optimal closed-loop poles are: (a) stable open-loop poles and (b) reflections of unstable open-loop poles about the imaginary axis.

Case II: Low Cost of Control ($r \to 0$)

When $r \to 0$, Equation 3.6.10 yields

$$a_2(s)a_2(-s) = 0 \tag{3.6.16}$$

Hence, m optimal closed-loop poles are the stable roots of $a_2(s)a_2(-s) = 0$. They are either open-loop left-half zeros or the reflections of open-loop right-half zeros about the imaginary axis. The remaining $(n-m)$ optimal closed-loop poles are located near infinity. To find their locations, Equation 3.6.10 is written as follows for large s by ignoring lower powers of s:

$$s^n(-s)^n + r^{-1}s^m(-s)^m(b_0)^2 = 0 \tag{3.6.17}$$

where

$$a_2(s) = b_0 s^m + \ldots + b_m \tag{3.6.18}$$

From (3.6.17),

$$s^{2(n-m)} = (-1)^{n-m+1} b_0^2 r^{-1} \tag{3.6.19}$$

or

$$s = [(-1)^{(n-m+1)}]^{\frac{1}{2(n-m)}} (b_0^2 r^{-1})^{\frac{1}{2(n-m)}} \tag{3.6.20}$$

$n - m + 1$ odd

$$s = (-1)^{\frac{1}{2(n-m)}} (b_0^2 r^{-1})^{\frac{1}{2(n-m)}} \tag{3.6.21}$$

Recall that

$$-1 = e^{j\pi(2\ell+1)} \text{ , where } \ell \text{ is an integer}$$

Hence

$$(-1)^{\frac{1}{2(n-m)}} = e^{\frac{j\pi(2\ell+1)}{2(n-m)}} \text{ ; } \quad \ell = 0, 1, 2, \ldots, 2(n-m)-1 \tag{3.6.22}$$

Substitution of (3.6.22) into (3.6.21) yields $2(n-m)$ roots of Equation 3.6.19.

$n - m + 1$ even

$$s = (1)^{\frac{1}{2(n-m)}} (b_0^2 r^{-1})^{\frac{1}{2(n-m)}} \tag{3.6.23}$$

Recall that

$$1 = e^{j\pi 2\ell} \text{ , where } \ell \text{ is an integer} \tag{3.6.24}$$

Hence

$$(1)^{\frac{1}{2(n-m)}} = e^{\frac{j\pi 2\ell}{2(n-m)}} \text{ ; } \quad \ell = 0, 1, 2, \ldots\ldots, 2(n-m)-1 \tag{3.6.25}$$

Substitution of (3.6.25) into (3.6.23) yields $2(n-m)$ roots of Equation 3.6.19.

To summarize, the $2(n-m)$ roots of equation (3.6.19) lie on a circle of radius $(b_0^2 r^{-1})^{\frac{1}{2(n-m)}}$ in a pattern described by (3.6.21) and (3.6.25). This pattern is known as Butterworth configuration (Kailath, 1980).

EXAMPLE 3.12: NONCOLOCATED SENSOR AND ACTUATOR (BRYSON, 1979)

Consider the two-degree-of-freedom spring-mass system shown in Figure 3.11, where the force $u_0(t)$ is applied on the left mass and the position of the right mass $y(t)$ is the output. This model has been extensively used to simulate noncolocated sensor and actuator in the structural vibration control.

The governing differential equations of motion are

$$m_0 \ddot{x} + k(x-y) = u_0(t) \tag{3.6.26}$$

$$m_0 \ddot{y} + k(y-x) = 0 \tag{3.6.27}$$

Nondimensional time t' is defined as

$$t' = t\sqrt{\frac{k}{m_0}} \tag{3.6.28}$$

Therefore,

$$\frac{d(.)}{dt'} = \sqrt{\frac{m_0}{k}} \frac{d(.)}{dt} \quad \text{and} \quad \frac{d^2(.)}{dt'^2} = \frac{m_0}{k} \frac{d^2(.)}{dt^2} \tag{3.6.29a,b}$$

Hence, Equation 3.6.26 and Equation 3.6.27 can be written as

$$x'' + (x-y) = u(t') \quad \text{where} \quad u = \frac{u_0}{k} \tag{3.6.30}$$

$$y'' + (y-x) = 0 \tag{3.6.31}$$

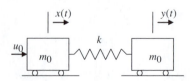

FIGURE 3.11 A two-degree-of-freedom system.

Taking the Laplace transform of (3.6.30) and (3.6.31) with zero initial conditions,

$$s^2 x(s) + (x(s) - y(s)) = u(s) \qquad (3.6.32)$$

$$s^2 y(s) + (y(s) - x(s)) = 0 \qquad (3.6.33)$$

Some simple algebra yields the SISO transfer function:

$$\frac{y(s)}{u(s)} = \frac{1}{s^2(s^2+2)} = \frac{a_2(s)}{a_1(s)} \qquad (3.6.34)$$

Hence, the root locus equation is

$$r^{-1} \frac{a_2(s)a_2(-s)}{a_1(s)a_1(-s)} = -1 \qquad (3.6.35)$$

or

$$r^{-1} \frac{1}{s^4(s^2+2)^2} = -1 \qquad (3.6.36)$$

The $180°$ root locus is shown in Figure 3.12. There are eight branches and all of them end at infinity. Angles of asymptotes are

$$(2l+1)\frac{\pi}{8} ; \quad l = 0, 1, 2, 3, 4, 5, 6, 7 \qquad (3.6.37)$$

All these asymptotes intersect the real axis at the origin of complex plane.

3.7 LINEAR QUADRATIC TRAJECTORY CONTROL

It is desired to find an optimal control law in such a way as to cause the output $\mathbf{y}(t)$ to track or follow a desired trajectory $\boldsymbol{\eta}(t)$. Hence, the objective function (3.5.2) is modified (Sage and White, 1977) to be

$$I = \frac{1}{2}\mathbf{z}^T(t_f)S_f\mathbf{z}(t_f) + \frac{1}{2}\int_0^{t_f} [\mathbf{z}^T(t)Q\mathbf{z}(t) + \mathbf{u}^T(t)R\mathbf{u}(t)]dt \qquad (3.7.1)$$

where $\mathbf{z}(t)$ is the trajectory error defined as

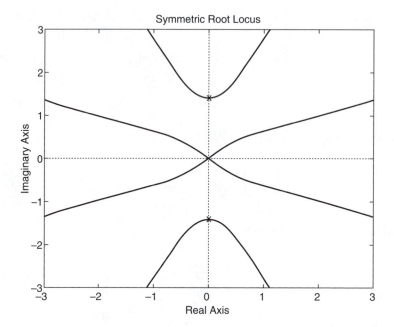

FIGURE 3.12 A symmetric root locus plot.

$$\mathbf{z}(t) = \boldsymbol{\eta}(t) - \mathbf{y}(t) \tag{3.7.2}$$

The state space equation (3.5.1) is modified (Sage and White, 1977) to include a deterministic external input or the plant *noise* vector $w(t)$:

$$\frac{d\mathbf{x}}{dt} = A\mathbf{x}(t) + B\mathbf{u}(t) + \mathbf{w}(t); \quad \mathbf{x}(0) = \mathbf{x}_0 \tag{3.7.3}$$

From Equation 3.4.11, the Hamiltonian H is defined as

$$H = \frac{1}{2}[\mathbf{z}^T(t)Q\mathbf{z}(t) + \mathbf{u}^T(t)R\mathbf{u}(t)] + \boldsymbol{\lambda}^T(t)[A\mathbf{x}(t) + B\mathbf{u}(t) + \mathbf{w}(t)] \tag{3.7.4}$$

The optimality condition (3.4.16) yields

$$\mathbf{u}(t) = -R^{-1}B^T\boldsymbol{\lambda}(t) \tag{3.7.5}$$

Equation 3.4.12 yields

$$\frac{d\boldsymbol{\lambda}}{dt} = -\{C^TQ[C\mathbf{x}(t) - \boldsymbol{\eta}(t)] + A^T\boldsymbol{\lambda}(t)\} \tag{3.7.6}$$

with the terminal condition

$$\lambda(t_f) = C^T S(t_f)[C\mathbf{x}(t_f) - \boldsymbol{\eta}(t_f)] \tag{3.7.7}$$

In order to determine the closed-loop control law, the transformation (3.5.14) is modified (Sage and White, 1977) to be

$$\lambda(t) = S(t)\mathbf{x}(t) - \boldsymbol{\xi}(t) \tag{3.7.8}$$

where $\boldsymbol{\xi}(t)$ is to be determined. Differentiating (3.7.8),

$$\frac{d\lambda}{dt} = \frac{dS}{dt}\mathbf{x}(t) + S(t)\frac{d\mathbf{x}}{dt} - \frac{d\boldsymbol{\xi}}{dt} \tag{3.7.9}$$

Using (3.7.3), (3.7.5), and (3.7.6),

$$\left(\frac{dS}{dt} + SA - SBR^{-1}B^T S + C^T QC + A^T S\right)\mathbf{x}(t)$$
$$+ \left(-\frac{d\boldsymbol{\xi}}{dt} + SBR^{-1}B^T \boldsymbol{\xi} + S\mathbf{w}(t) - C^T Q\boldsymbol{\eta}(t) - A^T \boldsymbol{\xi}(t)\right) = 0 \tag{3.7.10}$$

Using (3.7.8), the terminal condition (3.7.7) can be written as

$$S(t_f)\mathbf{x}(t_f) - \boldsymbol{\xi}(t_f) = C^T S(t_f)C\mathbf{x}(t_f) - C^T S(t_f)\boldsymbol{\eta}(t_f) \tag{3.7.11}$$

The solution of (3.7.10) can be obtained by solving it as two separate problems:

$$\frac{dS}{dt} + SA - SBR^{-1}B^T S + C^T QC + A^T S = 0$$

with

$$S(t_f) = C^T S(t_f)C \tag{3.7.12}$$

and

$$-\frac{d\boldsymbol{\xi}}{dt} + SBR^{-1}B^T \boldsymbol{\xi} + S\mathbf{w}(t) - C^T Q\boldsymbol{\eta}(t) - A^T \boldsymbol{\xi}(t) = 0$$

with

$$\xi(t_f) = C^T S(t_f) \eta(t_f) \tag{3.7.13}$$

Lastly, from (3.7.5) and (3.7.8), the control law is

$$\mathbf{u}(t) = -R^{-1} B^T [S(t)\mathbf{x}(t) - \xi(t)] = -K(t)\mathbf{x}(t) + R^{-1} B^T \xi(t) \tag{3.7.14}$$

The state feedback gain matrix $K(t)$ is the same as that given by (3.5.19). Hence, the solution of the linear quadratic trajectory control problem is composed of two parts: (1) a linear regulator part and (2) a correction term containing $\xi(t)$. The computation of $\xi(t)$ requires the solution of (3.7.13) backward in time. Hence, it is required that $\mathbf{w}(t)$ and $\eta(t)$ are exactly known *a priori* for all time t. From the disturbance rejection point of view, the control law can be described to be noncausal.

EXAMPLE 3.13: OPTIMAL CONTROL OF SUN TRACKING SOLAR CONCENTRATORS (HUGHES, 1979)

A solar collector consists of a concentrator and a receiver. As a concentrator, point focusing parabolic dishes have been used. It reflects the sun's energy towards its focal point and the receiver accepts the concentrated energy for further conversions. The axis of the paraboloid must be pointed at the sun in order to produce the required flux densities at the receiver aperture. Whenever there is a pointing error, energy is lost; hence, this energy loss is minimized by an appropriate control technique.

A linear model of a single axis of the concentrator, which is driven by an electric motor, is shown in Figure 3.13, where

$\theta_0(t)$: Collector's line of sight (LOS)
$\theta_i(t)$: Sun's position
$u(t)$: Command input to the motor
ξ: Damping ratio
K_s: System gain
ω_n: Natural frequency

The state space model of the system can be written as

$$\frac{d\mathbf{x}}{dt} = A\mathbf{x} + \mathbf{b}u(t) \tag{3.7.15}$$

$$\mathbf{y}(t) = C\mathbf{x}(t) \tag{3.7.16}$$

where

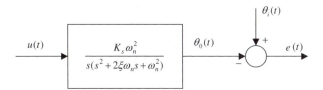

FIGURE 3.13 A model for sun tracking solar concentrator.

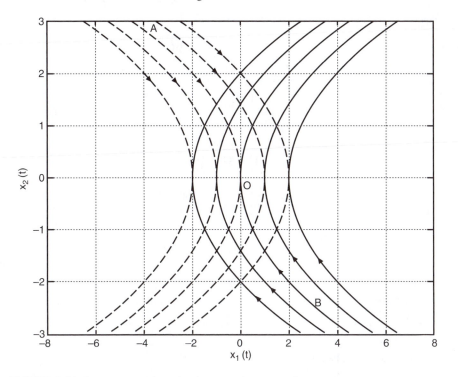

FIGURE 3.14 State space trajectories (____ $u = 1$, ---- $u = 1$).

$$\mathbf{x} = \begin{bmatrix} \theta_0 \\ \dot{\theta}_0 \\ \ddot{\theta}_0 \end{bmatrix}; \quad A = \begin{bmatrix} 0 & 1 & 0 \\ 0 & 0 & 1 \\ 0 & -\omega_n^2 & -2\xi\omega_n \end{bmatrix};$$

$$\mathbf{b} = \begin{bmatrix} 0 \\ 0 \\ K_s\omega_n^2 \end{bmatrix}; \quad \text{and} \quad C = \begin{bmatrix} 1 & 0 & 0 \\ 0 & 1 & 0 \\ 0 & 0 & 1 \end{bmatrix} \qquad (3.7.17)$$

Let $\mathbf{\eta}(t)$ be a vector of the sun's position, velocity, and acceleration, i.e.,

$$\boldsymbol{\eta}(t) = \begin{bmatrix} \theta_i \\ \dot{\theta}_i \\ \ddot{\theta}_i \end{bmatrix} \tag{3.7.18}$$

To minimize the energy loss, the output vector $\mathbf{y}(t)$ should follow the sun's trajectory $\boldsymbol{\eta}(t)$ as closely as possible. Hence, a linear servomechanism or LQ tracking problem is solved. The objective function is

$$J = \frac{1}{2} \int_0^{t_f - t_0} \mathbf{z}^T Q \mathbf{z} + r u^2 \, dt \tag{3.7.19}$$

where

$$\mathbf{z}(t) = \boldsymbol{\eta}(t) - \mathbf{y}(t) \tag{3.7.20}$$

t_0 is the time of sunrise, and t_f is the time of sunset. Using (3.7.14), the control law is given as

$$u(t) = -K(t)\mathbf{x}(t) + r^{-1} b^T \boldsymbol{\xi}(t) \tag{3.7.21}$$

The variable $\boldsymbol{\xi}(t)$ is obtained by solving Equation 3.7.13 with $\mathbf{w}(t) = 0$. Because the desired trajectory $\boldsymbol{\eta}(t)$ changes every day, the variable $\boldsymbol{\xi}(t)$ has to be computed every day.

3.8 FREQUENCY-SHAPED LQ CONTROL

Let the quadratic objective function be

$$I = \int_0^\infty \mathbf{x}(t)^T Q \mathbf{x}(t) + \mathbf{u}(t)^T R \mathbf{u}(t) dt \tag{3.8.1}$$

Using Parseval's Theorem (Appendix B),

$$I = \frac{1}{2\pi} \int_{-\infty}^\infty \mathbf{x}^T(-j\omega) Q \mathbf{x}(j\omega) + \mathbf{u}^T(-j\omega) R \mathbf{u}(j\omega) d\omega \tag{3.8.2}$$

Modify the objective function by making Q and R functions of the frequency ω (Gupta, 1980):

$$I = \frac{1}{2\pi} \int\limits_{-\infty}^{\infty} \mathbf{x}^T(-j\omega)Q(\omega)\mathbf{x}(j\omega) + \mathbf{u}^T(-j\omega)R(\omega)\mathbf{u}(j\omega)d\omega \qquad (3.8.3)$$

Assume that $Q(\omega)$ and $R(\omega)$ are rational functions of ω^2. Factoring these matrices,

$$Q(\omega) = N_q^T(-j\omega)N_q(j\omega) \qquad (3.8.4)$$

$$R(\omega) = N_r^T(-j\omega)N_r(j\omega) \qquad (3.8.5)$$

Let the rank of the matrix $Q(\omega)$ be p. Then, $N_q(j\omega)$ will be an $p \times n$ matrix. For the existence of the solution of an LQ problem, the matrix $R(\omega)$ has to be of full rank. Therefore, $N_r(j\omega)$ will be an $m \times m$ matrix
Define

$$\mathbf{x}^T(-j\omega)Q(\omega)\mathbf{x}(j\omega) = \mathbf{z}^T(-j\omega)\mathbf{z}(j\omega) \qquad (3.8.6)$$

where

$$N_q(j\omega)\mathbf{x}(j\omega) = \mathbf{z}(j\omega) \qquad (3.8.7)$$

Similarly,

$$\mathbf{u}^T(-j\omega)R(\omega)\mathbf{u}(j\omega) = \boldsymbol{\chi}^T(-j\omega)\boldsymbol{\chi}(j\omega) \qquad (3.8.8)$$

where

$$N_r(j\omega)\mathbf{u}(j\omega) = \boldsymbol{\chi}(j\omega) \qquad (3.8.9)$$

The objective function (3.8.3) can be expressed as

$$I = \frac{1}{2\pi} \int\limits_{-\infty}^{\infty} \mathbf{z}^T(-j\omega)\mathbf{z}(j\omega) + \boldsymbol{\chi}^T(-j\omega)\boldsymbol{\chi}(j\omega)d\omega \qquad (3.8.10)$$

Using Parseval's Theorem (Appendix B),

$$I = \int\limits_{0}^{\infty} \mathbf{z}^T(t)\mathbf{z}(t) + \boldsymbol{\chi}^T(t)\boldsymbol{\chi}(t)dt \qquad (3.8.11)$$

From (3.8.7), $\mathbf{z}(t)$ is the output of the linear system with the transfer function matrix $N_q(s)$ and input $\mathbf{x}(t)$; i.e.,

$$\mathbf{z}(s) = N_q(s)\mathbf{x}(s) \tag{3.8.12}$$

Similarly, from (3.8.9), $\boldsymbol{\chi}(t)$ is the output of the linear system with the transfer function matrix $N_r(s)$ and input $\mathbf{u}(t)$; i.e.,

$$\boldsymbol{\chi}(s) = N_r(s)\mathbf{u}(s) \tag{3.8.13}$$

Let the state space model of the MIMO system (3.8.12) be

$$\dot{\boldsymbol{\xi}} = A_q\boldsymbol{\xi} + B_q\mathbf{x}(t) \tag{3.8.14}$$

$$\mathbf{z}(t) = C_q\boldsymbol{\xi} + D_q\mathbf{x}(t) \tag{3.8.15}$$

Similarly, let the state space model of the MIMO system (3.8.13) be

$$\dot{\mathbf{v}} = A_r\mathbf{v} + B_r\mathbf{u}(t) \tag{3.8.16}$$

$$\boldsymbol{\chi}(t) = C_r\mathbf{v} + D_r\mathbf{u}(t) \tag{3.8.17}$$

Define the augmented state vector $\mathbf{x}_a(t)$:

$$\mathbf{x}_a(t) = \begin{bmatrix} \mathbf{x} \\ \boldsymbol{\xi} \\ \mathbf{v} \end{bmatrix} \tag{3.8.18}$$

Let the augmented state space model be

$$\dot{\mathbf{x}}_a(t) = A_a\mathbf{x}_a(t) + B_a\mathbf{u}(t) \tag{3.8.19}$$

where

$$A_a = \begin{bmatrix} A & 0 & 0 \\ B_q & A_q & 0 \\ 0 & 0 & A_r \end{bmatrix} \quad \text{and} \quad B_a = \begin{bmatrix} B \\ 0 \\ B_r \end{bmatrix} \tag{3.8.20}$$

Now, it can be shown that

$$\mathbf{z}^T\mathbf{z}+\boldsymbol{\chi}^T\boldsymbol{\chi}=\mathbf{x}_a^T Q_f \mathbf{x}_a+2\mathbf{x}_a^T N_f \mathbf{u}+\mathbf{u}^T R_f \mathbf{u} \tag{3.8.21}$$

where

$$Q_f = \begin{bmatrix} D_q^T D_q & D_q^T C_q & 0 \\ C_q^T D_q & C_q^T C_q & 0 \\ 0 & 0 & C_r^T C_r \end{bmatrix}, \quad N_f = \begin{bmatrix} 0 \\ 0 \\ C_r^T D_r \end{bmatrix}, \quad \text{and} \quad R_f = D_r^T D_r \tag{3.8.22}$$

Therefore, the objective function (3.8.11) is

$$I = \int_0^\infty \mathbf{x}_a^T Q_f \mathbf{x}_a + 2\mathbf{x}_a^T N_f \mathbf{u} + \mathbf{u}^T R_f \mathbf{u} \, dt \tag{3.8.23}$$

Using (3.5.29), the optimal control law will be

$$\mathbf{u}(t) = -K_a \mathbf{x}_a(t) \tag{3.8.24}$$

where

$$K_a = R_f^{-1}(PB_a + N_f)^T \tag{3.8.25}$$

and

$$P(A_a - B_a R_f^{-1} N_f^T) + (A_a - B_a R_f^{-1} N_f^T)^T P - PB_a R_f^{-1} B_a^T P + Q_f - N_f R_f^{-1} N_f^T = 0 \tag{3.8.26}$$

Representing K_a as

$$K_a = \begin{bmatrix} K_x & K_\xi & K_v \end{bmatrix} \tag{3.8.27}$$

$$\mathbf{u}(t) = -K_x \mathbf{x}(t) - K_\xi \boldsymbol{\xi}(t) - K_v \mathbf{v}(t) \tag{3.8.28}$$

EXAMPLE 3.14: A SIMPLE HARMONIC OSCILLATOR

Consider a second-order system:

$$\ddot{y} + \omega_n^2 y = u(t) \tag{3.8.29}$$

for which the state equations are

$$\dot{\mathbf{x}} = A\mathbf{x} + \mathbf{b}u(t) \tag{3.8.30}$$

$$y(t) = x_1(t) \tag{3.8.31}$$

where

$$\mathbf{x} = \begin{bmatrix} x_1 & x_2 \end{bmatrix}^T \tag{3.8.32}$$

$$A = \begin{bmatrix} 0 & 1 \\ -\omega_n^2 & 0 \end{bmatrix} \quad \text{and} \quad \mathbf{b} = \begin{bmatrix} 0 \\ 1 \end{bmatrix} \tag{3.8.33a,b}$$

Standard LQ Control

Let the objective function be

$$I = \int_0^\infty (y^2 + u^2)\,dt \tag{3.8.34}$$

The optimal state feedback control law is

$$u(t) = -\mathbf{k}\mathbf{x}(t) \tag{3.8.35}$$

where $\mathbf{k}$ is the state feedback gain vector.

For $\omega_n = 10$ rad/sec,

$$\mathbf{k} = \begin{bmatrix} 0.005 & 0.1 \end{bmatrix} \tag{3.8.36}$$

and the eigenvalues of the optimal closed-loop system are $-0.05 \pm 10j$. Therefore, if the disturbance to the system is a sinusoidal function with the frequency equal to 10 rad/sec, its effects will be extremely large.

Frequency-Shaped LQ Control

Let the objective function be

$$I = \frac{1}{2\pi} \int_{-\infty}^{\infty} [|y(j\omega)|^2 |Q(j\omega)|^2 + |u(j\omega)|^2]\,d\omega \tag{3.8.37}$$

where

$$Q(s) = \frac{\omega_n^2 s}{s^2 + 2\varsigma\omega_n s + \omega_n^2}$$ (3.8.38)

For a small value of the damping ratio ς, $|Q(j\omega_n)|$ will be extremely large, and as a result the response of the optimal closed-loop system is expected to be insensitive to external disturbance at the frequency $\omega = \omega_n$. Now, let

$$Q(s)y(s) = z(s)$$ (3.8.39)

The state space realization of the system (3.8.39) can be written as

$$\begin{bmatrix} \dot{\xi}_1 \\ \dot{\xi}_2 \end{bmatrix} = \begin{bmatrix} 0 & 1 \\ -\omega_n^2 & -2\varsigma\omega_n \end{bmatrix}\begin{bmatrix} \xi_1 \\ \xi_2 \end{bmatrix} + \begin{bmatrix} 0 \\ \omega_n^2 \end{bmatrix}y(t)$$ (3.8.40)

and

$$z(t) = \xi_2(t)$$ (3.8.41)

Combining (3.8.30) and (3.8.40),

$$\begin{bmatrix} \dot{x}_1 \\ \dot{x}_2 \\ \dot{\xi}_1 \\ \dot{\xi}_2 \end{bmatrix} = \begin{bmatrix} 0 & 1 & 0 & 0 \\ -\omega_n^2 & 0 & 0 & 0 \\ 0 & 0 & 0 & 1 \\ \omega_n^2 & 0 & -\omega_n^2 & -2\varsigma\omega_n \end{bmatrix}\begin{bmatrix} x_1 \\ x_2 \\ \xi_1 \\ \xi_2 \end{bmatrix} + \begin{bmatrix} 0 \\ 1 \\ 0 \\ 0 \end{bmatrix}u(t)$$ (3.8.42)

From (3.8.39), the objective function (3.8.37) can be written as

$$I = \int_0^\infty (z^2 + u^2)dt$$ (3.8.43)

Equation 3.8.42 and Equation 3.8.43 constitute a standard LQ control problem. The optimal state feedback control law is

$$u(t) = -\mathbf{k}_a[x_1 \quad x_2 \quad \xi_1 \quad \xi_2]^T$$ (3.8.44)

where $\mathbf{k}_a$ is the optimal state feedback gain vector. For $\omega_n = 10$ rad/sec and $\varsigma = 0.1$,

$$\mathbf{k}_a = [4.1717 \quad 2.8885 \quad -4.1717 \quad 0.0178]$$ (3.8.45)

And the eigenvalues of the optimal closed-loop system are $-1.217 \pm 10.9791j$ and $-1.2272 \pm 8.96921j$. Therefore, if the disturbance to the system is a sinusoidal function with frequency 10 rad/sec, its effect will be small.

3.9 MINIMUM-TIME CONTROL OF A LINEAR TIME-INVARIANT SYSTEM

Consider the linear system

$$\frac{d\mathbf{x}}{dt} = A\mathbf{x}(t) + B\mathbf{u}(t); \quad \mathbf{x}(0) = \mathbf{x}_0 \tag{3.9.1}$$

It is desired to apply a control $\mathbf{u}(t)$, such that the system reaches the origin of the state space, in a minimum time when inputs must satisfy the following constraints:

$$\left| u_i(t) \right| \le \alpha_i \ ; \quad i = 1, 2, \dots, m \tag{3.9.2}$$

If the final time is denoted by t_f, the objective is to minimize the following objective function:

$$I = t_f = \int_0^{t_f} 1 dt \tag{3.9.3}$$

Therefore, the Hamiltonian H is

$$H = 1 + \boldsymbol{\lambda}^T (A\mathbf{x} + B\mathbf{u}) \tag{3.9.4}$$

The adjoint equations are

$$\frac{d\boldsymbol{\lambda}}{dt} = -\frac{\partial H}{\partial \mathbf{x}} = -A^T \boldsymbol{\lambda} \tag{3.9.5}$$

The solution of (3.9.5) is

$$\boldsymbol{\lambda}(t) = e^{-A^T t} \boldsymbol{\lambda}(0) \tag{3.9.6}$$

Substituting (3.9.6) into (3.9.4),

$$H = 1 + \boldsymbol{\lambda}^T(0) e^{-At} A\mathbf{x} + \boldsymbol{\lambda}^T(0) e^{-At} B\mathbf{u} \tag{3.9.7}$$

Therefore,

$$\frac{\partial H}{\partial u_i} = p_i \qquad (3.9.8)$$

where

$$p_i = \boldsymbol{\lambda}^T(0)e^{-At}\mathbf{b}_i \qquad (3.9.9)$$

$$B = \begin{bmatrix} \mathbf{b}_1 & \mathbf{b}_2 & \cdot & \cdot & \mathbf{b}_m \end{bmatrix} \qquad (3.9.10)$$

$$\mathbf{u}^T = \begin{bmatrix} u_1 & u_2 & \cdot & \cdot & u_m \end{bmatrix} \qquad (3.9.11)$$

Using (3.4.17) and (3.4.18), the optimal control inputs are

$$u_{0i}(t) = \begin{cases} +\alpha_i & \text{if} & p_i < 0 \\ -\alpha_i & \text{if} & p_i > 0 \end{cases} \qquad (3.9.12)$$

NORMALITY OF LINEAR SYSTEMS

When $p_i(t) = 0$ over a finite interval $[t_1 \quad t_2]$, it is called a singular control case. This also implies that higher derivatives of $p_i(t)$, with respect to time, are zero over this finite interval. Therefore, for $t \in [t_1, t_2]$,

$$p_i = \boldsymbol{\lambda}^T(0)e^{-At}\mathbf{b}_i = 0$$

$$-\frac{dp_i}{dt} = \boldsymbol{\lambda}^T(0)e^{-At}A\mathbf{b}_i = 0$$

$$\frac{d^2 p_i}{dt^2} = \boldsymbol{\lambda}^T(0)e^{-At}A^2\mathbf{b}_i = 0 \qquad (3.9.13)$$

$$\vdots$$

$$(-1)^{n-1}\frac{d^{n-1}p_i}{dt^{n-1}} = \boldsymbol{\lambda}^T(0)e^{-At}A^{n-1}\mathbf{b}_i = 0$$

Equation 3.9.13 can be expressed as

$$\boldsymbol{\lambda}^T(0)e^{-At}C_i = 0 \qquad (3.9.14)$$

where

$$C_i = [\mathbf{b}_i \quad A\mathbf{b}_i \quad A^2\mathbf{b}_i \quad . \quad . \quad A^{n-1}\mathbf{b}_i] \tag{3.9.15}$$

is the controllability matrix with respect to u_i.

It should be noted that $\lambda(0) \neq 0$ because $H = 1$ otherwise; i.e., $H \neq 0$ and the solution will not be optimal (Section 3.4.2). Therefore, for Equation 3.9.14 to be true,

$$rank(C_i) < n \tag{3.9.16}$$

This analysis implies that there are no finite intervals on which $p_i(t) = 0$, provided C_i is not singular for any i, $i = 1, 2, \ldots m$. A system for which C_i is not singular for any i is called *normal* (Gopal, 1984).

EXISTENCE AND UNIQUENESS THEOREMS ON MINIMUM-TIME CONTROL (GOPAL, 1984)

1. If the linear time-invariant system is controllable, and if all the eigenvalues of A have nonpositive real parts, a time-optimal control exists that transfers any initial state to the origin of the state space in a minimum time.
2. If the linear time-invariant system is normal, and if the time-optimal control exists, it is unique.
3. If eigenvalues of the matrix A are real and a unique time-optimal exists, each control component can switch at the most $(n-1)$ times, where n is the dimension of the state space.

EXAMPLE 3.15: TIME-OPTIMAL CONTROL OF A RIGID BODY OR A DOUBLE INTEGRATOR SYSTEM

Consider the system

$$\frac{d\mathbf{x}}{dt} = A\mathbf{x}(t) + Bu(t) \tag{3.9.17}$$

where

$$A = \begin{bmatrix} 0 & 1 \\ 0 & 0 \end{bmatrix} \quad \text{and} \quad B = \begin{bmatrix} 0 \\ 1 \end{bmatrix} \tag{3.9.18}$$

and

$$|u| \leq 1 \tag{3.9.19}$$

The objective is to find the optimal control $u(t)$, such that the system reaches the origin of the state space in a minimum time from an initial state $\mathbf{x}(0)$.

First, note that eigenvalues of A are 0 and 0, which are real and nonpositive. Furthermore,

$$C = \begin{bmatrix} 0 & 1 \\ 1 & 0 \end{bmatrix} \tag{3.9.20}$$

Therefore, the system is controllable and normal as well. Application of the existence and uniqueness theorem indicates that a unique optimal control exists with at the most one switching.

$$\text{Hamiltonian } H = 1 + \lambda_1 x_2 + \lambda_2 u \tag{3.9.21}$$

$$\frac{\partial H}{\partial u} = \lambda_2 \tag{3.9.22}$$

Hence, the optimal control input is given by

$$u(t) = \begin{cases} -1 & if \quad \lambda_2 > 0 \\ +1 & if \quad \lambda_2 < 0 \end{cases} \tag{3.9.23}$$

Equation 3.9.23 can be expressed as

$$u(t) = -\operatorname{sgn}(\lambda_2(t)) \tag{3.9.24}$$

Adjoint equations are

$$\frac{d\lambda_1}{dt} = -\frac{\partial H}{\partial x_1} = 0 \tag{3.9.25}$$

$$\frac{d\lambda_2}{dt} = -\frac{\partial H}{\partial x_2} = -\lambda_1 \tag{3.9.26}$$

The solution of (3.9.25) is

$$\lambda_1(t) = \lambda_1(0) \tag{3.9.27}$$

Substituting (3.9.27) into (3.9.26),

$$\lambda_2(t) = -\lambda_1(0)t + \lambda_2(0) \tag{3.9.28}$$

Because states are specified at initial and final time, there are no constraints on adjoint variables. Hence, state equations must be used to determine adjoint variables.

If $u = 1$, the solution to (3.9.17) is

$$x_2(t) = t + x_2(0) \tag{3.9.29}$$

$$x_1(t) = \frac{t^2}{2} + x_2(0)t + x_1(0) \tag{3.9.30}$$

Eliminating t between (3.9.29) and (3.9.30),

$$x_1 = \frac{x_2^2}{2} + x_1(0) - \frac{(x_2(0))^2}{2} \tag{3.9.31}$$

Equation 3.9.31 describes state space trajectories (which are parabolas) when $u = 1$. These trajectories are shown as solid curves in Figure 3.14.

If $u = -1$, solution to (3.9.17) is

$$x_2(t) = -t + x_2(0) \tag{3.9.32}$$

$$x_1(t) = -\frac{t^2}{2} + x_2(0)t + x_1(0) \tag{3.9.33}$$

Eliminating t between (3.9.32) and (3.9.33),

$$x_1 = -\frac{x_2^2}{2} + x_1(0) + \frac{(x_2(0))^2}{2} \tag{3.9.34}$$

Equation 3.9.34 describes state space trajectories (which are parabolas) when $u = -1$. These trajectories are shown as dashed curves in Figure 3.14.

Because it is known that the optimal control input can only have one switching at the most, only four cases exist:

Case I:

$$u = 1, \ 0 \le t \le t_f \tag{3.9.35}$$

Initial states must be such that the system is on segment BO in Figure 3.14.

Case II:

$$u = -1, \quad 0 \le t \le t_f \tag{3.9.36}$$

Initial states must be such that the system is on segment AO in Figure 3.14.

Case III:

$$u = \begin{cases} +1 & 0 \le t \le t_s \\ -1 & t_s < t \le t_f \end{cases} \tag{3.9.37}$$

Initial states must be such that the system will follow one of solid parabolas, which intersect the segment AO. At the switching instant t_s, the system will reach the segment AO and go to the origin of the state space along the segment AO.

Case IV:

$$u = \begin{cases} -1 & 0 \le t \le t_s \\ +1 & t_s < t \le t_f \end{cases} \tag{3.9.38}$$

Initial states must be such that the system will follow one of dashed parabolas, which intersect the segment BO. At the switching instant t_s, the system will reach the segment BO and go to the origin of the state space along the segment BO.

The switching curve is composed of segments AO and BO (Figure 3.14). For segment AO,

$$x_1(0) + \frac{(x_2(0))^2}{2} = 0 \tag{3.9.39}$$

Equation 3.9.39 implies the following equation for the curve AO:

$$x_1(t) = -\frac{1}{2}(x_2(t))^2 \tag{3.9.40}$$

For segment BO,

$$x_1(0) - \frac{(x_2(0))^2}{2} = 0 \tag{3.9.41}$$

Equation 3.9.41 implies the following equation for the curve BO:

$$x_1(t) = \frac{1}{2}(x_2(t))^2 \tag{3.9.42}$$

Combining (3.9.40) and (3.9.42), the equation of the switching curve AOB can be expressed as

$$x_1(t) = -\frac{1}{2}x_2(t)|x_2(t)| \tag{3.9.43}$$

On the basis of (3.9.43), a switching function (Gopal, 1984) is defined as follows:

$$s(\mathbf{x}(t)) = x_1(t) + \frac{1}{2}x_2(t)|x_2(t)| \tag{3.9.44}$$

If $s > 0$, $\mathbf{x}(t)$ lies above the curve AOB. In this case, $u = -1$ because dotted
  parabolas are directed toward the switching curve.
If $s < 0$, $\mathbf{x}(t)$ lies below the curve AOB. In this case, $u = 1$ because solid
  parabolas are directed toward the switching curve.
If $s = 0$ and $x_2 > 0$, $\mathbf{x}(t)$ lies on the curve AO. In this case, $u = -1$.
If $s = 0$ and $x_2 < 0$, $\mathbf{x}(t)$ lies on the curve BO. In this case, $u = 1$.

In summary, the time-optimal control in feedback form is

$$u(t) = \begin{cases} -1 & when & s(\mathbf{x}(t)) > 0 \\ +1 & when & s(\mathbf{x}(t)) = 0 & and & x_2(t) < 0 \\ +1 & when & s(\mathbf{x}(t)) < 0 \\ -1 & when & s(\mathbf{x}(t)) = 0 & and & x_2(t) > 0 \end{cases} \tag{3.9.45}$$

The structure of the optimal feedback control system is shown in Figure 3.15.

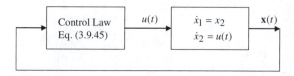

FIGURE 3.15 Implementation of bang-bang control via state feedback.

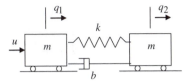

FIGURE 3.16 A system with rigid and flexible modes.

EXAMPLE 3.16: MINIMUM-TIME CONTROL OF A SYSTEM WITH RIGID AND FLEXIBLE MODES

The differential equations of motion of the mechanical system shown in Figure 3.16 are

$$\begin{bmatrix} m & 0 \\ 0 & m \end{bmatrix}\begin{bmatrix} \ddot{q}_1 \\ \ddot{q}_2 \end{bmatrix} + \begin{bmatrix} b & -b \\ -b & b \end{bmatrix}\begin{bmatrix} \dot{q}_1 \\ \dot{q}_2 \end{bmatrix} + \begin{bmatrix} k & -k \\ -k & k \end{bmatrix}\begin{bmatrix} q_1 \\ q_2 \end{bmatrix} = \begin{bmatrix} u(t) \\ 0 \end{bmatrix} \quad (3.9.46)$$

Natural frequencies are found to be

$$\omega_0^2 = 0 \quad (3.9.47)$$

and

$$\omega_1^2 = \frac{2k}{m} \quad (3.9.48)$$

Modal vectors are as follows:

$$\begin{bmatrix} 1 & 1 \end{bmatrix}^T \text{ corresponding to } \omega_0^2 = 0 \text{ (Rigid mode)} \quad (3.9.49)$$

and

$$\begin{bmatrix} 1 & -1 \end{bmatrix}^T \text{ corresponding to } \omega_1^2 = 2k/m \text{ (Flexible mode)} \quad (3.9.50)$$

Now, the displacement vector can be represented as a linear combination of these modal vectors:

$$\begin{bmatrix} q_1 \\ q_2 \end{bmatrix} = \Phi \begin{bmatrix} r_1 \\ r_2 \end{bmatrix} \quad (3.9.51)$$

where

$$\Phi = \begin{bmatrix} 1 & 1 \\ 1 & -1 \end{bmatrix} \tag{3.9.52}$$

Substituting (3.9.51) into (3.9.46), and premultiplying both sides by Φ^T,

$$\ddot{r}_1 = \frac{1}{2m} u \tag{3.9.53}$$

and

$$\ddot{r}_2 + 2\varsigma\omega_1\dot{r}_2 + \omega_1^2 r_2 = \frac{1}{2m} u \tag{3.9.54}$$

where $\varsigma = 2b/(2m\omega_1)$. Defining state variables as

$$x_1 = r_1, \ x_2 = \dot{r}_1, \ x_3 = r_2, \ \text{and} \ x_4 = \dot{r}_2 \tag{3.9.55}$$

State equations are

$$\dot{x} = Ax(t) + bu(t) \tag{3.9.56}$$

where

$$A = \begin{bmatrix} 0 & 1 & 0 & 0 \\ 0 & 0 & 0 & 0 \\ 0 & 0 & 0 & 1 \\ 0 & 0 & -\omega_1^2 & -2\varsigma\omega_1 \end{bmatrix} \quad \text{and} \quad b = \begin{bmatrix} 0 \\ 0.5m \\ 0 \\ 0.5m \end{bmatrix} \tag{3.9.57}$$

Let the initial and final states be

$$x(0) = \begin{bmatrix} -L & 0 & 0 & 0 \end{bmatrix}^T \quad \text{and} \quad x(t_f) = 0 \quad \text{(3.9.58a and b)}$$

and

$$|u(t)| \leq \alpha \qquad (3.9.59)$$

The minimum time control will be bang-bang and unique. However, because all eigenvalues are not real, the maximum number of switching is not necessarily $(n-1)$. Let switching instants be t_i, $i = 1, 2, \ldots, k$, and initially $u(t) = \alpha$ as the position of the system has to be increased. Solving state equations with this input and initial condition (3.9.58a), and then imposing the final condition (3.9.58b), the following nonlinear equations (Pao and Singhose, 1998; Singhose and Pao, 1997) are obtained:

$$(-1)^k t_f + 2 \sum_{i=1}^{k} (-1)^{i-1} t_i \qquad (3.9.60)$$

$$(-1)^{k+1} t_f^2 + 2 \sum_{i=1}^{k} (-1)^i (t_i)^2 = \frac{4mL}{\alpha} \qquad (3.9.61)$$

$$1 + (-1)^{k+1} e^{\varsigma \omega_1 t_f} \cos(\omega_d t_f) + 2 \sum_{i=1}^{k} (-1)^i e^{\varsigma \omega_1 t_i} \cos(\omega_d t_i) = 0 \qquad (3.9.62)$$

$$(-1)^{k+1} e^{\varsigma \omega_1 t_f} \sin(\omega_d t_f) + 2 \sum_{i=1}^{k} (-1)^i e^{\varsigma \omega_1 t_i} \sin(\omega_d t_i) = 0 \qquad (3.9.63)$$

where $\omega_d = \omega_n \sqrt{1 - \varsigma^2}$. There are many solutions of t_i, $i = 1, 2, \ldots, k$, that satisfy (3.9.60) to (3.9.63). However, only one of them will satisfy Equation 3.9.12.

EXAMPLE 3.17: INPUT SHAPING

Consider the prototype second-order system (Kuo, 1995):

$$\ddot{y} + 2\xi\omega_n \dot{y} + \omega_n^2 y = \omega_n^2 u(t) \qquad (3.9.64)$$

where ξ, ω_n, and $u(t)$ are the damping ratio, undamped natural frequency, and the reference input, respectively. The reference input is often a unit step function for which the response of an underdamped system is

$$y(t) = 1 - e^{-\xi\omega_n t} \cos \omega_d t - \chi e^{-\xi\omega_n t} \sin \omega_d t; \quad t \geq 0 \qquad (3.9.65)$$

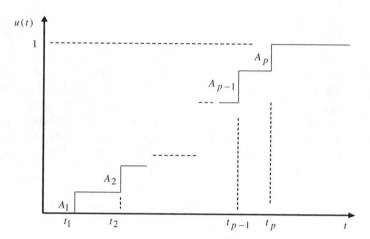

FIGURE 3.17 A staircase input.

where

$$\chi = \frac{\xi}{\sqrt{1-\xi^2}} \quad \text{and} \quad \omega_d = \omega_n\sqrt{1-\xi^2} \qquad (3.9.66\text{a,b})$$

The value of the steady state error is guaranteed to be zero. However, there is a large amount of oscillation, and the settling time can be large for a small damping level.

The concept of input shaping is to modify the command input so that that the system reaches the final position with zero oscillation. For the staircase input shown in Figure 3.17, the response can be written as

$$y(t) = \sum_{i=1}^{p} A_i [1 - e^{-\xi\omega_n(t-t_i)}(\cos\omega_d(t-t_i) + \chi\sin\omega_d(t-t_i))]u_s(t-t_i) \quad (3.9.67)$$

where $u_s(t-t_i)$ is a unit step function applied at $t = t_i$.

For $t > t_p$,

$$y(t) = g(t) + \sum_{i=1}^{p} A_i \qquad (3.9.68)$$

where

$$g(t) = -e^{-\xi\omega_n t}[\alpha \sin \omega_d t + \beta \cos \omega_d t] \qquad (3.9.69)$$

$$\alpha = \sum_{i=1}^{p} e^{\xi\omega_n t_i} A_i (\sin \omega_d t_i + \chi \cos \omega_d t_i) \qquad (3.9.70)$$

$$\beta = \sum_{i=1}^{p} e^{\xi\omega_n t_i} A_i (\cos \omega_d t_i - \chi \sin \omega_d t_i) \qquad (3.9.71)$$

The conditions for $g(t)$ to be zero for $t > t_p$ are

$$\alpha = \sum_{i=1}^{p} e^{\xi\omega_n t_i} A_i (\sin \omega_d t_i + \chi \cos \omega_d t_i) = 0 \qquad (3.9.72)$$

$$\beta = \sum_{i=1}^{p} e^{\xi\omega_n t_i} A_i (\cos \omega_d t_i - \chi \sin \omega_d t_i) = 0 \qquad (3.9.73)$$

Equation 3.9.72 and Equation 3.9.73 are equivalent to the following well-known conditions (Singer and Seering, 1990):

$$\sum_{i=1}^{p} e^{\xi\omega_n t_i} A_i \sin \omega_d t_i = 0 \qquad (3.9.74)$$

$$\sum_{i=1}^{p} e^{\xi\omega_n t_i} A_i \cos \omega_d t_i = 0 \qquad (3.9.75)$$

Consider the case of $p = 2$. Without any loss of generality, assume that

$$t_1 = 0 \quad \text{and} \quad A_1 + A_2 = 1 \qquad (3.9.76a \text{ and } b)$$

In this case, Equation 3.9.74 and Equation 3.9.75 can be solved for two unknowns. The results are

$$t_2 = \frac{\pi}{\omega_d}, \quad A_1 = \frac{1}{1+K}, \quad \text{and} \quad K = e^{-\frac{\xi\pi}{\sqrt{1-\xi^2}}} \qquad (3.9.77a, b, \text{ and } c)$$

Therefore, from (3.9.68),

$$y(t) = 1 \quad \text{for } t > t_2 \qquad (3.9.78)$$

In other words, the system reaches the desired value without any oscillation for $t > t_2$, when the input $u(t)$ is as follows:

$$u(t) = \begin{cases} (1+K)^{-1}; & 0 < t \le \pi/\omega_d \\ 1; & t > \pi/\omega_d \end{cases} \qquad (3.9.79)$$

Time Optimality of the Shaped Input

Define

$$x_1 = y - 1, \quad x_2 = \dot{y}, \quad v = u - 1 \qquad (3.9.80)$$

The state equations are

$$\dot{\mathbf{x}} = A\mathbf{x} + \mathbf{b}v(t) \qquad (3.9.81)$$

where

$$\mathbf{x} = \begin{bmatrix} x_1 & x_2 \end{bmatrix}^T \qquad (3.9.82)$$

$$A = \begin{bmatrix} 0 & 1 \\ -\omega_n^2 & -2\xi\omega_n \end{bmatrix} \quad \text{and} \quad \mathbf{b} = \begin{bmatrix} 0 \\ \omega_n^2 \end{bmatrix} \qquad (3.9.83a,b)$$

Now, initial and final states are

$$\mathbf{x}(0) = \begin{bmatrix} -1 & 0 \end{bmatrix}^T \quad \text{and} \quad \mathbf{x}(t_f) = \begin{bmatrix} 0 & 0 \end{bmatrix}^T \qquad (3.9.84a,b)$$

where t_f is the final time. Another constraint for a positive shaper is that the input $u(t)$ should be between 0 and 1. This implies the following constraint on $v(t)$:

$$-1 \le v(t) \le 0 \qquad (3.9.85)$$

For minimum-time control, the Hamiltonian H is defined as follows:

$$H = 1 + \lambda_1 x_2 + \lambda_2 (-\omega_n^2 x_1 - 2\xi\omega_n x_2 + \omega_n^2 v) \tag{3.9.86}$$

where adjoint variables $\lambda_1(t)$ and $\lambda_2(t)$ satisfy the following equations:

$$\dot{\lambda}_1 = -\frac{\partial H}{\partial x_1} = \omega_n^2 \lambda_2 \tag{3.9.87}$$

$$\dot{\lambda}_2 = -\frac{\partial H}{\partial x_2} = -\lambda_1 + 2\xi\omega_n \lambda_2 \tag{3.9.88}$$

and

$$\frac{\partial H}{\partial v} = \omega_n^2 \lambda_2 \tag{3.9.89}$$

Therefore, the minimum-time control will depend on the sign of $\lambda_2(t)$. Differentiating (3.9.88) and using (3.9.87),

$$\ddot{\lambda}_2 - 2\xi\omega_n \dot{\lambda}_2 + \omega_n^2 \lambda_2 = 0 \tag{3.9.90}$$

Solving (3.9.90),

$$\lambda_2(t) = e^{\xi\omega_n t} \left[\frac{\dot{\lambda}_2(0) - \xi\omega_n \lambda_2(0)}{\omega_d} \sin\omega_d t + \lambda_2(0)\cos\omega_d t \right] \tag{3.9.91}$$

From (3.9.88),

$$\dot{\lambda}_2(0) = -\lambda_1(0) + 2\xi\omega_n \lambda_2(0) \tag{3.9.92}$$

Substituting (3.9.92) into (3.9.91),

$$\lambda_2(t) = e^{\xi\omega_n t} \left[\frac{-\lambda_1(0) + \xi\omega_n \lambda_2(0)}{\omega_d} \sin\omega_d t + \lambda_2(0)\cos\omega_d t \right] \tag{3.9.93}$$

Because both $\mathbf{x}(0)$ and $\mathbf{x}(t_f)$ are specified, $\lambda_1(0)$ and $\lambda_2(0)$ are not specified, and have to be determined so that the resulting input leads to satisfaction of initial and final conditions of states. To achieve the control input (3.9.79) that leads to satisfaction of initial and final conditions of states,

$$\lambda_2(0) = 0 \quad \text{and} \quad \lambda_1(0) < 0 \tag{3.9.94}$$

Then, from (3.9.93),

$$\lambda_2(t) = -\lambda_1(0)e^{\xi\omega_n t} \sin\omega_d t > 0 \quad \text{for } 0 < t < \frac{\pi}{\omega_d} \tag{3.9.95}$$

and

$$\lambda_2(t) = -\lambda_1(0)e^{\xi\omega_n t} \sin\omega_d t < 0 \quad \text{for } \frac{\pi}{\omega_d} < t < \frac{2\pi}{\omega_d} \tag{3.9.96}$$

The control input (3.9.79) is minimum-time and bang-bang if the constraint (3.9.85) is modified as follows:

$$-1 + (1 + K)^{-1} \le v(t) \le 0 \tag{3.9.97}$$

EXERCISE PROBLEMS

P3.1   Consider a linear system described by the following transfer function:

$$\frac{y(s)}{u(s)} = \frac{10}{s(s+1)}$$

Design a state feedback controller such that the eigenvalues of the closed-loop system are at −2 and −3.

P3.2   Consider a linear system described by the following state space equation:

$$\frac{dx_1}{dt} = x_2$$

$$\frac{dx_2}{dt} = -x_2 + u(t)$$

$$y(t) = x_1(t) + x_2(t)$$

i. Find the open-loop eigenvalues.

ii. Find the state feedback gain vector such that both eigenvalues for the closed-loop system are located at -2.

iii. Let $u(t) = -k_1 x_1(t) - k_2 x_2(t)$, and k_1 and k_2 be as calculated in part ii. Furthermore, $y(0) = 0$ and $\dfrac{dy}{dt}(0) = 1$.

   a. Determine $x_1(0)$ and $x_2(0)$

   b. Determine the response $x_1(t)$ and $x_2(t)$ via the matrix exponential.

P3.3   Consider a linear system described by the transfer function

$$\frac{y(s)}{u(s)} = \frac{10(s+3)}{s(s+1)(s+2)}$$

   a. Develop a state space realization and construct the block diagram.

   b. Find the state feedback gain vector such that eigenvalues of the closed-loop system are located at -2, -3, and -4.

   c. Examine the controllability and observability of the closed-loop system.

P3.4   Consider the single-link flexible robot manipulator (Figure P3.4) which is driven at one end by a DC motor. The input torque is represented by $u(t)$. Assume that the response of the system can be adequately represented by considering only two modes of vibration. Parameters for the manipulator are given as follows:

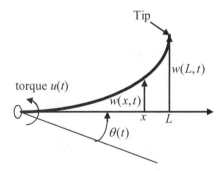

FIGURE P3.4 A single-link flexible manipulator.

$$EI = 2.04146 \text{ N-m}^2 \text{ , } L = 1.05 \text{ m}$$

$$m = 0.4252 \text{ kg, } I_\alpha = 0.15628 \text{ kg-m}^2$$

where $E =$ Young's modulus of elasticity, $I =$ area moment of inertia of the beam cross-section, $L =$ length of the beam, $\rho =$ mass of the beam per unit length, $m =$ mass of the beam, and $I_\alpha =$ mass moment of inertia of the beam about the torque axis $= mL^2/3$.

The governing partial differential equation (PDE) of motion is

$$EI \frac{\partial^4 y}{\partial x^4} + \rho \frac{\partial^2 y}{\partial t^2} = 0 \quad \text{where} \quad y(x,t) = w(x,t) + x\theta(t)$$

with the following boundary conditions:

$$EI \frac{\partial^2 y}{\partial x^2}\Big|_{x=0} + u(t) = 0$$

$$w(0,t) = 0$$

$$EI \frac{\partial^2 y}{\partial x^2}\Big|_{x=L} = 0$$

$$EI \frac{\partial^3 y}{\partial x^3}\Big|_{x=L} = 0$$

The solution of this PDE can be expressed as

$$y(x,t) = \sum_{i=0}^{\infty} \psi_i(x) q_i(t)$$

where $i = 0$ corresponds to the rigid body mode with

$$\psi_0(x) = x \quad \text{and} \quad q_0(t) = \theta(t)$$

$\psi_i(x)$ with $i \geq 1$ are mode shapes of a pinned-free beam, which are expressed as

$$\psi_i(x) = \sin \alpha_i x + \frac{\sin \alpha_i L}{\sinh \alpha_i L} \sinh \alpha_i x$$

where α_i is the solution of the following transcendental equation:

$$\tan \alpha_i L = \tanh \alpha_i L$$

Natural frequencies ω_i of the pinned-free beam are related to α_i as follows:

$$\omega_i^2 = \frac{EI\alpha_i^4}{\rho}$$

Finally, it can be shown that

$$\frac{d^2 q_0}{dt^2} = \frac{u(t)}{I_\alpha}$$

$$\frac{d^2 q_i}{dt^2} + \omega_i^2 q_i = \frac{\psi_i'(0)}{\rho a_i} u(t); \quad i = 1, 2, \ldots, \infty$$

Design a full state feedback controller such that all the closed-loop poles are located at $-\sigma$, $-\sigma$, $-\sigma \pm j\omega_1$, and $-\sigma \pm j\omega_2$. Choose σ such that the damping ratio in each vibratory mode is higher than 0.1.

P3.5   Consider a linear system for which the transfer function is given as

$$\frac{y(s)}{u(s)} = \frac{s}{(s-1)(s+1)}$$

i. Construct a suitable state space realization and find the state feedback gain vector such that the closed-loop poles are located at -2 and -3.

ii. Consider the control law $u(t) = v(t) - k_1 x_1(t) - k_2 x_2(t)$, where $x_1(t)$ and $x_2(t)$ are states. Can a nonzero value of the steady state output be obtained? Explain your answer.

iii. Let $u(t) = v(t) - k_1 x_1(t) - k_2 x_2(t)$. Determine the zeros of the closed-loop transfer function.

P3.6   Consider the following system:

$$\frac{dx}{dt} = -x(t) + u(t) + w$$

where w is a scalar constant disturbance of *unknown* magnitude and $u(t)$ is the control input.

a. Let $u(t) = -k\, x(t)$. Find an expression for the steady state value of x and show that the steady state value of x cannot be zero for a finite value of k.

b. Develop a suitable state feedback control law such that $x(t) \to x_d \neq 0$ as $t \to \infty$ for finite values of state feedback gains. Justify your answer.

P3.7   Consider a system with the following transfer function matrix:

$$G(s) = \begin{bmatrix} 0 & 1 \end{bmatrix} \begin{bmatrix} -\dfrac{s-2}{2} & \dfrac{s}{2} \\ s & 0 \end{bmatrix}^{-1}$$

   i. Find a state space realization in controller form.

   ii. Determine the state feedback gain matrix corresponding to the controller form state space realization such that the closed-loop eigenvalues are located at −1 and −2 and the closed-loop eigenvectors are $[1 \quad 0]^T$ and $[0 \quad 1]^T$.

P3.8   Consider the flexible tetrahedral truss structure (Appendix F) with collocated sensors and actuators. The controller is to be designed on the basis of the first four modes of vibration.

   Design a full state feedback controller such that all the closed-loop poles are located at $-0.1\omega_i \pm j\omega_i$; $i = 1, 2, 3$, and 4. Find at least two state feedback gain matrices.

P3.9   Find the open-loop optimal control inputs $u_1(t)$ (and $u_2(t)$) for the system

$$\frac{dx_1}{dt} = x_2 + u_1$$

$$\frac{dx_2}{dt} = u_2(t)$$

which minimizes

$$I = \frac{1}{2} \int_0^2 [u_1^2(t) + u_2^2(t)]dt$$

Given: $x_1(0) = x_2(0) = 1$ and $x_1(2) = 0$.

P3.10 The system

$$\frac{dx}{dt} = -x(t) + u(t)$$

is to be transferred from $x(0) = 5$ to $x(4) = 0$ such that

$$I = \frac{1}{2} \int_0^4 u^2(t)dt$$

is minimized. Find the optimal control input $u(t)$.

P3.11 Consider the Example 3.9. It is desired to move the load shaft with inertia 1 kg-m² by 30° in 2 sec and minimize the work done by the DC motor. Furthermore, this load shaft must start from rest and end at rest. The parameters for the DC motor are as follows: $k_b = 0.04297$ V-sec/rad, $R_a = 1.025$ Ω, and $k_i = 0.04297$ N-m/A.

   i. Set up a suitable objective function such that the optimal controller can be implemented in a closed-loop fashion. Find the optimal feedback gain vector.

  ii. Find the open-loop optimal input voltage for the DC motor such that the constraints on final states are exactly satisfied. Compare the optimal control input to that for item i.

P3.12 Consider the following system:

$$\frac{dx}{dt} = x(t) + u(t)$$

It is desired to develop an optimal state feedback law such that the following objective function is minimized:

$$I = \frac{1}{2}\int_0^{t_f} [x^2(t) + u^2(t)]dt$$

  a. For a finite time t_f, determine the equation for finding the optimal state feedback gain.

  b. If $t_f \to \infty$, what is the optimal state feedback gain?

P3.13 Consider the second-order system

$$\frac{dx_1}{dt} = x_2$$

$$\frac{dx_2}{dt} = -x_1 - 2x_2 + u(t)$$

and the performance index

$$I = 20x_1^2(5) + \int_0^5 x_1^2(t) + 2x_2^2(t) + x_1(t)x_2(t) + u^2(t)dt$$

  a. Set up the Riccati differential equations with proper boundary conditions.

b. Solve the Riccati differential equation using the Matlab routine ode23 or ode45, and find the time-varying state feedback gain vector and the closed-loop response. Also, find the optimal value of I.

P3.14   Consider a plant consisting of a DC motor, the shaft of which has the angular velocity $\omega(t)$. The DC motor is driven by the input voltage $u(t)$. The dynamics of the deviation in the angular speed from its constant nominal value $\bar{\omega}$ is given by

$$\frac{d\omega}{dt} = -0.5(\omega(t) - \bar{\omega}) + (u(t) - \bar{u})$$

where $\bar{u}$ is the voltage corresponding to $\bar{\omega}$. A feedback control system is to be designed such that the following objective function is minimized:

$$I = \int_0^\infty [(\omega(t) - \bar{\omega})^2 + r(u(t) - \bar{u})^2]dt \; ; \quad r > 0$$

i. Find the optimal state feedback gain by solving the algebraic Riccati equation.

ii. Using the gain found in item i, find the response of the closed-loop system for a specified $\omega(0) - \bar{\omega}$.

iii. Assuming that $\omega(0) - \bar{\omega} = 1$ unit, determine the value of r such that the magnitude of optimal $u(t) - \bar{u}$ never exceeds 1 unit.

P3.15   An unstable spring-mass-damper is described by the following differential equation:

$$\frac{d^2x}{dt^2} - \frac{dx}{dt} + x = u(t)$$

where $u(t)$ is the control input. It is desired to design a state feedback control system such that

$$I = \int_0^\infty x^2 + ru^2 \, dt$$

is minimized. Assuming $r \to \infty$, find the optimal location of poles and the corresponding state feedback gain vector.

P3.16   A simple spring-mass-damper system is shown in Figure P3.16a. Because of some external disturbances, the initial displacement $x(0)$ and veloc-

ity $\dfrac{dx}{dt}(0)$ are not equal to zero. It is desired to introduce additional spring (stiffness $= k_1$) and viscous damper (damping coefficient $= c_1$) such that the following objective function is minimized:

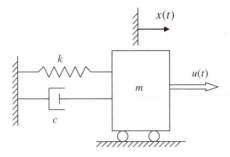

FIGURE P3.16A A spring-mass system.

$$I = \int_0^\infty x^2 + ru^2 \, dt \; ; \quad r \to 0$$

where $u(t)$ is the force acting on the unit mass because of additional spring and viscous damper (Figure P3.16b). Calculate the optimal values of k_1 and c_1.

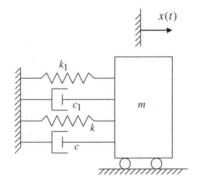

FIGURE P3.16B Additional spring and damper.

P3.17 Consider a plant with the following transfer function:

$$\frac{y(s)}{u(s)} = \frac{(s+1)(s-1)}{(s^2 - s + 1)(s+2)}$$

It is desired to develop a state feedback law such that the following objective function is minimized:

$$I = \frac{1}{2}\int_0^\infty y^2(t) + ru^2(t)dt \quad \text{where } r > 0$$

a. Draw the symmetric root locus plot.
b. Find the optimal closed-loop poles when $r \to 0$.
c. Find the optimal closed-loop poles when $r \to \infty$.

P3.18 Consider the system shown in Figure 5.11.4 in which the torque $T(t)$ is the control input. Design a full state feedback controller such that the following objective function is minimized:

$$I = \frac{1}{2}\int_0^\infty \theta_1^2 + rT^2 dt \quad \text{where } r > 0$$

Construct the symmetric root locus plot to determine optimal closed-loop pole locations. Show the performance of your controller as r changes from 0 to ∞.

Parameters (ECP Manual):

$$J_1 = 0.0024 \text{ kg-m}^2, J_2 = 0.0019 \text{ kg-m}^2, J_3 = 0.0019 \text{ kg-m}^2$$

$$k_1 = k_2 = 2.8 \text{ N-m/rad}$$

$$c_1 = 0.007 \text{ N-m/rad/sec}, c_2 = c_3 = 0.001 \text{ N-m/rad/sec}$$

$$k_{hw} = \frac{100}{6} \text{ units}$$

P3.19 Refer to Example 3.11. Select $q_1 = 10$ and $q_2 = 1$.
a. Find optimal state feedback gain for three different values of ρ: 10^{-7}, 10^{-8}, and 10^{-9}. In each case, assume that the initial condition on state vector is $-[1 \quad 1 \quad 0 \quad 0]^T$ and plots $x_1(t)$, $x_2(t)$, and $u(t)$ vs. time. Discuss your results.
b. Refer to Equation 3.5.88 and Figure 3.10. For all the three aforementioned values of ρ, find λ_2 and β_2. Also, plot $f(t)$ vs. time for initial state vector chosen in part a. Discuss your results.

P3.20 The following model is developed for longitudinal pressure oscillation in a uniform chamber (Yang et al., 1992):

$$\ddot{\eta}_i + \omega_i^2 \eta_i + \sum_{\ell=1}^n (D_{i\ell}\dot{\eta}_\ell + E_{i\ell}\eta_\ell) = v_i(t) ; \, i = 1, 2, ..., n$$

$$v_i(t) = \frac{2\bar{a}^2}{\alpha} \sum_{\ell=1}^{m} \psi_i(z_{a\ell}) \frac{\bar{u}_\ell(t)}{\bar{p}}$$

where $\bar{p}$ and $\bar{u}_\ell(t)$ are the mean pressure and the pressure excitation supplied by the actuator# ℓ located at $z_{a\ell}$, respectively. The normal mode# i of the chamber is described by

$$\psi_i(z) = \cos(i\pi z / \alpha); \quad 0 \leq z \leq \alpha$$

where α is the length of the chamber.

$D_{i\ell}$	$\ell = 1$	$\ell = 2$	$\ell = 3$	$\ell = 4$
$i = 1$	−0.01	0.007	−0.001	0.007
$i = 2$	0.01	0.1	0.007	−0.001
$i = 3$	−0.01	0.01	0.75	0.008
$i = 4$	0.02	−0.005	0.01	1.50

$E_{i\ell}$	$\ell = 1$	$\ell = 2$	$\ell = 3$	$\ell = 4$
$i = 1$	−0.005	−0.005	0.0025	0.0016
$i = 2$	−0.0025	−0.015	0.01	0.01
$i = 3$	−0.005	0.0	−0.02	0.02
$i = 4$	0.01	0.02	0.02	−0.025

a. Develop a state space realization and find the open-loop poles with $n = 4$.
b. Select a suitable location of a single actuator with $n = 4$, and find the state feedback gain vector to locate the poles such that closed-loop frequencies are same as those of the open loop, and there is at least a 5% damping ratio in each mode.
c. With the location of the actuator in part b, draw the symmetric root locus for the following objective function:

$$\int_0^\infty \left(\sum_{i=1}^{n} \eta_i \right)^2 + r u_1^2 dt$$

d. Select suitable locations for two actuators, and find a state feedback gain matrix to locate the poles, such that closed-loop frequencies are same as those of open loop, and there is at least a 5% damping ratio in each mode.
e. With the location of the actuator in part b, draw the symmetric root locus for the following objective function:

$$\int_0^\infty \left[\left(\sum_{i=1}^n \eta_i \right)^2 + u_1^2 + u_2^2 \right] dt$$

P3.21 Consider the flexible tetrahedral truss structure (Appendix F). The controller is to be designed on the basis of first four modes of vibration. Design a full state feedback LQR controller such that

$$\frac{1}{2} \int_0^\infty \mathbf{y}^T(t)\mathbf{y}(t) + \rho \mathbf{u}^T(t)\mathbf{u}(t) dt$$

is minimized. Demonstrate your controller performance for two values of $\rho = 0.1$ and 1.

P3.22 Consider the single-link flexible manipulator described in Problem P3.4. Consider two vibratory modes for the link. It is given that $u < 1$ N-m. If the link is to be rotated by 30°, find the switching instants and move time for the bang-bang (minimum-time) control. The link starts from the rest and there should not be any vibration when the arm reaches its final position.

4 Control with Estimated States

In Chapter 3, a state feedback control system was developed under the assumption that all states are available to us. In general, it will be impractical to measure all the states. Therefore, methods to estimate states are developed on the basis of inputs and outputs of a deterministic and a stochastic system as well. Then, theories and examples of optimal state estimation and linear quadratic Gaussian control are presented.

4.1 OPEN-LOOP OBSERVER

Consider the system

$$\dot{\mathbf{x}} = A\mathbf{x}(t) + B\mathbf{u}(t) \tag{4.1.1}$$

If the initial conditions are estimated to be

$$\hat{\mathbf{x}}(0) = \mathbf{x}(0) + \boldsymbol{\varepsilon} \tag{4.1.2}$$

where $\boldsymbol{\varepsilon}$ is the error in the estimate of the initial state vector. Then, states can be computed by the solution of the following equation (Kailath, 1980):

$$\dot{\hat{\mathbf{x}}} = A\hat{\mathbf{x}} + B\mathbf{u}(t) \tag{4.1.3}$$

Define

$$\tilde{\mathbf{x}} = \mathbf{x}(t) - \hat{\mathbf{x}}(t) \tag{4.1.4}$$

Errors in the state vector can be obtained as the solution of

$$\dot{\tilde{\mathbf{x}}} = A\tilde{\mathbf{x}}(t); \quad \tilde{\mathbf{x}}(0) = -\boldsymbol{\varepsilon} \tag{4.1.5}$$

Hence, if the matrix A is stable, $\tilde{\mathbf{x}}(t) \to 0$ as $t \to \infty$. On the other hand, if the matrix A is unstable, $\tilde{\mathbf{x}}(t) \to \infty$ as $t \to \infty$.

This strategy of estimating states will be useful only when all the eigenvalues of the matrix are far inside the left half of the s-plane because only in this case will errors die out quickly. But, in most cases that need control, the eigenvalues of the matrix A will be close to the imaginary axis, or the matrix A will be unstable. In these cases, initial state errors will either persist for a long time or lead to unbounded state errors.

4.2 CLOSED-LOOP OBSERVER

The algorithm presented in Section 4.1 works in an open-loop fashion. Measured outputs, $\mathbf{y}(t)$, are not used at all. At any instant, estimated outputs are

$$\hat{\mathbf{y}}(t) = C\hat{\mathbf{x}}(t) \tag{4.2.1}$$

Define an error signal,

$$\delta\mathbf{y}(t) = \mathbf{y}(t) - \hat{\mathbf{y}}(t) \tag{4.2.2}$$

If there is no estimation error, $\delta\mathbf{y}(t) = 0$. The signal $\delta\mathbf{y}(t)$ can be used as feedback variables to influence the dynamics of estimated states as follows (Kailath, 1980):

$$\dot{\hat{\mathbf{x}}} = A\hat{\mathbf{x}} + B\mathbf{u}(t) + L(\mathbf{y}(t) - \hat{\mathbf{y}}(t)) \; ; \quad \hat{\mathbf{x}}(0) = \hat{\mathbf{x}}_0 \tag{4.2.3}$$

where L is an observer gain matrix (Figure 4.1). Here, the following questions arise:

1. What is the role of L in having estimated states $\hat{\mathbf{x}}(t)$ converge to true states $\mathbf{x}(t)$?
2. Is it possible to select L for a guaranteed convergence of estimated states $\hat{\mathbf{x}}(t)$ to true states $\mathbf{x}(t)$?
3. Can we select L to achieve a desired rate of convergence of estimated states $\hat{\mathbf{x}}(t)$ to true states $\mathbf{x}(t)$?
4. What should be the value of $\hat{\mathbf{x}}(0)$?

To answer these questions, the estimator equation (4.2.3) is subtracted from the state equation (4.1.1):

$$\dot{\tilde{\mathbf{x}}} = A\tilde{\mathbf{x}} - L(\mathbf{y}(t) - C\hat{\mathbf{x}}(t))$$

$$= (A - LC)\tilde{\mathbf{x}} \; ; \quad \tilde{\mathbf{x}}(0) = \mathbf{x}(0) - \hat{\mathbf{x}}(0) \tag{4.2.4}$$

If the matrix $(A - LC)$ is stable, $\lim \tilde{\mathbf{x}}(t) \to 0$ as $t \to \infty$, or $\mathbf{x}(t) \to \hat{\mathbf{x}}(t)$ as $t \to \infty$. Equation 4.2.3 is also called the Luenberger observer. Usually, $\hat{\mathbf{x}}(0) = 0$.

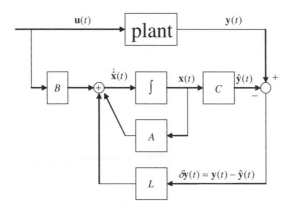

FIGURE 4.1 Schematic drawing of the closed-loop observer.

Theorem

Eigenvalues of $(A - LC)$ can be arbitrarily located in the complex plane, provided the realization (A, C) is observable.

Proof

As eigenvalues of a square matrix and its transpose are identical,

$$\det(sI - (A - LC)) = \det(sI - (A - LC)^T) = \det(sI - (A^T - C^T L^T)) \quad (4.2.5)$$

With the interpretation (Kailath, 1980)

$$A \rightarrow A^T \; ; \quad B \rightarrow C^T \; ; \quad \text{and} \quad K \rightarrow L^T \quad (4.2.6)$$

the problem of finding L is dual to that of finding K (see Chapter 3). Therefore, eigenvalues of $(A^T - C^T L^T)$ can be arbitrarily placed in the complex plane, provided (A^T, C^T) is controllable. It is easy to see that the controllability of (A^T, C^T) is equivalent to the observability of (A, C). This completes the proof.

Definition 4.1.1: Detectability

The detectability of a system is a weaker condition than the observability. It only requires that all the unstable modes of the system are observable. The system (A, C) is called detectable if there exists a matrix L such that the matrix $(A\text{-}LC)$ is stable (Zhou et al., 1996).

4.2.1 DETERMINATION OF OBSERVER GAIN VECTOR *L* FOR A SINGLE-OUTPUT SYSTEM

Let $\boldsymbol{\alpha}$ be the vector of the coefficients of desired characteristic polynomial of the matrix $(A - L\mathbf{c})$; i.e.,

$$\boldsymbol{\alpha} = [\alpha_1 \quad \alpha_2 \quad . \quad . \quad \alpha_n] \tag{4.2.7}$$

$$\det(sI - (A - L\mathbf{c})) = s^n + \alpha_1 s^{n-1} + \ldots + \alpha_{n-1}s + \alpha_n \tag{4.2.8}$$

Therefore,

$$(\boldsymbol{\alpha} - \mathbf{a}) = L^T C_* U_t \tag{4.2.9}$$

where

$$C_* = [\mathbf{c}^T \quad A^T \mathbf{c}^T \quad . \quad . \quad (A^T)^{n-1} \mathbf{c}^T] = O^T \tag{4.2.10}$$

where O is the observability matrix. Hence, from Equation 3.1.18,

$$L^T = (\boldsymbol{\alpha} - \mathbf{a})(O^T U_t)^{-1} = (\boldsymbol{\alpha} - a)(U_t)^{-1}(O^T)^{-1} \tag{4.2.11}$$

Because the matrix U_t is nonsingular, the solution of L can be obtained, provided O^T or, equivalently, O is nonsingular. In other words, the existence of L is guaranteed for an arbitrary choice of $\boldsymbol{\alpha}$, provided the realization (A, C) is observable.

4.2.2 DETERMINATION OF OBSERVER GAIN MATRIX L FOR A MULTIPLE-OUTPUT SYSTEM

For a multioutput system, the gain matrix L is not unique for specified locations of observer poles. A method, dual to that presented in Chapter 3, Section 3.2, can be developed to determine the gain matrix L.

4.2.3 LOCATIONS OF OBSERVER POLES

When observer poles are located farther into the left half of the s-plane, errors in estimated states go to zero at a faster rate. In general, this will also imply higher values of observer gains, which do not cost anything in terms of the actuator size or the magnitude of the control input. However, the observer starts behaving like a differentiator as the poles are pushed farther into the left half of the s-plane. As a result, high-frequency stochastic noises in the system can get amplified, and the system performance can be reduced. Therefore, the optimal selection of observer poles requires considerations of stochastic disturbances in the system, and will be presented in Section 4.6 as the *Kalman filter* (Kalman and Bucy, 1961).

EXAMPLE 4.1: AN AORTIC PRESSURE OBSERVER FOR AN ARTIFICIAL HEART

The Penn State electric ventricular assist device (EVAD) consists of a high torque brushless DC motor coupled to a pusher plate by a roller-screw mechanism (Tsach

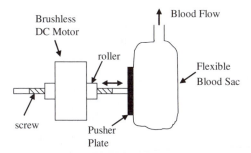

FIGURE 4.2 Schematic drawing for Penn State electric ventricular assist device (EVAD).

et al., 1990). The blood flow is generated by moving the pusher plate against a flexible blood sac (Figure 4.2).

The state space equations can be written as

$$\dot{\mathbf{x}} = A\mathbf{x}(t) + \mathbf{b}u(t) \tag{4.2.12}$$

$$\mathbf{x}(t) = \begin{bmatrix} x_1(t) \\ x_2(t) \\ x_3(t) \end{bmatrix} \quad A = \begin{bmatrix} 0 & 1 & 0 \\ 0 & -68.3 & -7.2 \\ 0 & 3.2 & -0.7 \end{bmatrix} \quad \mathbf{b} = \begin{bmatrix} 0 \\ 425.4 \\ 0 \end{bmatrix} \quad (4.2.13\text{a,b,c})$$

where states x_1, x_2, and x_3 are pusher plate's position (mm), velocity (mm/sec), and aortic pressure (mm Hg), respectively. The input, $u(t)$, is the DC motor voltage.

The position of the pusher plate is measured by a sensor; i.e., the output is x_1, and the output equation is

$$y(t) = \mathbf{c}\mathbf{x}(t) \tag{4.2.14}$$

where

$$\mathbf{c} = \begin{bmatrix} 1 & 0 & 0 \end{bmatrix} \tag{4.2.15}$$

In this case, eigenvalues of A are 0, -1.0426, and -67.9574. The open-loop characteristic polynomial is

$$\det(sI - A) = s^3 + 69s^2 + 70.85s \tag{4.2.16}$$

The observability matrix is

$$O = \begin{bmatrix} 1 & 0 & 0 \\ 0 & 1 & 0 \\ 0 & -68.3 & -7.2 \end{bmatrix}$$ (4.2.17)

As $\det O \neq 0$, the state space realization is observable. Now,

$$\det(sI - A + L\mathbf{c}) = s^3 + (69 + \ell_1)s^2 + (70.85 + 69\ell_1 + \ell_2)s + \\ 70.85\ell_1 + 0.7\ell_2 - 7.2\ell_3$$ (4.2.18)

where

$$L = \begin{bmatrix} \ell_1 & \ell_2 & \ell_3 \end{bmatrix}^T$$ (4.2.19)

Let the eigenvalues of the observer be –3, –27, and –67.9574. Then, the desired characteristic polynomial is

$$\det(sI - A + L\mathbf{c}) = s^3 + 98s^2 + 2119.7s + 5504.5$$ (4.2.20)

Comparing coefficients,

$$69 + \ell_1 = 98$$ (4.2.21)

$$70.85 + 69\ell_1 + \ell_2 = 2119.7$$ (4.2.22)

$$70.85\ell_1 + 0.7\ell_2 - 7.2\ell_3 = 5504.5$$ (4.2.23)

Therefore,

$$\ell_1 = 29, \ \ell_2 = 47.5, \text{ and } \ell_3 = -474.527$$ (4.2.24)

EXAMPLE 4.2: A DISTURBANCE OBSERVER

Consider a spring-mass-damper system (Chapter 2, Figure 2.4) subjected to a sinusoidal excitation. Then, the state equation (2.5.16) will be modified to be

$$\dot{\mathbf{x}} = A\mathbf{x}(t) + \mathbf{b}u(t) + \mathbf{b}w(t)$$ (4.2.25)

where

$$w(t) = w_0 \sin(\omega t + \phi) \tag{4.2.26}$$

It is assumed that the excitation frequency ω is known. However, the amplitude w_0 and phase ϕ are not known. The objective is to design an observer to determine $w(t)$ or, equivalently, the amplitude w_0 and the phase ϕ.

The excitation $w(t)$ satisfies the following second-order equation:

$$\ddot{w} + \omega^2 w = 0 \tag{4.2.27}$$

Converting (4.2.27) into a state space model,

$$\begin{bmatrix} \dot{w}_1 \\ \dot{w}_2 \end{bmatrix} = \begin{bmatrix} 0 & 1 \\ -\omega^2 & 0 \end{bmatrix} \begin{bmatrix} w_1 \\ w_2 \end{bmatrix} ; \quad w_1 = w \tag{4.2.28}$$

Combining (4.2.25) and (4.2.28),

$$\dot{\mathbf{x}}_a = A_a \mathbf{x}_a(t) + \mathbf{b}_a u(t) \tag{4.2.29}$$

where

$$\mathbf{x}_a = \begin{bmatrix} x_1 \\ x_2 \\ w_1 \\ w_2 \end{bmatrix} ; \quad A_a = \begin{bmatrix} 0 & 1 & 0 & 0 \\ -\beta/m & -\alpha/m & 1 & 0 \\ 0 & 0 & 0 & 1 \\ 0 & 0 & -\omega^2 & 0 \end{bmatrix} ; \quad \mathbf{b}_a = \begin{bmatrix} 0 \\ 1 \\ 0 \\ 0 \end{bmatrix} \tag{4.2.30a,b,c}$$

If the position of the mass is measured by a sensor, the output equation is

$$y(t) = \mathbf{c}\mathbf{x}_a(t) \tag{4.2.31}$$

where

$$\mathbf{c} = [1 \quad 0 \quad 0 \quad 0] \tag{4.2.32}$$

The observability matrix is

$$O = \begin{bmatrix} 1 & 0 & 0 & 0 \\ 0 & 1 & 0 & 0 \\ -\beta/m & -\alpha/m & 1 & 0 \\ \alpha\beta/m^2 & -(\beta/m)+(\alpha^2/m^2) & -\alpha/m & 1 \end{bmatrix} \tag{4.2.33}$$

Therefore,

$$\det O = 1 \tag{4.2.34}$$

And the realization is observable for all system parameters. Let $m = 1$ kg, $\alpha = 1$ N-sec/m, $\beta = 1 N/m$, and $\omega = 1$ rad/sec. In this case, eigenvalues of A_a are $-0.5 \pm 0.86603j$ and $\pm 1j$. The open-loop characteristic polynomial is

$$\det(sI - A_a) = s^4 + s^3 + 2s^2 + s + 1 \tag{4.2.35}$$

$$\mathbf{a} = [1 \quad 1 \quad 1 \quad 1] \tag{4.2.36}$$

Let the eigenvalues of the observer be $-0.5 \pm 0.86603j$ and $-0.5 \pm 1j$. Then, the desired characteristic polynomial is

$$\det(sI - A_a + L\mathbf{c}) = s^4 + 2s^3 + 3.25s^2 + 2.25s + 1.25 \tag{4.2.37}$$

and

$$\boldsymbol{\alpha} = [2 \quad 3.25 \quad 2.25 \quad 1.25] \tag{4.2.38}$$

Using Equation 4.2.11,

$$L^T = [1 \quad 0.25 \quad 0.25 \quad -1]$$

Simulation results are shown in Figure 4.3, in which the estimated disturbance is shown by the dotted line.

4.3 COMBINED OBSERVER–CONTROLLER

In Chapter 3, we designed the control system with the full state feedback. Because all states are not available, we can only feed back estimated states. Let the control law be defined (Figure 4.4) as

$$u(t) = v(t) - K\hat{\mathbf{x}}(t) \tag{4.3.1}$$

where

$$\dot{\hat{\mathbf{x}}} = A\hat{\mathbf{x}} + B\mathbf{u}(t) + L(\mathbf{y}(t) - \hat{\mathbf{y}}(t)) \tag{4.3.2}$$

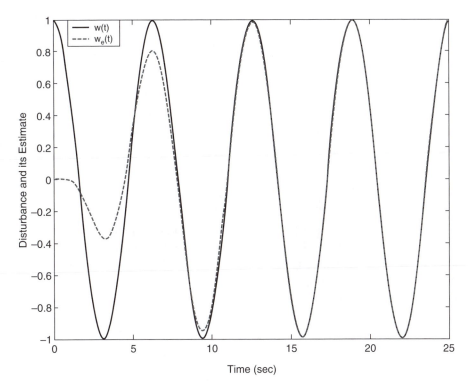

FIGURE 4.3 Estimation of the disturbance.

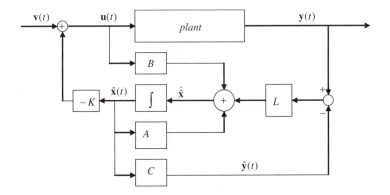

FIGURE 4.4 Combined observer–controller system.

Recall the plant dynamics

$$\dot{\mathbf{x}} = A\mathbf{x} + B\mathbf{u}(t) \tag{4.3.3}$$

Subtracting (4.3.2) from (4.3.3), the state error $\tilde{\mathbf{x}} = \mathbf{x} - \hat{\mathbf{x}}$ is still governed by

$$\dot{\tilde{\mathbf{x}}} = (A - LC)\tilde{\mathbf{x}} \tag{4.3.4}$$

From (4.3.1) and (4.3.3),

$$\dot{\mathbf{x}} = (A - BK)\mathbf{x} + BK\tilde{\mathbf{x}} + B\mathbf{v}(t) \tag{4.3.5}$$

Representing Equation 4.3.4 and Equation 4.3.5 in the matrix form,

$$\begin{bmatrix} \dot{\mathbf{x}} \\ \dot{\tilde{\mathbf{x}}} \end{bmatrix} = \begin{bmatrix} (A - BK) & BK \\ 0 & (A - LC) \end{bmatrix} \begin{bmatrix} \mathbf{x} \\ \tilde{\mathbf{x}} \end{bmatrix} + \begin{bmatrix} B \\ 0 \end{bmatrix} \mathbf{v}(t) \tag{4.3.6}$$

Equation 4.3.6 represents the dynamics of the closed-loop system (Figure 4.4). Note that the order of the closed-loop system is $2n$ where the dynamics of the plant is of the order n and observer dynamics is also of the order n. The system matrix for the closed-loop system is

$$A_{sys} = \begin{bmatrix} (A - BK) & BK \\ 0 & A - LC \end{bmatrix} \tag{4.3.7}$$

Therefore,

$$\det(sI_{2n} - A_{sys}) = \det \begin{bmatrix} sI_n - (A - BK) & -BK \\ 0 & sI_n - (A - LC) \end{bmatrix} \tag{4.3.8}$$

$$= \det(sI_n - (A - BK)) \det(sI_n - (A - LC))$$

The eigenvalues of the closed-loop system, A_{sys}, are same as those of $(A - BK)$ and $(A - LC)$. Hence, K and L are to be chosen such that eigenvalues of $(A - BK)$ and $(A - LC)$ are located at desired places in the left half of the complex plane, respectively. In other words, K can be chosen as if all states are available for feedback, and L can be chosen as if only an observer as described in Section 4.2 is to be designed. This is called the *separation property*, which implies that the observer and controller can be designed independent of each other and the stability of the closed-loop system is guaranteed. However, as far as the performance of the closed-loop system is concerned, there is certainly a coupling between selection of K and L. A rule of thumb that has often been used is to have eigenvalues of $(A - LC)$ at least four times faster than those of $(A - BK)$. But, more precisely, they can be selected to minimize certain performance criteria; e.g., LQG, H_∞ control, etc.

The control law, (4.3.1) and (4.3.2), can be represented as a classical output feedback control system shown in Figure 4.5. Assuming $\mathbf{v}(t) = 0$, Equation 4.3.1 and Equation 4.3.2 can also be written as

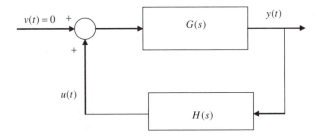

FIGURE 4.5 Combined observer–controller system as an output feedback compensator.

$$\dot{\hat{\mathbf{x}}} = (A - BK - LC)\hat{\mathbf{x}} + L\mathbf{y}(t) \tag{4.3.9}$$

and

$$u(t) = -K\hat{\mathbf{x}}(t) \tag{4.3.10}$$

Hence, the controller transfer function $H(s)$ can be expressed as

$$H(s) = -K(sI - (A - BK - LC))^{-1}L \tag{4.3.11}$$

It is important to realize that the stability of $H(s)$ or, equivalently, $(A - BK - LC)$ is not guaranteed. It is possible to have an unstable $(A - BK - LC)$ while $(A - BK)$ and $(A - LC)$ are stable.

The transfer function of the closed-loop system is

$$\mathbf{y}(s) = \begin{bmatrix} C & 0 \end{bmatrix} (sI_{2n} - A_{sys})^{-1} \begin{bmatrix} B \\ 0 \end{bmatrix} \mathbf{v}(s) \tag{4.3.12}$$

Because of the block-diagonal structure of A_{sys},

$$(sI_{2n} - A_{sys})^{-1} = \begin{bmatrix} (sI_n - (A - BK))^{-1} & ? \\ 0 & (sI_n - (A - Lc))^{-1} \end{bmatrix} \tag{4.3.13}$$

Substituting (4.3.13) into (4.3.12),

$$\mathbf{y}(s) = C(sI_n - (A - BK))^{-1}B\mathbf{v}(s) \tag{4.3.14}$$

This transfer function does not depend on the observer dynamics (Kailath, 1980), and it is exactly the same as that obtained with the full state feedback in Chapter 3.

This can be understood by realizing that a transfer function is defined for zero initial conditions. In this case, $\tilde{\mathbf{x}}(0) = 0$, and the solution of Equation 4.3.4 yields

$$\tilde{\mathbf{x}}(t) = 0 \quad \text{for all } t \geq 0 \tag{4.3.15}$$

Therefore, an observer is not needed at all to estimate states. This result also highlights what an observer really guarantees. It basically drives the state errors caused by errors in initial values of states to zero. It may not compensate for modeling errors and unknown external disturbances.

EXAMPLE 4.3: BALANCING A POINTER

Consider Example 3.1, Balancing a Pointer, in Chapter 3. If the length of the pointer is 0.98 m,

$$A = \begin{bmatrix} 0 & 1 \\ 10 & 0 \end{bmatrix} \quad \text{and} \quad \mathbf{b} = \begin{bmatrix} 0 \\ -1 \end{bmatrix} \tag{4.3.16}$$

Let the output $y(t)$ be the angular position of the pointer $\phi(t)$; i.e.,

$$\mathbf{c} = \begin{bmatrix} 1 & 0 \end{bmatrix} \tag{4.3.17}$$

To locate the controller poles at -1 and -1, Equation 3.1.32 yields

$$\mathbf{k} = \begin{bmatrix} -11 & -2 \end{bmatrix} \tag{4.3.18}$$

Next,

$$\det(sI - (A - Lc)) = s^2 + \ell_1 s + (\ell_2 - 10) \tag{4.3.19}$$

Let the observer poles be at -4 and -4. Then, the desired characteristic equation will be

$$\det(sI - (A - Lc)) = (s + 4)^2 = s^2 + 8s + 16 \tag{4.3.20}$$

Comparing coefficients of polynomials (4.3.19) and (4.3.20),

$$\ell_1 = 8 \quad \text{and} \quad \ell_2 = 16 \tag{4.3.21}$$

The controller equations (4.3.9 and 4.3.10), which are to be solved in real time, are as follows:

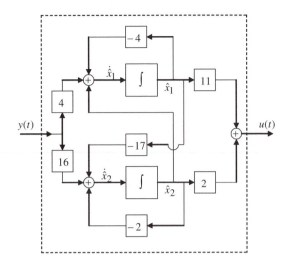

FIGURE 4.6 Observer-based state feedback controller to balance a pointer.

$$u(t) = 11\hat{x}_1(t) + 2\hat{x}_2(t) \tag{4.3.22}$$

where

$$\dot{\hat{x}}_1(t) = -4\hat{x}_1 + \hat{x}_2(t) + 4y(t) ; \quad \hat{x}_1(0) = 0 \tag{4.3.23}$$

$$\dot{\hat{x}}_2(t) = -17\hat{x}_1 - 2\hat{x}_2(t) + 16y(t) ; \quad \hat{x}_2(0) = 0 \tag{4.3.24}$$

The analog computer simulation diagram for the observer-based state feedback controller is shown in Figure 4.6. The controller transfer function is

$$H(s) = -\mathbf{k}(sI - (A - \mathbf{bk} - L\mathbf{c}))^{-1}L = \frac{76s + 256}{s^2 + 6s + 25} \tag{4.3.25}$$

4.4 REDUCED-ORDER OBSERVER

Because p outputs $\mathbf{y}(t)$ are linear combinations of n states $\mathbf{x}(t)$, output equations $\mathbf{y}(t) = C\mathbf{x}(t)$ represent p equations in n unknowns where $p < n$. Therefore, there is no need to estimate all the states, and the order of the observer dynamics can be reduced to be $n–p$. Here, as an example, a single-output system (Kailath, 1980) is considered. Let

$$\mathbf{c} = \begin{bmatrix} 0 & 0 & . & . & 0 & 1 \end{bmatrix} \tag{4.4.1}$$

Then, $y = x_n$; i.e., x_n is directly being measured. Therefore, we have only to observe or estimate

$$\mathbf{x}_r = \begin{bmatrix} x_1 & x_2 & . & . & x_{n-1} \end{bmatrix}^T \tag{4.4.2}$$

Now, the state space equation is partitioned as

$$\begin{bmatrix} \dot{x}_r \\ \dot{x}_n \end{bmatrix} = \begin{bmatrix} A_r & \mathbf{b}_r \\ \mathbf{c}_r & a_{nn} \end{bmatrix} \begin{bmatrix} \mathbf{x}_r \\ x_n \end{bmatrix} + \begin{bmatrix} \mathbf{g}_r \\ g_n \end{bmatrix} u \tag{4.4.3}$$

or

$$\dot{x}_r = A_r \mathbf{x}_r + \mathbf{b}_r y + \mathbf{g}_r u \tag{4.4.4}$$

$$\dot{x}_n = \mathbf{c}_r \mathbf{x}_r + a_{nn} y + g_n u \tag{4.4.5}$$

Define

$$y_r = \mathbf{c}_r \mathbf{x}_r \tag{4.4.6}$$

Then, Equation 4.4.5 can be written as

$$y_r = \dot{y} - a_{nn} y - g_n u \tag{4.4.7}$$

Because the right-hand side of Equation 4.4.7 is known, y_r can be viewed as the output, and c_r as the output vector. Also, from Equation 4.4.4, $\mathbf{b}_r y + \mathbf{g}_r u$ can be considered as the input for the $\mathbf{x}_r$ dynamics. Analogous to the equation for the full-order observer dynamics, the reduced-order observer can be now described by

$$\dot{\hat{x}}_r = A_r \hat{x}_r + \mathbf{b}_r y + \mathbf{g}_r u + \ell_r (y_r - \mathbf{c}_r \hat{x}_r) \tag{4.4.8}$$

where ℓ_r is an $(n-1)$ dimensional vector which has to be determined. Define the error in the estimated state vector as

$$\tilde{\mathbf{x}}_r = \mathbf{x}_r - \hat{\mathbf{x}}_r \tag{4.4.9}$$

Subtracting (4.4.9) from (4.4.4),

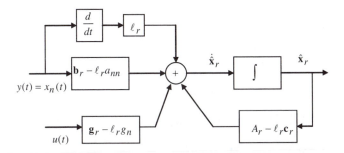

FIGURE 4.7 Reduced-order observer with a differentiator.

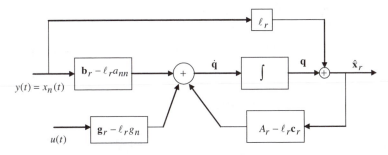

FIGURE 4.8 Reduced-order observer without a differentiator.

$$\dot{\tilde{\mathbf{x}}}_\mathbf{r} = (A_r - \ell_r\mathbf{c}_r)\tilde{\mathbf{x}}_\mathbf{r} \tag{4.4.10}$$

It has been shown that $\{\mathbf{c}_r, A_r\}$ is observable, provided $\{\mathbf{c}, A\}$ is observable (Kailath, 1980). Therefore, ℓ_r can be calculated for any choice of the desired characteristic polynomial for (4.4.10) when the state space realization is observable. We can then guarantee that $\tilde{\mathbf{x}}_r(t) \to 0$ as $t \to \infty$ irrespective of the value of $\tilde{\mathbf{x}}_r(0)$.

The problem of observing unmeasured states has been solved only in theory because the knowledge of y_r via (4.4.7) implies differentiating the output signal to obtain $\dot{y}$. The resulting system is shown in Figure 4.7.

From (4.4.7) and (4.4.8),

$$\begin{aligned}\dot{\hat{\mathbf{x}}}_r &= (A_r - \ell_r\mathbf{c}_r)\hat{\mathbf{x}}_r + \mathbf{b}_r y + \mathbf{g}_r u + \ell_r(\dot{y} - a_{nn}y - g_n u) \\ &= (A_r - \ell_r\mathbf{c}_r)\hat{x}_r + (\mathbf{b}_r - \ell_r a_{nn})y + (\mathbf{g}_r - \ell_r g_n)u + \ell_r\dot{y}\end{aligned} \tag{4.4.11}$$

In order to get rid of the differentiator, the system in Figure 4.7 is modified by changing the path with the block ℓ_r from the input to the output side of the integrator (Figure 4.8). The output of the integrator is now

$$\mathbf{q}(t) = \hat{\mathbf{x}}_r(t) - \ell_r y(t) \tag{4.4.12}$$

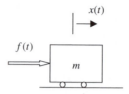

FIGURE 4.9 A simple mass.

From (4.4.11) and (4.4.12),

$$\dot{\mathbf{q}} = (A_r - \ell_r \mathbf{c}_r)\mathbf{q}(t) + (A_r \ell_r - \ell_r \mathbf{c}_r \ell_r + \mathbf{b}_r - \ell_r a_{nn})y(t) + (\mathbf{g}_r - \ell_r g_n)u(t) \quad (4.4.13)$$

With $y(t)$ and $u(t)$ known, Equation 4.4.13 has to be solved in real time to determine $\mathbf{q}(t)$. Then the states are estimated by

$$\hat{\mathbf{x}}_r(t) = \mathbf{q}(t) + \ell_r y(t) \quad (4.4.14)$$

EXAMPLE 4.4: A SIMPLE MASS

Consider a simple mass m which is subjected to a control force $f(t)$ (Figure 4.9). The differential equation of motion is

$$\ddot{x} = u \quad (4.4.15)$$

where

$$u(t) = \frac{f(t)}{m} \quad (4.4.16)$$

Define states as

$$x_1(t) = \dot{x} \quad \text{and} \quad x_2 = x \quad (4.4.17)$$

Then, the state equations are

$$\begin{bmatrix} \dot{x}_1 \\ \dot{x}_2 \end{bmatrix} = \begin{bmatrix} 0 & 0 \\ 1 & 0 \end{bmatrix} \begin{bmatrix} x_1 \\ x_2 \end{bmatrix} + \begin{bmatrix} 1 \\ 0 \end{bmatrix} u \quad (4.4.18)$$

Let us assume that we have only one sensor that can measure the position of the mass. Then, the output equation is

$$y(t) = \begin{bmatrix} 0 & 1 \end{bmatrix} \begin{bmatrix} x_1 \\ x_2 \end{bmatrix} \tag{4.4.19}$$

The objective is to design a reduced-order observer to estimate the velocity of the mass. This is interesting because the velocity will be estimated from the position of the mass via an integrator, i.e., without using a differentiator. In view of the partitioning shown in Equation 4.4.3,

$$A = \begin{bmatrix} 0 & 0 \\ 1 & 0 \end{bmatrix} = \begin{bmatrix} A_r & b_r \\ c_r & a_{nn} \end{bmatrix} \tag{4.4.20}$$

$$\mathbf{b} = \begin{bmatrix} 1 \\ 0 \end{bmatrix} = \begin{bmatrix} g_r \\ g_n \end{bmatrix} \tag{4.4.21}$$

Hence, the state error dynamics is written as

$$\dot{\tilde{x}}_1 = (A_r - l_r c_r)\tilde{x}_1 = -l_r \tilde{x}_1 \tag{4.4.22}$$

Solving (4.4.22),

$$\tilde{x}_1(t) = \tilde{x}_1(0)e^{-l_r t} \tag{4.4.23}$$

Equation 4.4.13 is written as

$$\dot{q} = -l_r q - l_r^2 y + u = -l_r(q + l_r y) + u = -l_r \hat{x}_1 + u \tag{4.4.24}$$

where

$$\hat{x}_1(t) = q(t) + l_r y(t) \tag{4.4.25}$$

The simulation diagram for the velocity estimator is shown in Figure 4.10.

4.5   RESPONSE OF A LINEAR CONTINUOUS-TIME SYSTEM TO WHITE NOISE

Consider the following linear system:

$$\dot{\mathbf{x}} = A\mathbf{x}(t) + B\mathbf{w}(t) \tag{4.5.1}$$

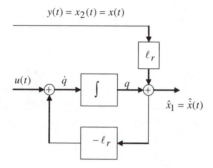

FIGURE 4.10 A reduced-order observer to estimate velocity.

where $\mathbf{w}(t)$ is a zero-mean white noise vector (Appendix D) with the intensity Q; i.e.,

$$E(\mathbf{w}(t)) = 0 \tag{4.5.2}$$

$$E[\mathbf{w}(t)\mathbf{w}^T(t+\tau)] = Q\delta(\tau) \tag{4.5.3}$$

Initial condition $\mathbf{x}(0)$ is assumed to be a random variable vector independent of $\mathbf{w}(t)$; i.e.,

$$E[\mathbf{x}(0)\mathbf{w}^T(\tau)] = E[\mathbf{x}(0)]E[\mathbf{w}^T(\tau)] = \mathbf{0} \tag{4.5.4a}$$

and

$$E(\mathbf{x}(0)) = \overline{\mathbf{x}}_0 \tag{4.5.4b}$$

$$E[(\mathbf{x}(0)\mathbf{x}^T(0)] = P_0 \tag{4.5.5}$$

The solution of (4.5.1) is

$$\mathbf{x}(t) = e^{At}\mathbf{x}(0) + \int_0^t e^{A(t-\tau)}B\mathbf{w}(\tau)d\tau \tag{4.5.6}$$

Mean of x(t)

$$E(\mathbf{x}(t)) = E(e^{At}\mathbf{x}(0)) + E\left(\int_0^t e^{A(t-\tau)}B\mathbf{w}(\tau)d\tau\right)$$

$$= e^{At}E(\mathbf{x}(0)) + \int_0^t e^{A(t-\tau)}BE(\mathbf{w}(\tau))d\tau = e^{At}\overline{\mathbf{x}}_0 \qquad (4.5.7)$$

Variance of x(t)

Define the covariance matrix:

$$P(t) = E(\mathbf{x}(t)\mathbf{x}^T(t)] \qquad (4.5.8)$$

Using (4.5.6),

$$\mathbf{x}(t)\mathbf{x}^T(t) = e^{At}\mathbf{x}(0)\mathbf{x}^T(0)e^{A^T t} + e^{At}\int_0^t \mathbf{x}(0)\mathbf{w}^T(\tau)B^T e^{A^T(t-\tau)}d\tau$$
$$\qquad (4.5.9)$$
$$+ \int_0^t e^{A(t-\tau)}B\mathbf{w}(\tau)\mathbf{x}^T(0)e^{A^T \tau}d\tau + \int_0^t \int_0^t e^{A(t-\tau)}B\mathbf{w}(\tau)\mathbf{w}^T(\nu)B^T e^{A^T(t-\nu)}d\tau d\nu$$

Taking expected values on both sides,

$$P(t) = e^{At}E(\mathbf{x}(0)\mathbf{x}^T(0))e^{A^T t} + \int_0^t \int_0^t e^{A(t-\tau)}BQ\delta(\nu-\tau)B^T e^{A^T(t-\nu)}d\nu d\tau \quad (4.5.10)$$

$$P(t) = e^{At}P(0)e^{A^T t} + \int_0^t e^{A(t-\tau)}BQB^T e^{A^T(t-\tau)}d\tau \qquad (4.5.11)$$

Differentiating (4.5.11) with respect to t,

$$\frac{dP}{dt} = AP + PA^T + BQB^T; \qquad P(0) = P_0 \qquad (4.5.12)$$

For a stable system,

$$\frac{dP}{dt} \to 0 \quad \text{as} \quad t \to \infty \qquad (4.5.13)$$

Therefore, in steady state, the covariance matrix P is governed by the Lyapunov equation (Chapter 2, Section 2.15):

$$AP + PA^T + BQB^T = 0 \qquad (4.5.14)$$

EXAMPLE 4.5: A SPRING-MASS DAMPER SUBJECTED TO A STOCHASTIC FORCE

Consider the spring-mass-damper system (Chapter 2, Figure 2.4) where the force $f(t)$ is a zero-mean Gaussian white noise with the following statistics:

$$E(f(t)f(t + \tau)) = 0.1\delta(\tau) \qquad (4.5.15)$$

Let the system have the following parameters: $m = 1$ kg, $\alpha = 0.2\,N - s/m$, and $\beta = 1\,N/m$. Then,

$$A = \begin{bmatrix} 0 & 1 \\ -1 & -0.2 \end{bmatrix} \quad \text{and} \quad BQB^T = 0.1\begin{bmatrix} 0 & 0 \\ 0 & 1 \end{bmatrix} \qquad (4.5.16a,b)$$

Solving Equation 4.5.14 via the Matlab routine "lyap,"

$$P = \begin{bmatrix} 0.25 & 0 \\ 0 & 0.25 \end{bmatrix} \qquad (4.5.17)$$

The response (Figure 4.11) is also generated by numerical solution of the state equations (2.5.16) in Chapter 2. The Matlab program is provided (Program 4.1). Note that the Gaussian white noise is generated by the method presented in Appendix D. Numerical results provide the following value of P:

$$P_{sim} = \begin{bmatrix} 0.2442 & 4.53e - 5 \\ 4.53e - 5 & 0.2516 \end{bmatrix} \qquad (4.5.18)$$

which is close to the analytical solution.

MATLAB PROGRAM 4.1: LINEAR SYSTEM SUBJECTED TO WHITE NOISE

```
%
clear all
close all
global wC
%
wone=randn(1,10000);
w=sqrt(2)*wone;
```

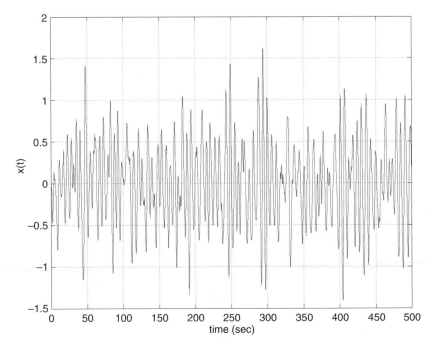

FIGURE 4.11 Response obtained from Matlab Program 4.1.

```
%
%
disp(1)=0;
Vel(1)=0;
tv(1)=0;
tinit=0;
delt=0.05;
%
%PSD of White Noise=2*delt=0.1;
for i=2:10000
tf=tinit+delt;
wC=w(1,i-1);
[t,Y]=ode45(@ex4p6,[tinit tf],[disp(i-1) Vel(i-1)]);
yud=flipud(Y);
tv(i)=tf;
```

```
disp(i)=yud(1,1);
Vel(i)=yud(1,2);
tinit=tf;
end
-------------------------------------------
function dx=ex4p6(t,x)
%
global wC
dx=zeros(2,1);
dx(1)=x(2);
dx(2)=-0.2*x(2)-x(1)+wC;
```

4.6 KALMAN FILTER: OPTIMAL STATE ESTIMATION

The Kalman filter deals with optimal selection of observer poles, and is dual to the linear quadratic control presented in Chapter 3.

STATE DYNAMICS

$$\dot{\mathbf{x}}(t) = A\mathbf{x}(t) + B\mathbf{u}(t) + \mathbf{w}(t) \tag{4.6.1}$$

The vector $\mathbf{w}(t)$ is a stochastic process called *process* (or *plant*) *noise*. It is assumed that $\mathbf{w}(t)$ is a continuous-time Gaussian white noise vector (Appendix D). Its mathematical characterization is

$$E(\mathbf{w}(t)) = 0 \tag{4.6.2}$$

and

$$E(\mathbf{w}(t)\mathbf{w}^T(t+\tau)) = W\delta(\tau) \tag{4.6.3}$$

The matrix W is called the *intensity matrix* with the property $W = W^T > 0$.

MEASUREMENT EQUATION

$$\mathbf{y}(t) = C\mathbf{x}(t) + \boldsymbol{\theta}(t) \tag{4.6.4}$$

where $\mathbf{y}(t)$ is the sensor noise vector and $\mathbf{\theta}(t)$ is a continuous-time Gaussian white noise vector. It is assumed that

$$E(\mathbf{\theta}(t)) = 0 \qquad (4.6.5)$$

and

$$E(\mathbf{\theta}(t)\mathbf{\theta}^T(t+\tau)) = \Theta\delta(\tau) \qquad (4.6.6)$$

where $\Theta = \Theta^T > 0$.

It is also assumed that process and measurement noise vectors are uncorrelated; i.e.,

$$E(\mathbf{\theta}(t)\mathbf{w}^T(t+\tau)) = E(\mathbf{\theta}(t))E(\mathbf{w}^T(t+\tau)) = 0 \qquad (4.6.7)$$

OBSERVER DYNAMICS

From Equation 4.2.3,

$$\dot{\hat{\mathbf{x}}} = A\hat{\mathbf{x}} + B\mathbf{u}(t) + L(\mathbf{y}(t) - C\hat{\mathbf{x}}(t)) \qquad (4.6.8)$$

Define the state error vector as

$$\tilde{\mathbf{x}} = \mathbf{x}(t) - \hat{\mathbf{x}}(t) \qquad (4.6.9)$$

Subtracting (4.6.8) from (4.6.1),

$$\dot{\tilde{\mathbf{x}}} = (A - LC)\tilde{\mathbf{x}} + \mathbf{\psi}(t) \qquad (4.6.10)$$

where

$$\mathbf{\psi}(t) = \mathbf{w}(t) - L\mathbf{\theta}(t) \qquad (4.6.11)$$

$$E(\mathbf{\psi}(t)) = E(\mathbf{w}(t)) - LE(\mathbf{\theta}(t)) = 0 \qquad (4.6.12)$$

Consider

$$\mathbf{\psi}(t)\mathbf{\psi}^T(t+\tau) = [\mathbf{w}(t) - L\mathbf{\theta}(t)][\mathbf{w}(t+\tau) - L\mathbf{\theta}(t+\tau)]^T$$
$$= \mathbf{w}(t)\mathbf{w}^T(t+\tau) - L\mathbf{\theta}(t)\mathbf{w}^T(t+\tau) - \mathbf{w}(t)\mathbf{\theta}^T(t+\tau)L^T + L\mathbf{\theta}(t)\mathbf{\theta}^T(t+\tau)L^T \qquad (4.6.13)$$

Using (4.6.3), (4.6.6), and (4.6.7),

$$E(\boldsymbol{\psi}(t)\boldsymbol{\psi}^T(t+\tau)) = (W + L\Theta L^T)\delta(\tau) \tag{4.6.14}$$

The state error dynamics (Equation 4.6.10) is represented by a linear system subjected to white noise vector $\boldsymbol{\psi}(t)$ with the intensity described by (4.6.14). Assuming that the matrix $(A - LC)$ is stable,

$$E(\tilde{\mathbf{x}}(t)) \to 0 \quad \text{as} \quad t \to \infty \tag{4.6.15}$$

and

$$(A - LC)\Sigma + \Sigma(A - LC)^T + W + L\Theta L^T = 0 \tag{4.6.16}$$

where Σ is the steady state error covariance matrix defined as

$$\Sigma = \lim_{t \to \infty} E(\tilde{\mathbf{x}}(t)\tilde{\mathbf{x}}^T(t)) \tag{4.6.17}$$

Problem

Find L such that

$$J = tr[\Sigma] = E(\tilde{x}_1^2(t)) + E(\tilde{x}_2^2(t)) + \ldots + E(\tilde{x}_n^2(t)) \tag{4.6.18}$$

is minimized subject to constraints (4.6.16).

Solution

Introducing symmetric matrix P as the Lagrange multiplier,

$$g(\Sigma, L, P) = tr(\Sigma) + tr(U(\Sigma, L)P) \tag{4.6.19}$$

where $U(\Sigma, L)$ is the left-hand side of Equation 4.6.16. As a result,

$$g(\Sigma, L, P) =$$
$$tr(\Sigma) + tr((A - LC)\Sigma P) + tr(\Sigma(A - LC)^T P) + tr(WP) + tr(L\Theta L^T P) \tag{4.6.20}$$

Using the Lemma for the optimal solution (Appendix E),

$$\frac{\partial g}{\partial \Sigma} = I + (A - LC)^T P + P(A - LC) = 0 \tag{4.6.21}$$

$$\frac{\partial g}{\partial L} = -P\Sigma C^T - P\Sigma C^T + PL\Theta + PL\Theta = 2P(-\Sigma C^T + L\Theta) = 0 \quad (4.6.22)$$

$$\frac{\partial g}{\partial P} = \Sigma(A - LC)^T + (A - LC)\Sigma + W + L\Theta L^T = 0 \quad (4.6.23)$$

Assuming that $(A - LC)$ is stable, $P > 0$ from the well-known result of the solution of the Lyapunov equation (Chapter 2, Section 2.15). Hence, from Equation 4.6.22,

$$-\Sigma C^T + L\Theta = 0 \quad (4.6.24)$$

or

$$L = \Sigma C^T \Theta^{-1} \quad (4.6.25)$$

Substituting (4.6.25) into (4.6.16),

$$\Sigma A^T + A\Sigma - \Sigma C^T \Theta^{-1} C\Sigma + W = 0 \quad (4.6.26)$$

Equation 4.6.26 is called the *filter algebraic Riccati equation (FARE)*.

The important statistical properties of the Kalman filter (Anderson and Moore, 1990) are as follows:

1. The state error vector is orthogonal to estimate the state vector in the following sense:

$$E(\tilde{\mathbf{x}}\hat{\mathbf{x}}^T) = 0 \quad (4.6.27)$$

2. The signal

$$\mathbf{v}(t) = \mathbf{y}(t) - C\hat{\mathbf{x}}(t) \quad (4.6.28)$$

is white noise with zero mean, and is called the *innovation process*. Furthermore,

$$E(\mathbf{v}(t)\mathbf{v}^T(t + \tau)) = \Theta\delta(\tau) \quad (4.6.29)$$

EXAMPLE 4.6: A SIMPLE MASS (FIGURE 4.9) OR A DOUBLE INTEGRATOR SYSTEM WITH STOCHASTIC DISTURBANCES

Consider the problem of finding an optimal state estimator for the system

$$\begin{bmatrix} \dot{x}_1 \\ \dot{x}_2 \end{bmatrix} = \begin{bmatrix} 0 & 1 \\ 0 & 0 \end{bmatrix} \begin{bmatrix} x_1 \\ x_2 \end{bmatrix} + \begin{bmatrix} 0 \\ 1 \end{bmatrix} u(t) + \begin{bmatrix} 0 \\ 1 \end{bmatrix} g(t) \qquad (4.6.30)$$

and

$$y(t) = \begin{bmatrix} 1 & 0 \end{bmatrix} \begin{bmatrix} x_1(t) \\ x_2(t) \end{bmatrix} + \theta(t) \qquad (4.6.31)$$

Let

$$E(g(t)g(t-\tau)) = \delta(t-\tau) \qquad (4.6.32)$$

and

$$E(\theta(t)\theta(t-\tau)) = \rho\delta(t-\tau) \qquad (4.6.33)$$

Therefore,

$$W = \begin{bmatrix} 0 & 0 \\ 0 & 1 \end{bmatrix} \quad \text{and} \quad \Theta = \rho \qquad (4.6.34)$$

Define

$$\Sigma = \begin{bmatrix} \Sigma_{11} & \Sigma_{12} \\ \Sigma_{12} & \Sigma_{22} \end{bmatrix} \qquad (4.6.35)$$

Then,

$$A\Sigma = \begin{bmatrix} \Sigma_{12} & \Sigma_{22} \\ 0 & 0 \end{bmatrix}; \quad \Sigma A^T = (A\Sigma)^T = \begin{bmatrix} \Sigma_{12} & 0 \\ \Sigma_{22} & 0 \end{bmatrix} \qquad (4.6.36a,b)$$

$$\Sigma C^T \frac{1}{\rho} C\Sigma = \frac{1}{\rho} \begin{bmatrix} \Sigma_{11}^2 & \Sigma_{11}\Sigma_{12} \\ \Sigma_{11}\Sigma_{12} & \Sigma_{12}^2 \end{bmatrix} \qquad (4.6.37)$$

Equation 4.6.26 yields the following three algebraic equations:

$$2\Sigma_{12} - \frac{\Sigma_{11}^2}{\rho} = 0 \tag{4.6.38}$$

$$\Sigma_{22} - \frac{\Sigma_{11}\Sigma_{12}}{\rho} = 0 \tag{4.6.39}$$

$$1 - \frac{\Sigma_{12}^2}{\rho} = 0 \tag{4.6.40}$$

Solving these three equations simultaneously,

$$\Sigma = \begin{bmatrix} \sqrt{2}\rho^{0.75} & \sqrt{\rho} \\ \sqrt{\rho} & \sqrt{2}\rho^{0.25} \end{bmatrix} > 0 \tag{4.6.41}$$

From (4.6.25),

$$L = \begin{bmatrix} \sqrt{2}/\rho^{0.25} \\ 1/\sqrt{\rho} \end{bmatrix} \tag{4.6.42}$$

For $\rho = 0.01$, Equation 4.6.41 yields

$$\Sigma = \begin{bmatrix} 0.0447 & 0.1 \\ 0.1 & 0.4472 \end{bmatrix} \tag{4.6.43}$$

The errors in states from numerical simulations are presented in Figure 4.12 and Figure 4.13. Also,

$$\Sigma_{sim} = \begin{bmatrix} 0.0461 & 0.1012 \\ 0.1012 & 0.4603 \end{bmatrix} \tag{4.6.44}$$

MATLAB PROGRAM 4.2: KALMAN FILTER

```
%
clear all
close all
```

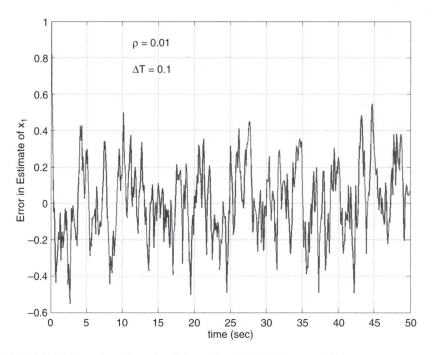

FIGURE 4.12 Error $\tilde{x}_1(t)$ from the Kalman filter (Matlab Program 4.2).

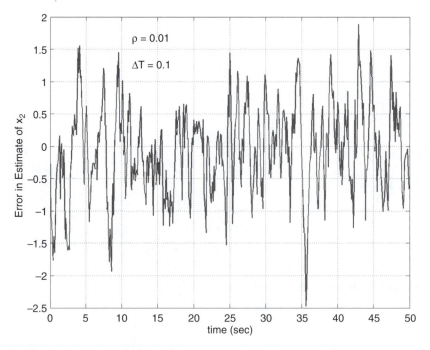

FIGURE 4.13 Error $\tilde{x}_2(t)$ from the Kalman filter (Matlab Program 4.2).

```
global wthC Aobs
%
rho=0.01;
SIGMA=[sqrt(2)*(rho^0.75) sqrt(rho);sqrt(rho)
sqrt(2)*(rho^0.25)]
delt=0.1;
%
load rand_chap4
w=wone/sqrt(delt);
load th_rand_chap4
th=sqrt(rho)*thone/sqrt(delt);
%
A=[0 1;0 0];
L=[sqrt(2)/(rho^0.25) sqrt(1/rho)]';
C=[1 0];
Aobs=A-L*C;
%
%
disp_er(1)=1;
Vel_er(1)=0.0;
tv(1)=0;
tinit=0;
%
for i=2:5000
tf=tinit+delt;
wC=w(i-1);
thC=th(i-1);
wthC=[0;wC]-L*thC;
[t,Y]=ode45(@ex4p6KF,[tinit tf],[disp_er(i-1)
Vel_er(i-1)]);
yud=flipud(Y);
```

```
tv(i)=tf;

disp_er(i)=yud(1,1);

Vel_er(i)=yud(1,2);

tinit=tf;

end

%

-------------------------------------------

function dx=ex4p6KF(t,x)

%

global wthC Aobs

%

dx=zeros(2,1);

dx=Aobs*x+wthC;
```

4.7 STOCHASTIC OPTIMAL REGULATOR IN STEADY STATE

It is assumed that all states are known. However, the plant dynamics is subjected to stochastic disturbances.

STATE DYNAMICS

$$\dot{\mathbf{x}}(t) = A\mathbf{x}(t) + B\mathbf{u}(t) + \mathbf{w}(t) \tag{4.7.1}$$

The vector $\mathbf{w}(t)$ is a stochastic process called *process* (or *plant*) *noise*. It is assumed that $\mathbf{w}(t)$ is a continuous-time Gaussian white noise vector (Appendix D). Its mathematical characterization is

$$E(\mathbf{w}(t)) = 0 \tag{4.7.2}$$

and

$$E(\mathbf{w}(t)\mathbf{w}^T(t+\tau)) = W\delta(\tau) \tag{4.7.3}$$

The matrix W is called the *intensity matrix* with the property $W = W^T > 0$.

OBJECTIVE FUNCTION

Minimize

$$J = \lim_{t_f \to \infty} E\left\{ \frac{1}{t_f} \int_0^{t_f} [\mathbf{x}^T(t)Q\mathbf{x}(t) + \mathbf{u}^T(t)R\mathbf{u}(t)dt] \right\} \qquad (4.7.4)$$

Let the optimal control be

$$\mathbf{u}(t) = -K\mathbf{x}(t) \qquad (4.7.5)$$

Then, from (4.7.1),

$$\dot{\mathbf{x}}(t) = (A - BK)\mathbf{x}(t) + \mathbf{w}(t) \qquad (4.7.6)$$

Assuming that the matrix $(A - BK)$ is stable,

$$E(\mathbf{x}(t)) \to 0 \quad \text{as} \quad t \to \infty \qquad (4.7.7)$$

and

$$(A - BK)M + M(A - BK)^T + W = 0 \qquad (4.7.8)$$

where M is the steady state covariance matrix defined as

$$M = \lim_{t \to \infty} E(\mathbf{x}(t)\mathbf{x}^T(t)) \qquad (4.7.9)$$

Assuming the asymptotic stability of the closed-loop system, Equation 4.7.4 can also be written as

$$J = \lim_{t \to \infty} E\left\{ \mathbf{x}^T(t)Q\mathbf{x}(t) + \mathbf{u}^T(t)R\mathbf{u}(t) \right\} \qquad (4.7.10)$$

Equation 4.7.10 can be rewritten as

$$J = \lim_{t \to \infty} E\left\{ tr(\mathbf{x}(t)\mathbf{x}^T(t)Q + \mathbf{u}(t)\mathbf{u}^T(t)R) \right\} \qquad (4.7.11)$$

Using (4.7.5) and interchanging $E(.)$ and tr operations,

$$J = tr(MQ + KMK^T R) \qquad (4.7.12)$$

Problem

Find K such that J Equation 4.7.12 is minimized subject to constraints (4.7.8).

Solution

First, Equation 4.7.12 is rewritten as

$$J = tr(MQ + RKMK^T) \qquad (4.7.13)$$

Introducing a symmetric matrix P as the Lagrange multiplier,

$$g(M,K,P) = tr(MQ + RKMK^T) + tr(U(M,K)P) \qquad (4.7.14)$$

where $U(M,K)$ is the left-hand side of Equation 4.7.8. As a result,

$$g(M,K,P) =$$
$$tr(MQ + RKMK^T) + tr((A-BK)MP) + tr(M(A-BK)^T P) + tr(WP) \qquad (4.7.15)$$

Using the Lemma for the optimal solution (Appendix E),

$$\frac{\partial g}{\partial P} = M(A-BK)^T + (A-BK)M + W = 0 \qquad (4.7.16)$$

$$\frac{\partial g}{\partial K} = RKM + RKM - B^T PM - B^T PM = 2(RK - B^T P)M = 0 \qquad (4.7.17)$$

$$\frac{\partial g}{\partial M} = Q + K^T RK + (A-BK)^T P + P(A-BK) = 0 \qquad (4.7.18)$$

For a stable $(A-BK)$, $M > 0$ from the well-known result of the solution of the Lyapunov equation (Chapter 2, Section 2.15). Hence, from Equation 4.7.17,

$$K = R^{-1}B^T P \qquad (4.7.19)$$

Substituting (4.7.19) into (4.7.18), the Riccati equation is obtained:

$$PA + A^T P - PBR^{-1}B^T P + Q = 0 \qquad (4.7.20)$$

To find the optimal value of the objective function, Equation 4.7.13 is written as

$$J = tr(MQ + MK^T RK) = tr(M(Q + K^T RK))$$ (4.7.21)

Then, using (4.7.18) and (4.7.16),

$$J = tr(WP) = tr(PW)$$ (4.7.22)

Summary

The optimal control law is

$$\mathbf{u} = -K\mathbf{x}(t)$$ (4.7.23)

$$K = R^{-1}B^T P$$ (4.7.24)

$$PA + A^T P - PBR^{-1}B^T P + Q = 0$$ (4.7.25)

The minimum value of *J* is

$$J = tr(PW)$$ (4.7.26)

EXAMPLE 4.7: ACTIVE SUSPENSION SYSTEM

Consider that the road profile, $x_0(t)$, is a white noise (Chapter 3, Example 3.11) with the following statistics:

$$E(x_0(t)x_0(t+\tau)) = v\delta(\tau)$$ (4.7.27)

In this case, the state equations are written as

$$
\begin{bmatrix} \dot{x}_1 \\ \dot{x}_2 \\ \dot{x}_3 \\ \dot{x}_4 \end{bmatrix} = \begin{bmatrix} 0 & 0 & 1 & 0 \\ 0 & 0 & 0 & 1 \\ -\dfrac{\lambda_1}{m_1} & 0 & 0 & 0 \\ 0 & 0 & 0 & 0 \end{bmatrix} \begin{bmatrix} x_1 \\ x_2 \\ x_3 \\ x_4 \end{bmatrix} + \begin{bmatrix} 0 \\ 0 \\ -\dfrac{1}{m_1} \\ \dfrac{1}{m_2} \end{bmatrix} u(t) + \mathbf{w}(t)
$$ (4.7.28)

where

$$\mathbf{w}(t) = \mathbf{b}_w x_0(t) \tag{4.7.29a}$$

$$\mathbf{b}_w^T = [0 \quad 0 \quad \frac{\lambda_1}{m_1} \quad 0] \tag{4.7.29b}$$

From (4.7.27) and (4.7.29),

$$W = \mathbf{b}_w \mathbf{b}_w^T v^2 \tag{4.7.30}$$

Then the controller

$$u(t) = -\mathbf{k}\mathbf{x} \tag{4.7.31}$$

where **k** computed by (3.5.84) minimizes the following objective function:

$$J = \lim_{t \to \infty} E\left\{\mathbf{x}^T(t)Q\mathbf{x}(t) + \rho u^2\right\} \tag{4.7.32}$$

From (4.7.26) and (4.7.30), the minimum value of the objective function will be

$$J = (P_{11} + P_{22} + P_{33} + P_{44})v^2 \frac{\lambda_1^2}{m_1^2} \tag{4.7.33}$$

EXAMPLE 4.8: A SIMPLE MASS OR A DOUBLE INTEGRATOR SYSTEM

Consider the system (4.6.30) again. Weighting matrices Q and R are chosen as follows:

$$Q = \begin{bmatrix} 1 & 0 \\ 0 & 0 \end{bmatrix} \quad \text{and} \quad R = \rho_c \tag{4.7.34a,b}$$

Define

$$P = \begin{bmatrix} p_{11} & p_{12} \\ p_{21} & p_{22} \end{bmatrix} \quad \text{with} \quad p_{12} = p_{21} \tag{4.7.35}$$

From Equation 4.7.25, the following nonlinear algebraic equations are obtained:

$$-\frac{p_{12}^2}{\rho_c} + 1 = 0 \qquad (4.7.36a)$$

$$2p_{12} - \frac{p_{22}^2}{\rho_c} = 0 \qquad (4.7.36b)$$

$$p_{11} - \frac{p_{12}p_{22}}{\rho_c} = 0 \qquad (4.7.36c)$$

Solving Equations 4.7.36a to Equation 4.7.36c simultaneously,

$$P = \begin{bmatrix} \sqrt{2\rho_c} & \sqrt{\rho_c} \\ \sqrt{\rho_c} & 2^{0.5}\rho_c^{0.75} \end{bmatrix} \qquad (4.7.37)$$

From (4.7.24),

$$K = [\sqrt{\rho_c} \quad 2^{0.5}\rho_c^{0.75}] \qquad (4.7.38)$$

From (4.7.37) and (4.6.34),

$$PW = \begin{bmatrix} 0 & \sqrt{\rho_c} \\ 0 & 2^{0.5}\rho_c^{0.75} \end{bmatrix} \qquad (4.7.39)$$

From (4.7.26),

$$J = tr(PW) = 2^{0.5}\rho_c^{0.75} \qquad (4.7.40)$$

For $\rho_c = 0.05$, Equation 4.7.40 yields

$$J = 0.1495$$

From numerical simulation (Matlab Program 4.3),

$$J_{sim} = 0.1581$$

The response $x_1(t)$ is shown in Figure 4.14.

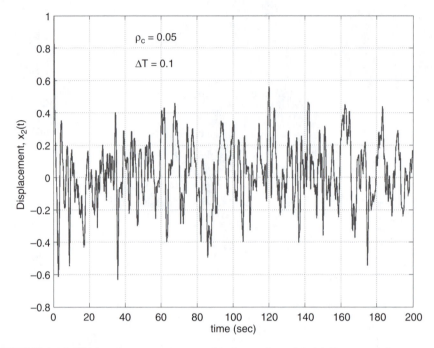

FIGURE 4.14 Displacement $x_1(t)$ from stochastic regulator (Matlab Program 4.3).

MATLAB PROGRAM 4.3: STOCHASTIC REGULATOR

```
%
clear all
close all
global wC Acl
%
rhoc=0.05;
delt=0.1;
%
load rand_chap4
w=wone/sqrt(delt);
%
A=[0 1;0 0];
K=[sqrt(1/rhoc) sqrt(2)/(rhoc^0.25)];
B=[0 1]';
```

```
Acl=A-B*K;
%
J=sqrt(2)*(rhoc)^0.75
%
disp(1)=1;
Vel(1)=0.0;
tv(1)=0;
tinit=0;
%
for i=2:5000
tf=tinit+delt;
wC=w(i-1);
[t,Y]=ode45(@ex4stochR,[tinit tf],[disp(i-1) Vel(i-
1)]);
yud=flipud(Y);
tv(i)=tf;
disp(i)=yud(1,1);
Vel(i)=yud(1,2);
uin(i)=-K(1)*disp(i)-K(2)*Vel(i);
tinit=tf;
end
-----------------------------------------
function dx=ex4stochR(t,x)
%
global wC Acl
%
dx=zeros(2,1);
dx=Acl*x+[0;1]*wC;
```

4.8 LINEAR QUADRATIC GAUSSIAN (LQG) CONTROL

It is assumed that all states are not known. The plant dynamics and output measurements are subjected to stochastic disturbances.

STATE DYNAMICS

$$\dot{\mathbf{x}}(t) = A\mathbf{x}(t) + B\mathbf{u}(t) + \mathbf{w}(t) \tag{4.8.1}$$

The vector $\mathbf{w}(t)$ is a stochastic process called *process* (or *plant*) *noise*. It is assumed that $\mathbf{w}(t)$ is a continuous-time Gaussian white noise vector. Its mathematical characterization is

$$E(\mathbf{w}(t)) = 0 \tag{4.8.2}$$

and

$$E(\mathbf{w}(t)\mathbf{w}^T(t+\tau)) = W\delta(\tau) \tag{4.8.3}$$

The matrix W is called the *intensity matrix* with the property $W = W^T \geq 0$.

MEASUREMENT EQUATION

$$\mathbf{y}(t) = C\mathbf{x}(t) + \boldsymbol{\theta}(t) \tag{4.8.4}$$

where $\mathbf{y}(t)$ is the sensor noise vector and $\boldsymbol{\theta}(t)$ is a continuous-time Gaussian white noise vector. It is assumed that

$$E(\boldsymbol{\theta}(t)) = 0 \tag{4.8.5}$$

and

$$E(\boldsymbol{\theta}(t)\boldsymbol{\theta}^T(t+\tau)) = \Theta\delta(\tau) \tag{4.8.6}$$

where $\Theta = \Theta^T > 0$.

It is also assumed that process and measurement noise vectors are uncorrelated; i.e.,

$$E(\boldsymbol{\theta}(t)\mathbf{w}^T(t+\tau)) = E(\boldsymbol{\theta}(t))E(\mathbf{w}^T(t+\tau)) = 0 \tag{4.8.7}$$

OBJECTIVE FUNCTION

Minimize

$$J = \lim_{t_f \to \infty} E\left\{ \frac{1}{t_f} \int_0^{t_f} [\mathbf{x}^T(t)Q\mathbf{x}(t) + \mathbf{u}^T(t)R\mathbf{u}(t)dt] \right\} \tag{4.8.8}$$

Assuming the asymptotic stability of the closed-loop system, Equation 4.8.8 can also be written as

$$J = \lim_{t \to \infty} E\{\mathbf{x}^T(t)Q\mathbf{x}(t) + \mathbf{u}^T(t)R\mathbf{u}(t)\} \qquad (4.8.9)$$

Let the states be estimated by the Kalman filter (4.6.8) with the observer gain L given by (4.6.25). Substituting $\mathbf{x} = \hat{\mathbf{x}} + \tilde{\mathbf{x}}$,

$$\mathbf{x}^T Q\mathbf{x} = \hat{\mathbf{x}}^T Q\hat{\mathbf{x}} + \tilde{\mathbf{x}}^T Q\tilde{\mathbf{x}} + 2\hat{\mathbf{x}}^T Q\tilde{\mathbf{x}} = \hat{\mathbf{x}}^T Q\hat{\mathbf{x}} + tr(\tilde{\mathbf{x}}\tilde{\mathbf{x}}^T Q) + 2tr(\tilde{\mathbf{x}}\hat{\mathbf{x}}^T Q) \quad (4.8.10)$$

Substituting (4.8.10) into (4.8.19), interchanging tr and $E(.)$ operations, and using the orthogonality property (4.6.27),

$$J = \lim_{t \to \infty} E\{\hat{\mathbf{x}}^T(t)Q\hat{\mathbf{x}}(t) + \mathbf{u}^T(t)R\mathbf{u}(t)\} + tr(\Sigma Q) \qquad (4.8.11)$$

where

$$\dot{\hat{\mathbf{x}}} = A\hat{\mathbf{x}} + B\mathbf{u}(t) + L\mathbf{v}(t) \qquad (4.8.12)$$

and $\mathbf{v}(t)$ is a white noise defined by (4.6.28). Because the second term $tr(\Sigma Q)$ in (4.8.11) is a constant, the minimization of J (Equation 4.8.11) subject to constraints (4.8.12) is the stochastic regulator problem solved in Section 4.6. Therefore, the optimal control law is

$$\mathbf{u}(t) = -K\hat{\mathbf{x}}(t) \qquad (4.8.13)$$

where K is defined by (4.7.24) and (4.7.25).

In summary, the optimal control law is obtained by separately solving the optimal estimation problem and the deterministic LQ control problem. This is known as the *separation theorem*. The LQG control system is developed by completing the following steps (Athans, 1992):

Step I: The optimal state estimator

$$\dot{\hat{\mathbf{x}}} = A\hat{\mathbf{x}} + B\mathbf{u}(t) + L(\mathbf{y}(t) - C\hat{\mathbf{x}}) \qquad (4.8.14)$$

where

$$L = \Sigma C^T \Theta^{-1} \qquad (4.8.15)$$

and

$$A\Sigma + \Sigma A^T + W - \Sigma C^T \Theta^{-1} C \Sigma = 0 \qquad (4.8.16)$$

Equation 4.8.16 is known as the *filter algebraic Riccati equation (FARE)*.

Step II: The optimal control law is

$$\mathbf{u}(t) = -K\hat{\mathbf{x}}(t) \qquad (4.8.17)$$

where

$$K = R^{-1}B^T P \qquad (4.8.18)$$

and

$$PA + A^T P - PBR^{-1}B^T P + Q = 0 \qquad (4.8.19)$$

Equation 4.8.19 is known as the *control algebraic Riccati equation (CARE)*.

CLOSED-LOOP SYSTEM DYNAMICS

Define the state error vector as

$$\tilde{\mathbf{x}} = \mathbf{x} - \hat{\mathbf{x}} \qquad (4.8.20)$$

Substituting (4.8.18) and (4.8.20) into (4.8.1),

$$\dot{\mathbf{x}} = (A - BK)\mathbf{x}(t) + BK\tilde{\mathbf{x}}(t) + \mathbf{w}(t) \qquad (4.8.21)$$

Subtracting (4.8.14) from (4.8.1),

$$\dot{\tilde{\mathbf{x}}} = (A - LC)\tilde{\mathbf{x}} - L\boldsymbol{\theta}(t) + \mathbf{w}(t) \qquad (4.8.22)$$

Putting (4.8.21) and (4.8.22) in the matrix form,

$$\begin{bmatrix} \dot{\mathbf{x}} \\ \dot{\tilde{\mathbf{x}}} \end{bmatrix} = \begin{bmatrix} A - BK & BK \\ 0 & A - LC \end{bmatrix} \begin{bmatrix} \mathbf{x} \\ \tilde{\mathbf{x}} \end{bmatrix} + \begin{bmatrix} \mathbf{w}(t) \\ \mathbf{w}(t) - L\boldsymbol{\theta}(t) \end{bmatrix} \qquad (4.8.23)$$

Equation 4.8.23 is equivalent to

$$\begin{bmatrix} \dot{\mathbf{x}} \\ \dot{\hat{\mathbf{x}}} \end{bmatrix} = \begin{bmatrix} A & -BK \\ LC & A-BK-LC \end{bmatrix} \begin{bmatrix} \mathbf{x} \\ \hat{\mathbf{x}} \end{bmatrix} + \begin{bmatrix} \mathbf{w}(t) \\ L\boldsymbol{\theta}(t) \end{bmatrix} \tag{4.8.24}$$

MINIMUM VALUE OF THE OBJECTIVE FUNCTION

Drawing analogy with the stochastic regulator problem,

$$\lim_{t \to \infty} E\{\hat{\mathbf{x}}^T(t)Q\hat{\mathbf{x}}(t) + \mathbf{u}^T(t)R\mathbf{u}(t)\} = tr(PL\Theta L^T) \tag{4.8.25}$$

where

$$E(L\mathbf{v}(t)\mathbf{v}^T(t+\tau)L^T) = L\Theta L^T \delta(\tau) \tag{4.8.26}$$

Therefore, using (4.8.11), the minimum value of the objective function is given by

$$J = tr\{PL\Theta L^T\} + tr\{\Sigma Q\} \tag{4.8.27}$$

EXAMPLE 4.9: A SIMPLE MASS OR A DOUBLE INTEGRATOR SYSTEM (LQG CONTROL)

Consider the system (4.6.30) again. The weighting matrices Q and R in (4.8.8) are also defined by Equation 4.7.34a and Equation 4.7.34b. Then, from (4.8.27),

$$J = \rho[\ell_1^2 \sqrt{2\rho_c} + 2\ell_1\ell_2\sqrt{\rho_c} + \sqrt{2}\ell_2^2\rho_c^{0.75}] + \sqrt{2}\rho^{0.75} \tag{4.8.28}$$

where from Equation 4.6.42,

$$\ell_1 = \frac{\sqrt{2}}{\rho^{0.25}} \quad \text{and} \quad \ell_2 = \frac{1}{\sqrt{\rho}} \tag{4.8.29}$$

Substituting (4.8.29) into (4.8.28),

$$J = 2\sqrt{2\rho_c}(\sqrt{\rho} + \rho^{0.25}) + \sqrt{2}(\rho_c^{0.75} + \rho^{0.75}) \tag{4.8.30}$$

For $\rho = 0.01$ and $\rho_c = 0.05$,

$$J = 0.5357 \tag{4.8.31}$$

From numerical simulations (Matlab Program 4.4),

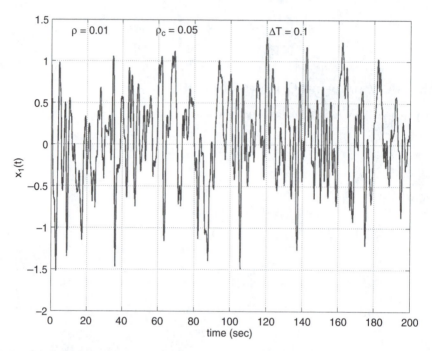

FIGURE 4.15 Displacement $x_1(t)$ from the linear quadratic Gaussian (LQG) control (Matlab Program 4.4).

$$J_{sim} = 0.5529 \tag{4.8.32}$$

The response $x_1(t)$ is shown in Figure 4.15.

MATLAB PROGRAM 4.4: LQG CONTROL

```
%
clear all
close all
global wC thC K L
%
rho=0.01;
rhoc=0.05;
delt=0.1;
%
load rand_chap4
```

```
w=wone/sqrt(delt);
load th_rand_chap4
th=sqrt(rho)*thone/sqrt(delt);
%
A=[0  1;0  0];
B=[0  1]';
C=[1  0];
%
L=[sqrt(2)/(rho^0.25)  sqrt(1/rho)]';
K=[sqrt(1/rhoc)  sqrt(2)/(rhoc^0.25)];
%
J=2*sqrt(2*rhoc)*((rhoc)^0.5+(rho)^0.25)+sqrt(2)*((r
hoc)^0.75+(rho)^0.75)
%
disp(1)=1;
Vel(1)=0.0;
disph(1)=0;
Velh(1)=0;
tv(1)=0;
tinit=0;
%
for  i=2:5000
tf=tinit+delt;
wC=w(i-1);
thC=th(i-1);
[t,Y]=ode45(@ex4p6lqg,[tinit tf],[disp(i-1) Vel(i-1)
disph(i-1)  Velh(i-1)]);
yud=flipud(Y);
tv(i)=tf;
disp(i)=yud(1,1);
Vel(i)=yud(1,2);
```

```
disph(i)=yud(1,3);

Velh(i)=yud(1,4);

uin(i)=-K(1)*disph(i)-K(2)*Velh(i);

tinit=tf;

end

%

function dx=ex4p6lqg(t,x)

%

global  wC  thC  K  L

%

dx=zeros(4,1);

u=-K(1)*x(3)-K(2)*x(4);

dx(1)=x(2);

dx(2)=u+wC;

dx(3)=x(4)+L(1)*(x(1)+thC-x(3));

dx(4)=u+L(2)*(x(1)+thC-x(3));
```

4.9 IMPACT OF MODELING ERRORS ON OBSERVER-BASED CONTROL

4.9.1 STRUCTURED PARAMETRIC UNCERTAINTIES

Consider the state space model of the plant:

$$\dot{\mathbf{x}} = A\mathbf{x}(t) + B\mathbf{u}(t) \qquad (4.9.1)$$

$$\mathbf{y}(t) = C\mathbf{x}(t) \qquad (4.9.2)$$

The elements of matrices A, B, and C are assumed to be uncertain. This type of modeling error is known as a *structured parametric error* because the structure of the model is considered accurate, only the parameters of the model are not well known. Let $\hat{A}$, $\hat{B}$, and $\hat{C}$ be estimates of A, B, and C, respectively. Then, the observer-based controller can be defined as

$$\mathbf{u} = -K\hat{\mathbf{x}} \qquad (4.9.3)$$

$$\dot{\hat{\mathbf{x}}} = \hat{A}\hat{\mathbf{x}} + \hat{B}\mathbf{u} + L(\mathbf{y} - \hat{C}\hat{\mathbf{x}}) \tag{4.9.4}$$

Combining (4.9.1) through (4.9.4), the closed-loop system dynamics is defined as

$$\begin{bmatrix} \dot{\mathbf{x}} \\ \dot{\hat{\mathbf{x}}} \end{bmatrix} = A_{sysp} \begin{bmatrix} \mathbf{x} \\ \hat{\mathbf{x}} \end{bmatrix} \tag{4.9.5}$$

where

$$A_{sysp} = \begin{bmatrix} A & -BK \\ LC & \hat{A} - \hat{B}K - L\hat{C} \end{bmatrix} \tag{4.9.6}$$

Now, eigenvalues of A_{sysp} are not those of $(\hat{A} - \hat{B}K)$ and $(\hat{A} - L\hat{C})$. In fact, it is possible that the closed-loop system is unstable under parametric errors.

EXAMPLE 4.10: BENCHMARK ROBUST CONTROL PROBLEM (WIE AND BERNSTEIN, 1992)
The differential equations of motion for the system in Figure 4.16 are

$$m_1\ddot{q}_1 + k(q_1 - q_2) = u \tag{4.9.7}$$

$$m_2\ddot{q}_2 + k(q_2 - q_1) = 0 \tag{4.9.8}$$

Defining states as

$$x_1 = q_1, \quad x_2 = q_2, \quad x_3 = \dot{q}_1, \quad \text{and} \quad x_4 = \dot{q}_2 \tag{4.9.9}$$

The state space model is

$$\dot{\mathbf{x}} = A\mathbf{x} + Bu \tag{4.9.10}$$

$$y = C\mathbf{x} \tag{4.9.11}$$

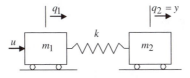

FIGURE 4.16 A two-mass system.

where

$$A = \begin{bmatrix} 0 & 0 & 1 & 0 \\ 0 & 0 & 0 & 1 \\ -k/m_1 & k/m_1 & 0 & 0 \\ k/m_2 & -k/m_2 & 0 & 0 \end{bmatrix}; \quad B = \begin{bmatrix} 0 \\ 0 \\ 1/m_1 \\ 0 \end{bmatrix} \qquad (4.9.12)$$

$$C = \begin{bmatrix} 0 & 1 & 0 & 0 \end{bmatrix} \qquad (4.9.13)$$

Let estimates of parameters k, m_1, and m_2 be $\hat{k}$, $\hat{m}_1$, and $\hat{m}_2$, respectively. Then,

$$\hat{A} = \begin{bmatrix} 0 & 0 & 1 & 0 \\ 0 & 0 & 0 & 1 \\ -\hat{k}/\hat{m}_1 & \hat{k}/\hat{m}_1 & 0 & 0 \\ \hat{k}/\hat{m}_2 & -\hat{k}/\hat{m}_2 & 0 & 0 \end{bmatrix}; \quad \hat{B} = \begin{bmatrix} 0 \\ 0 \\ 1/\hat{m}_1 \\ 0 \end{bmatrix} \qquad (4.9.14)$$

Let $\hat{k} = 1$, $\hat{m}_1 = 1$, and $\hat{m}_2 = 1$. When there is no modeling errors, eigenvalues of A_{sysp} are controller poles ($-0.1 \pm 1.414j$, 0.1, and 0.1) and observer poles ($-0.4 \pm 1.414j$, -0.4, and -0.4). Associated vectors K and L are as follows:

$$L^T = [0.256 \quad 1.6 \quad -0.6144 \quad 0.96] \qquad (4.9.15)$$

and

$$K = [0.06 \quad -0.0399 \quad 0.4 \quad 0.004] \qquad (4.9.16)$$

Let $k = 0.5$, $\hat{k} = 1$, $\hat{m}_1 = m_1 = 1$, and $\hat{m}_2 = m_2 = 1$. The observer and controller gain vectors L and K are again chosen by Equation 4.9.15 and Equation 4.9.16, respectively. In this case, $\hat{k} \neq k$, and the eigenvalues of A_{sysp} are 1.0085, -0.0881, -0.1231, -1.1567, $-0.4989 \pm 1.4805j$, and $-0.3214 \pm 0.0955j$. That is, the closed-loop system has become unstable.

4.9.2 UNMODELED DYNAMICS

Let the dynamics of a system be given as

$$\dot{\mathbf{x}} = A\mathbf{x} + B\mathbf{u} \qquad (4.9.17)$$

$$\dot{\mathbf{x}}_R = A_R \mathbf{x}_R + B_R \mathbf{u} \tag{4.9.18}$$

and

$$\mathbf{y}(t) = C\mathbf{x} + C_R \mathbf{x}_R \tag{4.9.19}$$

Here, the controller will be designed on the basis of state equations (4.9.17) only, and state equations (4.9.18) will become *residual or unmodeled* dynamics. The observer-based controller is

$$\mathbf{u} = -K\hat{\mathbf{x}} \tag{4.9.20}$$

$$\dot{\hat{\mathbf{x}}} = \hat{A}\hat{\mathbf{x}} + \hat{B}\mathbf{u} + L(\mathbf{y} - \hat{C}\hat{\mathbf{x}}) \tag{4.9.21}$$

Because of nonzero B_R, the input designed to control the system (4.9.17) will also influence unmodeled dynamics. The term B_R is also known as the *control spillover*. Similarly, because of nonzero C_R, the measurement data (Equation 4.9.19) will contain unmodeled dynamics also. The term C_R is also known as the *observation spillover* (Balas, 1978).

Define estimated state error vector $\tilde{\mathbf{x}}$ as

$$\tilde{\mathbf{x}}(t) = \mathbf{x}(t) - \hat{\mathbf{x}}(t) \tag{4.9.22}$$

The complete closed-loop system dynamics is represented by

$$\begin{bmatrix} \dot{\mathbf{x}} \\ \dot{\tilde{\mathbf{x}}} \\ \dot{\mathbf{x}}_R \end{bmatrix} = A_{sys} \begin{bmatrix} \mathbf{x} \\ \tilde{\mathbf{x}} \\ \mathbf{x}_R \end{bmatrix} \tag{4.9.23}$$

where

$$A_{sys} = \begin{bmatrix} A - BK & BK & 0 \\ 0 & A - LC & LC_R \\ -B_R K & B_R K & A_R \end{bmatrix} \tag{4.9.24}$$

When $C_R = 0$, eigenvalues of A_{sys} are same as those of $(A - BK)$, $(A - LC)$, and A_R. However, when $C_R \neq 0$, eigenvalues of A_{sys} are not same as those of $(A - BK)$, $(A - LC)$, and A_R, and it is possible to get an unstable closed-loop system in spite of $(A - BK)$, $(A - LC)$, and A_R being stable.

EXAMPLE 4.11: FLEXIBLE MODE AS THE UNMODELED DYNAMICS IN THE BENCHMARK ROBUST CONTROL PROBLEM

Assume that $m_1 = m_2 = m$. Without an input, Equation 4.9.7 and Equation 4.9.8 can be written in matrix form as

$$M\ddot{\mathbf{q}} + K\mathbf{q} = 0 \qquad (4.9.25)$$

where

$$M = \begin{bmatrix} m & 0 \\ 0 & m \end{bmatrix} \quad \text{and} \quad K = \begin{bmatrix} k & -k \\ -k & k \end{bmatrix} \qquad (4.9.26a,b)$$

Natural frequencies and modal vectors are computed by solving the following eigenvalue problem:

$$(-\omega^2 M + K)\mathbf{q}_0 = 0 \qquad (4.9.27)$$

Natural frequencies are found to be

$$\omega_1^2 = 0 \quad \text{and} \quad \omega_2^2 = \frac{2k}{m} \qquad (4.9.28)$$

and modal vectors are as follows:

Rigid mode:

$$\begin{bmatrix} 1 & 1 \end{bmatrix}^T \text{ corresponding to } \omega_1^2 = 0 \qquad (4.9.29)$$

and

Flexible mode:

$$\begin{bmatrix} 1 & -1 \end{bmatrix}^T \text{ corresponding to } \omega_2^2 = 2k / m \qquad (4.9.30)$$

Now, a displacement vector can be represented as a linear combination of these modal vectors:

$$\begin{bmatrix} q_1 \\ q_2 \end{bmatrix} = \Phi \begin{bmatrix} r_1 \\ r_2 \end{bmatrix} \qquad (4.9.31)$$

where

$$\Phi = \begin{bmatrix} 1 & 1 \\ 1 & -1 \end{bmatrix} \tag{4.9.32}$$

Substituting (4.9.31) into (4.9.25) with the input and premultiplying both sides by Φ^T,

$$\ddot{r}_1 = \frac{1}{2m} u \tag{4.9.33}$$

and

$$\ddot{r}_2 + \omega_2^2 r_2 = \frac{1}{2m} u \tag{4.9.34}$$

Equation 4.9.33 represents the rigid body mode, whereas Equation 4.9.34 represents the vibratory mode. The output equation from (4.9.11) is

$$y(t) = q_2 = r_1 - r_2 \tag{4.9.35}$$

Let the controller be designed on the basis of the rigid mode only. Then, the state space model for the controlled mode will be

$$\begin{bmatrix} \dot{p}_1 \\ \dot{p}_2 \end{bmatrix} = \begin{bmatrix} 0 & 1 \\ 0 & 0 \end{bmatrix} \begin{bmatrix} p_1 \\ p_2 \end{bmatrix} + \begin{bmatrix} 0 \\ 1/2m \end{bmatrix} u \tag{4.9.36}$$

where

$$p_1 = r_1 \quad \text{and} \quad p_2 = \dot{r}_1 \tag{4.9.37}$$

The state space model for the residual mode (4.9.34) will be

$$\begin{bmatrix} \dot{z}_1 \\ \dot{z}_2 \end{bmatrix} = \begin{bmatrix} 0 & 1 \\ -\omega_2^2 & 0 \end{bmatrix} \begin{bmatrix} z_1 \\ z_2 \end{bmatrix} + \begin{bmatrix} 0 \\ 1/2m \end{bmatrix} u \tag{4.9.38}$$

where

$$z_1 = r_2 \quad \text{and} \quad z_2 = \dot{r}_2 \tag{4.9.39}$$

Equation 4.9.35, Equation 4.9.36, and Equation 4.9.38 can be cast as Equation 4.9.17 to Equation 4.9.19 where

$$A = \begin{bmatrix} 0 & 1 \\ 0 & 0 \end{bmatrix} \quad B = \begin{bmatrix} 0 \\ 1/2m \end{bmatrix}$$

$$A_R = \begin{bmatrix} 0 & 1 \\ -\omega_2^2 & 0 \end{bmatrix} \quad B_R = \begin{bmatrix} 0 \\ 1/2m \end{bmatrix}$$

$$C = \begin{bmatrix} 1 & 0 \end{bmatrix} \quad C_R = -\begin{bmatrix} 1 & 0 \end{bmatrix}$$

EXERCISE PROBLEMS

P4.1   Consider a simple electromagnetic suspension system shown in Figure P4.1. The electromagnetic force f_m is given by

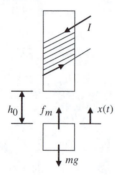

FIGURE P4.1 An electromagnetic suspension system.

$$f_m = \alpha \frac{I^2}{h^2} \qquad (P4.1.1)$$

where I and h are coil current and air gap, respectively. The constant $\alpha = \mu_0 N A_p^2$ where μ_0, N, and A_p are air permeability, number of coil turns, and the face area per single pole of the magnet, respectively. Let h_0 be the desired air gap. Then, the current I_0 is calculated from the following static equilibrium condition:

$$\alpha \frac{I_0^2}{h_0^2} = mg \qquad (P4.1.2)$$

Let $x(t)$ be the dynamic displacement of the mass with respect to the static equilibrium position. The nonlinear differential equation of motion is

$$m\ddot{x} = -mg + \alpha \frac{(I_0 + i)^2}{(h_0 - x)^2} \qquad \text{(P4.1.3)}$$

where i is the incremental current. Linearizing (P4.1.3) around the static equilibrium condition,

$$m\ddot{x} = k_x x + k_i i(t) \qquad \text{(P4.1.4)}$$

where

$$k_x = 2\alpha \frac{I_0^2}{h_0^3} \quad \text{and} \quad k_i = 2\alpha \frac{I_0}{h_0^2} \qquad \text{(P4.1.5a,b)}$$

The system parameters for simulation (Fujita et al., 1995) are as follows: $\mu_0 = 0.000001258N / A^2$, $A_p = 0.000146m^2$, $h_0 = 0.000508m$, $m = 0.3\,\text{kg}$.

a. Assuming that the output is $x(t)$, design a full-order observer. Present simulation results.

b. Assuming that the output is $x(t)$, design a reduced-order observer. Present simulation results.

c. Develop a combined observer-controller system with the full-order observer. Your answer should contain a combined observer-controller transfer function. Present simulation results.

d. Develop a combined observer-controller system with the reduced-order observer. Your answer should contain a combined observer-controller transfer function. Present simulation results.

P4.2 Consider a plant with the following state space realization:

$$\frac{d\mathbf{x}}{dt} = A\mathbf{x}(t) + \mathbf{b}u(t) \; ; \quad y(t) = \mathbf{c}x(t)$$

where

$$A = \begin{bmatrix} 0 & 1 \\ -3 & 4 \end{bmatrix}, \; \mathbf{b} = \begin{bmatrix} 0 \\ 1 \end{bmatrix}, \; \text{and } \mathbf{c} = [1\ 1]$$

a. Develop a combined observer-controller system. Place the state feedback controller poles at -1 and -2, and let both the observer poles be at -4. Give all the equations necessary for implementing your controller in real time.

b. Find the observer-controller transfer function relating $u(s)$ and $y(s)$.

c. Demonstrate the performance of the controller via numerical simulations.

P4.3   Consider the model for longitudinal pressure oscillation in a uniform chamber (Problem P3.19). The outputs $y_i(t)$ (unsteady pressures) are obtained by p sensors located at z_{si}:

$$y_i(t) = \bar{p}\sum_{\ell=1}^{n} \psi_\ell(z_{si})\eta_\ell(t) ; \quad i = 1, 2, ..., p$$

a. Design a full-order observer with $n = 4$ and $p = 1$. Assume that the sensor is located at $\alpha / 7.5$.

b. Assuming that sensors and actuators are collocated at $\alpha / 7.5$, design a combined observer-controller system with $n = 4$ and $p = 1$. Demonstrate the performance of your controller via numerical simulations.

P4.4   Consider the system

$$\frac{d\mathbf{x}}{dt} = \begin{bmatrix} 1 & 1 \\ 0 & 1 \end{bmatrix}\mathbf{x} + \begin{bmatrix} 0 \\ 1 \end{bmatrix}u(t) + \xi(t)$$

$$y(t) = [1 \quad 0]\mathbf{x}(t) + \theta(t)$$

where $\xi(t)$ and $\theta(t)$ are zero mean Gaussian white noise processes with

$$E[\xi(t)\xi^T(t+\tau)] = \sigma\begin{bmatrix} 1 & 0 \\ 0 & 1 \end{bmatrix}\delta(\tau)$$

$$E[\theta(t)\theta(t+\tau)] = 1\delta(\tau)$$

a. Find the steady state Kalman filter gain matrix for two different values of σ, 2 and 0.2.

b. Demonstrate the performance of the Kalman filter via numerical simulations for both the values of σ. Discuss your results.

Hint: Use Matlab routines: randn and ode23 or ode45.

P4.5   Consider the system

$$\frac{d\mathbf{x}}{dt} = \begin{bmatrix} 1 & 1 \\ 0 & 1 \end{bmatrix}\mathbf{x} + \begin{bmatrix} 0 \\ 1 \end{bmatrix}u(t) + \xi(t)$$

$$y(t) = \begin{bmatrix} 1 & 0 \end{bmatrix}\mathbf{x} + \theta(t)$$

where $\xi(t)$ and $\theta(t))$ are zero mean Gaussian white noise processes with

$$E[\xi(t)\xi^T(t+\tau)] = \begin{bmatrix} 0.5 & 0 \\ 0 & 0.5 \end{bmatrix}\delta(\tau)$$

$$E[\theta(t)\theta(t+\tau)] = \delta(\tau)$$

a. Design the LQG controller to minimize the following objective function:

$$\lim_{t\to\infty} E(y^2(t)+u^2(t))$$

b. Demonstrate the performance of your controller via numerical simulations.

Hint: Use Matlab routines: ODE45 and randn.

P4.6 Consider the flexible tetrahedral truss structure (Appendix F). The controller is to be designed on the basis of the first four modes of vibration. Design an LQG controller with collocated sensors and actuators such that

$$J = \lim_{t_f\to\infty} E\left\{\frac{1}{t_f}\int_0^{t_f}[\mathbf{y}^T(t)\mathbf{y}(t)+0.1\mathbf{u}^T(t)\mathbf{u}(t)dt]\right\}$$

is minimized. Introduce fictitious noises $\mathbf{w}(t)$ and $\theta(t)$ (Equation 4.6.1 and Equation 4.6.4), where $W = I_8$ and $\Theta = I_3$.

a. Find the controller transfer matrix and determine its stability.

b. Demonstrate the performance of your controller via numerical simulations.

P4.7 For the control of thermoacoustic instability (Annaswamy et al., 2000), a microphone and a loudspeaker are used as a sensor (Figure P4.7) and the actuator, respectively. The plant transfer function is derived to be

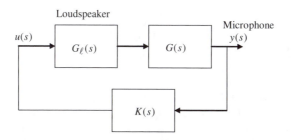

FIGURE P4.7 Control of thermoacoustic instability.

$$G(s) = 2.15 \times 10^5 \frac{(s+440)}{(s+14.9)(s^2+407s+1.03x10^6)} \frac{s^2-83s+7.70\times10^6}{s^2-63s+9.41\times10^6}$$

The transfer function of the loudspeaker is given by

$$G_\ell(s) = \frac{35.5s^2}{s^2+364s+3.320\times10^6}$$

a. Design an LQG controller such that

$$\lim_{t\to\infty} E(y^2(t)+u^2(t))$$

is minimized. Introduce fictitious noises $\mathbf{w}(t)$ and $\mathbf{\theta}(t)$ (Equation 4.6.1 and Equation 4.6.4), where $W = I$ and $\Theta = 1$.

b. Demonstrate the performance of your controller via numerical simulations.

P4.8   Consider the single-link manipulator (Figure P4.8a) modeled by Cannon and Schmitz (1984). This model is the same as that in P3.4 except that the mass-moment of inertia of the hub, I_h, is included here. Defining $I_T = I_h + I_\alpha$,

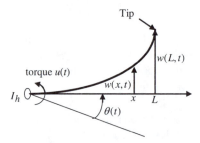

FIGURE P4.8A A single-link flexible manipulator.

$$\frac{y(L,s)}{u(s)} = \frac{L}{I_T s^2} + \frac{1}{I_T} \sum_{i=1}^{\infty} \frac{\beta_i}{s^2+2\xi_i\omega_i s+\omega_i^2}$$

where I_T = 0.44 kg-m² and L = 1 m. Parameters β_i, ξ_i, and ω_i are experimentally identified and listed in Table P4.8.

TABLE P4.8
Parameter Values

Mode i	ω_i, Hz	ξ_i	β_i
1	1.8	0.05	−2.8830
2	3.5	0.02	1.4950
3	8.4	0.02	−0.5940

a. On the basis of the rigid and the first mode, design an LQG controller $k_{lqg}(s)$ (Banavar and Dominic, 1995) to minimize the following objective function:

$$E(0.25y^2(L,t) + 3000u^2(t))$$

Introduce fictitious noises $\mathbf{w}(t)$ and $\mathbf{\theta}(t)$ (Equation 4.6.1 and Equation 4.6.4), where $W = 0.1I_4$ and $\Theta = 100$.

b. Find the feedforward function k_f in Figure P4.8b such that the unit step response has a zero steady state error.

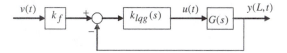

FIGURE P4.8B Linear quadratic Gaussian (LQG) control system.

5 Robust Control: Fundamental Concepts and H_2, H_∞, and μ Techniques

First, fundamental concepts for robust control are developed. Then, the robustness of liner quadratic (LQ) and liner quadratic Gaussian (LQG) control techniques developed in Chapter 3 and Chapter 4 are examined. Lastly, theories for H_2, H_∞, and μ techniques are presented along with Bode's sensitivity integrals and illustrative examples.

5.1 IMPORTANT ASPECTS OF SINGULAR VALUE ANALYSIS

5.1.1 SIGNIFICANCE OF MINIMUM AND MAXIMUM SINGULAR VALUES OF THE TRANSFER FUNCTION MATRIX AT A FREQUENCY

Consider a multiinput/multioutput (MIMO) system with the $p \times m$ transfer function matrix $G(s)$; i.e.,

$$\mathbf{y}(s) = G(s)\mathbf{u}(s) \tag{5.1.1}$$

where $\mathbf{y}(s)$ and $\mathbf{u}(s)$ are output and input vectors, respectively. Let the input vector be sinusoidal with the frequency ω; i.e.,

$$\mathbf{u}(t) = e^{j\omega t}\mathbf{a}_u \tag{5.1.2}$$

where

$$\mathbf{a}_u = \begin{bmatrix} a_{u1} & a_{u2} & . & . & a_{um} \end{bmatrix}^T \tag{5.1.3}$$

and the corresponding steady state output vector be

$$\mathbf{y}(t) = e^{j\omega t}\mathbf{a}_y \tag{5.1.4}$$

where

$$\mathbf{a}_y = \begin{bmatrix} a_{y1} & a_{y2} & . & . & a_{yp} \end{bmatrix}^T \tag{5.1.5}$$

It is well known that

$$\mathbf{a}_y = G(j\omega)\mathbf{a}_u \tag{5.1.6}$$

Bode magnitude plots for a single input/single output (SISO) system is generalized to a MIMO system by plotting all the singular values (Appendix C) of the matrix $G(j\omega)$ as a function of the frequency ω. The plots of maximum ($\bar{\sigma}(G(j\omega))$) and minimum ($\underline{\sigma}(G(j\omega))$) singular values have a special significance because of the following relationship (Maciejowski, 1989):

$$\underline{\sigma}(G(j\omega)) \le \frac{\|G(j\omega)\mathbf{a}_u\|}{\|\mathbf{a}_u\|} = \frac{\|\mathbf{a}_y(\omega)\|}{\|\mathbf{a}_u\|} \le \bar{\sigma}(G(j\omega)) \tag{5.1.7}$$

Therefore, the maximum and the minimum singular values of $G(j\omega)$ are upper and lower bounds of the ratio of 2-norms of amplitude vectors of the steady state output and the sinusoidal input of frequency ω.

EXAMPLE 5.1.1: SPRING-MASS-DAMPER SYSTEM

Consider the spring-mass-damper system shown in Figure 5.1.1. Let the outputs of the system be displacements x_1 and x_2 of both masses, and the inputs be the forces u_1 and u_2 acting on masses. The state space model of the system is

$$\dot{\mathbf{x}} = A\mathbf{x} + B\mathbf{u} \tag{5.1.8}$$

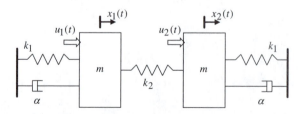

FIGURE 5.1.1 A spring-mass-damper system.

$$\mathbf{y} = C\mathbf{x} + D\mathbf{u} \tag{5.1.9}$$

where

$$\mathbf{y} = \begin{bmatrix} x_1 & x_2 \end{bmatrix}^T ; \quad \mathbf{x} = \begin{bmatrix} x_1 & x_2 & \dot{x}_1 & \dot{x}_2 \end{bmatrix}^T ; \quad \mathbf{u} = \begin{bmatrix} u_1 & u_2 \end{bmatrix}^T \tag{5.1.10}$$

$$A = \begin{bmatrix} 0 & I \\ -M^{-1}K & -M^{-1}C_d \end{bmatrix} ; \quad B = \begin{bmatrix} 0 \\ I \end{bmatrix} ; \quad C = \begin{bmatrix} I & 0 \end{bmatrix} ; \quad D = 0 \tag{5.1.11}$$

$$K = \begin{bmatrix} k_1 + k_2 & -k_2 \\ -k_2 & k_1 + k_2 \end{bmatrix} ; \quad C_d = \begin{bmatrix} \alpha & 0 \\ 0 & \alpha \end{bmatrix} ; \quad M = \begin{bmatrix} m & 0 \\ 0 & m \end{bmatrix} \tag{5.1.12}$$

The transfer function matrix relating inputs and outputs is described as follows:

$$\mathbf{y}(s) = G(s)\mathbf{u}(s) \tag{5.1.13}$$

where

$$G(s) = C(sI - A)^{-1}B + D \tag{5.1.14}$$

is a square matrix of order 2. At any frequency ω, there will be two singular values of $G(j\omega)$. These singular values provide bounds of the magnitude of the frequency response (Equation 5.1.7). Using the Matlab command "sigma," singular values are plotted in Figure 5.1.2 as a function of the frequency ω for $m = 1$ kg, $k_1 = 100 \ N/m$, $k_2 = 500 \ N/m$, and $\alpha = 1 \ N-sec/m$. It is interesting to observe that peaks of singular values are located near the natural frequencies of the system.

5.1.2 An Inequality for Robustness Test

Let A be a nonsingular matrix; i.e.,

$$\underline{\sigma}(A) > 0 \tag{5.1.15}$$

where $\underline{\sigma}(A)$ is the minimum singular value of the matrix A. Let us try to find a perturbation matrix L of the smallest size in the sense of a 2-norm such that the matrix $(A+L)$ is singular or rank deficient. In this case, there exists a nonzero vector x such that

$$\|\mathbf{x}\|_2 = 1 \quad \text{and} \quad (A + L)\mathbf{x} = 0 \tag{5.1.16a,b}$$

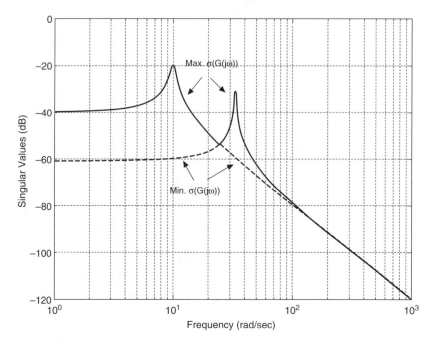

FIGURE 5.1.2 Singular values of the mechanical system in Figure 5.1.1.

Therefore, using properties of singular values (Appendix C),

$$\underline{\sigma}(A) \le \left\| A\mathbf{x} \right\|_2 = \left\| L\mathbf{x} \right\|_2 \le \left\| L \right\|_2 = \overline{\sigma}(L) \tag{5.1.17}$$

where $\overline{\sigma}(L)$ is the maximum singular value of the matrix L. In other words, the singularity of the matrix $(A+L)$ implies $\underline{\sigma}(A) \le \overline{\sigma}(L)$; i.e., the size of the perturbation matrix must be at least $\underline{\sigma}(A)$. This analysis can also be viewed as the following condition for the nonsingularity of the matrix $(A+L)$:

$$\underline{\sigma}(A) > \overline{\sigma}(L) \tag{5.1.18}$$

The condition (5.1.18) serves as an inequality for the development of various robustness criteria.

5.2 ROBUSTNESS: SENSITIVITY AND COMPLEMENTARY SENSITIVITY

5.2.1 BASIC DEFINITIONS

In Figure 5.2.1, the transfer function matrix for a MIMO system is denoted by $G(s)$. The controller transfer function matrix is denoted by $K(s)$. Other variables are described as follows:

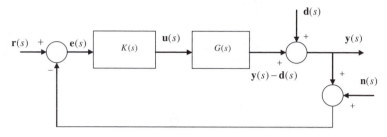

FIGURE 5.2.1 A multiinput/multioutput system with disturbances.

$\mathbf{n}(s)$: Measurement or sensor noise vector
$\mathbf{d}(s)$: Disturbance vector at the plant output
$\mathbf{r}(s)$: Reference (external) input vector
$\mathbf{e}(s)$: Tracking error vector

Analyzing the Figure 5.2.1 (Anderson and Moore, 1990), it is clear that

$$\mathbf{y}(s) - \mathbf{d}(s) = G(s)K(s)\mathbf{e}(s) \tag{5.2.1}$$

and

$$\mathbf{e}(s) = \mathbf{r}(s) - (\mathbf{y}(s) + \mathbf{n}(s)) \tag{5.2.2}$$

Substituting (5.2.2) into (5.2.1),

$$\mathbf{y}(s) = T_o(s)[\mathbf{r}(s) - \mathbf{n}(s)] + S_o(s)\mathbf{d}(s) \tag{5.2.3}$$

where

$$T_o(s) = (I + G(s)K(s))^{-1}G(s)K(s) \tag{5.2.4}$$

and

$$S_o(s) = (I + G(s)K(s))^{-1} \tag{5.2.5}$$

In addition, from (5.2.2),

$$\mathbf{e}(s) = \mathbf{r}(s) - \mathbf{n}(s) - \mathbf{d}(s) - (\mathbf{y}(s) - \mathbf{d}(s)) \tag{5.2.6}$$

Using (5.2.1),

$$\mathbf{e}(s) = S_o(s)(\mathbf{r}(s) - \mathbf{d}(s)) - S_o(s)\mathbf{n}(s) \tag{5.2.7}$$

Also,

$$\mathbf{u}(s) = K(s)\mathbf{e}(s) = K(s)S_o(s)(\mathbf{r}(s) - \mathbf{d}(s)) - K(s)S_o(s)\mathbf{n}(s) \qquad (5.2.8)$$

Definitions 5.2.1

$$S_o(s) = (I + G(s)K(s))^{-1} \text{ is called the } output\ sensitivity\ function.\quad (5.2.9)$$

$T_o(s) = (I + G(s)K(s))^{-1}G(s)K(s)$ is called the *output complementary*
$$\qquad\qquad\qquad\qquad sensitivity\ function. \qquad\qquad\qquad (5.2.10)$$

Fact 5.2.1

$$T_o(s) = (I + G(s)K(s))^{-1}G(s)K(s) = G(s)K(s)(I + G(s)K(s))^{-1} \quad (5.2.11)$$

Proof

$$(I + GK)^{-1}GK = [(GK)^{-1}(I + GK)]^{-1}$$

$$= [(GK)^{-1} + I]^{-1} = [(I + GK)(GK)^{-1}]^{-1} = GK(I + GK)^{-1}$$

This completes the proof.

Fact 5.2.2

$$S_o(s) + T_o(s) = I \qquad (5.2.12)$$

Proof

$$S_o(s) + T_o(s) = (I + GK)^{-1} + (I + GK)^{-1}GK$$

$$= (I + GK)^{-1}(I + GK) = I$$

This completes the proof.

The relationship (5.2.12) represents a fundamental trade-off in the design of a control system. It implies that both S_o and $T_o(s)$ cannot be simultaneously small. Note the following observations (Anderson and Moore, 1990):

1. Assuming that $\mathbf{d}(s)$ and $\mathbf{n}(s)$ are zero, Equation 5.2.7 yields

$$\mathbf{e}(s) = S_o(s)\mathbf{r}(s) \quad \text{where} \quad \mathbf{e}(s) = \mathbf{r}(s) - \mathbf{y}(s) \tag{5.2.13}$$

The error signal $\mathbf{e}(s) = \mathbf{r}(s) - \mathbf{y}(s)$ is the tracking error. Therefore, the sensitivity function $S_o(s)$ should be small for a small tracking error. More specifically, the following condition should be satisfied for a good tracking:

$$\bar{\sigma}(S_o(j\omega)) \ll 1 \quad \text{at a frequency } \omega \tag{5.2.14}$$

2. From Equation 5.2.3, the effect of the disturbance (at the output) is small on the output if the sensitivity function $S_o(s)$ is small. Therefore, for a good disturbance rejection,

$$\bar{\sigma}(S_o(j\omega)) \ll 1 \quad \text{at a frequency } \omega \tag{5.2.15}$$

3. From Equation 5.2.3, the effect of the sensor noise is small on the output if the complementary sensitivity function $T_o(s)$ is small. Therefore, for a good noise rejection,

$$\bar{\sigma}(T_o(j\omega)) \ll 1 \quad \text{at a frequency } \omega \tag{5.2.16}$$

4. Assume that $\bar{\sigma}(S_o(j\omega)) \ll 1$ in a certain range of ω. In this case, $G(s)K(s)$ is large, and

$$S_o(s) = (I + G(s)K(s))^{-1} \approx (G(s)K(s))^{-1} \tag{5.2.17}$$

Assuming that the number of inputs equals the number of outputs, $K(s)$ and $G(s)$ will be square matrices and Equation 5.2.17 yields

$$S_o(s) \approx K^{-1}(s)G^{-1}(s) \tag{5.2.18}$$

Using (5.2.8) and (5.2.18),

$$\mathbf{u}(s) = G^{-1}(s)(\mathbf{r}(s) - \mathbf{n}(s) - \mathbf{d}(s)) \tag{5.2.19}$$

Therefore, the input is related to the inverse of the plant transfer function. If $S_o(s)$ is small outside the bandwidth of the plant, Equation 5.2.19 will be applicable to the frequency range where

$$\bar{\sigma}(G(j\omega)) \ll 1 \qquad (5.2.20)$$

This condition is equivalent to

$$\underline{\sigma}(G^{-1}(j\omega)) \gg 1 \qquad (5.2.21)$$

Therefore, Equation 5.2.19 suggests that the control input will be large.

5.2.2 ROBUSTNESS TO STRUCTURED UNCERTAINTIES

Structured uncertainties refer to parameter errors in a model of the plant. Because these parameters appear in the model according to its structure, parametric errors are called structured uncertainties. To discuss the robustness to structured uncertainties, consider the multivariable feedback system shown in Figure 5.2.2.

Here, from (5.2.3),

$$\mathbf{y}(s) = G(s, \chi)K(s)(I + G(s, \chi)K(s))^{-1}\mathbf{r}(s) \qquad (5.2.22)$$

where the plant transfer function $G(s, \chi)$ has a parameter χ which is uncertain. Equation 5.2.22 can also be viewed as an equivalent open-loop system (Anderson and Moore, 1990) shown in Figure 5.2.3 where

$$\mathbf{y}(s) = G(s, \chi)\tilde{K}(s)\mathbf{r}(s) \qquad (5.2.23)$$

and

$$\tilde{K}(s) = K(s)S_o(s) \qquad (5.2.24)$$

It should be noted that $\tilde{K}(s)$ is the equivalent open-loop controller and is not a function of χ.

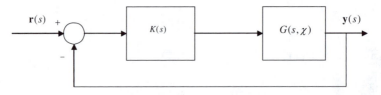

FIGURE 5.2.2 A multiinput/multioutput feedback system with structured uncertainties.

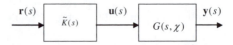

FIGURE 5.2.3 An equivalent multiinput/multioutput open-loop system.

Therefore,

$$\left(\frac{\partial \mathbf{y}}{\partial \chi}\right)_{OL} = \frac{\partial G}{\partial \chi} \tilde{K}(s)\mathbf{r}(s) \qquad (5.2.25)$$

From (5.2.22) and (5.2.11),

$$\mathbf{y}(s) = (I + G(s,\chi)K(s))^{-1} G(s,\chi)K(s)\mathbf{r}(s) \qquad (5.2.26)$$

Equation 5.2.26 can also be written as

$$(I + G(s,\chi)K(s))\mathbf{y}(s) = G(s,\chi)K(s)\mathbf{r}(s) \qquad (5.2.27)$$

Differentiating the closed-loop (CL) relationship (5.2.27) with respect to χ,

$$(I+GK)\left(\frac{\partial \mathbf{y}}{\partial \chi}\right)_{CL} + \frac{\partial G}{\partial \chi} K\mathbf{y}(s) = \frac{\partial G}{\partial \chi} K\mathbf{r}(s) \qquad (5.2.28)$$

or

$$(I+GK)\left(\frac{\partial \mathbf{y}}{\partial \chi}\right)_{CL} = \frac{\partial G}{\partial \chi} K(-\mathbf{y}(s)+\mathbf{r}(s)) \qquad (5.2.29)$$

From (5.2.26) and (5.2.12),

$$\mathbf{r}(s) - \mathbf{y}(s) = S_o(s)\mathbf{r}(s) \qquad (5.2.30)$$

From (5.2.29), (5.2.24), and (5.2.30),

$$\left(\frac{\partial \mathbf{y}}{\partial \chi}\right)_{CL} = S_o(s) \frac{\partial G}{\partial \chi} \tilde{K}(s)\mathbf{r}(s) \qquad (5.2.31)$$

Using (5.2.25) and (5.2.31),

$$(\frac{\partial \mathbf{y}}{\partial \chi})_{CL} = S_o(s)(\frac{\partial \mathbf{y}}{\partial \chi})_{OL} \qquad (5.2.32)$$

Therefore, the sensitivity to a structured plant parameter variation in a closed-loop system is smaller than that of the equivalent open-loop system when the sensitivity function $S_o(s)$ is small. Furthermore, from (5.2.31),

$$\left(\frac{\partial \mathbf{y}}{\partial \chi}\right)_{CL} = S_o(s)\frac{\partial G}{\partial \chi}K(s)S_o(s)\mathbf{r}(s) \qquad (5.2.33)$$

A small $S_o(s)$ usually implies a large feedback gain $K(s)$. Therefore, in this case for a square plant, Equation 5.2.33 yields

$$\left(\frac{\partial \mathbf{y}}{\partial \chi}\right)_{CL} \approx K^{-1}(s)G^{-1}(s)\frac{\partial G}{\partial \chi}G^{-1}(s)\mathbf{r}(s) \qquad (5.2.34)$$

Equation 5.2.34 indicates that a higher value of gain will reduce the sensitivity of the output to parameter errors. But outside the plant bandwidth where $G^{-1}(s)$ is large, the sensitivity of the output to parameter errors can be large.

5.2.3 ROBUSTNESS TO UNSTRUCTURED UNCERTAINTIES

Unstructured modeling errors usually refer to unmodeled dynamics in the range of high frequencies. They can be represented in the form of a multiplicative uncertainty $L(s)$ as

$$G_L(s) = (I + L(s))G(s) \qquad (5.2.35)$$

where $G_L(s)$ and $G(s)$ are actual and nominal transfer functions, respectively. Assume that the modeling error $L(s)$ is bounded in the following sense:

$$\bar{\sigma}(L(j\omega)) < \ell(j\omega) \qquad (5.2.36)$$

Referring to Figure 5.2.4, the closed-loop input/output relationship is

$$\mathbf{y}(s) = G_L(s)K(s)(I + G_L(s)K(s))^{-1}\mathbf{r}(s) \qquad (5.2.37)$$

For the stability of (5.2.37), all poles of the multivariable transfer function matrix must be in the left half of the s-plane. The stability can also be determined by the Nyquist stability criterion (Doyle and Stein, 1981):

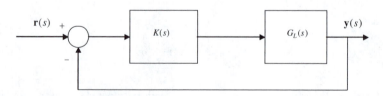

FIGURE 5.2.4 A multiinput/multioutput feedback system with unstructured uncertainties.

The encirclement count of the map $\det(I + G(s)K(s))$ *evaluated on the standard Nyquist path in the s-plane must be equal to the (negative) number of unstable open-loop poles of* $G(s)K(s)$.

The following assumptions are made:

1. The nominal feedback system $G(s)K(s)(I + G(s)K(s))^{-1}$ is stable.
2. The perturbed system $G_L(s)$ and the nominal system $G(s)$ have the same number of unstable poles. It should be noted that unstable poles of $G_L(s)$ and $G(s)$ are not assumed to be identical.

Applying the Nyquist stability criterion with these assumptions, the closed-loop system will be stable under all possible multiplicative perturbations provided the number of encirclements of the origin remains unchanged for $\det(I + G_L(s)K(s))$. This condition is guaranteed if and only if $\det(I + G_L(s)K(s))$ remains nonzero or, equivalently,

$$\underline{\sigma}(I + G_L(j\omega)K(j\omega)) > 0 \tag{5.2.38}$$

Using (5.2.35),

$$\begin{aligned} I + G_L(j\omega)K(j\omega) &= I + (I + L(j\omega))G(j\omega)K(j\omega) \\ &= I + G(j\omega)K(j\omega) + L(j\omega)G(j\omega)K(j\omega) \\ &= [I + LGK(I + GK)^{-1}](I + GK) \end{aligned} \tag{5.2.39}$$

From (5.2.39),

$$\begin{aligned} &\det(I + G_L(j\omega)K(j\omega)) = \\ &\det[I + L(j\omega)G(j\omega)K(j\omega)(I + G(j\omega)K(j\omega))^{-1}].\det(I + G(j\omega)K(j\omega)) \end{aligned} \tag{5.2.40}$$

Because $\det(I + G(j\omega)K(j\omega)) \neq 0$, the condition $\det(I + G_L(j\omega)K(j\omega)) \neq 0$ is equivalent to

$$\det(I + L(j\omega)G(j\omega)K(j\omega)(I + G(j\omega)K(j\omega))^{-1}) \neq 0 \tag{5.2.41}$$

Using (5.1.18), the condition (5.2.41) is satisfied when

$$\bar{\sigma}(L(j\omega)G(j\omega)K(j\omega)(I + G(j\omega)K(j\omega))^{-1}) < 1 \tag{5.2.42}$$

But

$$\bar{\sigma}(LGK(I+GK)^{-1}) \leq \bar{\sigma}(L)\bar{\sigma}(GK(I+GK)^{-1}) \qquad (5.2.43)$$

Therefore, the condition (5.2.42) will be satisfied when

$$\bar{\sigma}(L)\bar{\sigma}(GK(I+GK)^{-1}) < 1 \qquad (5.2.44)$$

The condition (5.2.44) can be written as

$$\bar{\sigma}(T_o(j\omega)) < \frac{1}{\bar{\sigma}(L(j\omega))} \qquad (5.2.45)$$

where $T_o(j\omega)$ is the output complementary sensitivity function. The condition (5.2.45) states that the output complementary sensitivity function should be smaller than the inverse of the largest size of the perturbation matrix $L(j\omega)$. Therefore, for a larger perturbation matrix $L(j\omega)$, the output complementary sensitivity function has to be smaller. A small complementary sensitivity function implies a large sensitivity function, which in turn means small feedback gains. This fact can be shown by rewriting Equation 5.2.44 as

$$\frac{1}{\bar{\sigma}(GK(I+GK)^{-1})} > \bar{\sigma}(L) \qquad (5.2.46)$$

Note that

$$[\bar{\sigma}(GK(I+GK)^{-1})]^{-1} = \underline{\sigma}[(I+GK)(GK)^{-1}] = \underline{\sigma}(I+(GK)^{-1}) \qquad (5.2.47)$$

Therefore, the condition (5.2.46) is equivalent to

$$\underline{\sigma}(I+(GK)^{-1}) > \bar{\sigma}(L) \qquad (5.2.48)$$

Hence, $G(j\omega)K(j\omega)$ or, equivalently, $K(j\omega)$ has to be small.

EXAMPLE 5.2.1: MULTIPLICATIVE UNCERTAINTY

Consider the following SISO system:

$$G_L(s) = \frac{\alpha}{s+\alpha} G(s) \qquad (5.2.49)$$

where $G(s)$ and $G_L(s)$ are nominal and actual transfer functions, respectively. Equation 5.2.49 can be rewritten in the form of Equation 5.2.35:

$$G_L(s) = (1 + L(s))G(s) \tag{5.2.50}$$

where

$$L(s) = -\frac{s}{s+\alpha} \tag{5.2.51}$$

Therefore,

$$\bar{\sigma}(L(j\omega)) = |L(j\omega)| = \frac{1}{\sqrt{1 + \dfrac{\alpha^2}{\omega^2}}} < 1 + \varepsilon \tag{5.2.52}$$

where ε is a small positive number. Therefore, the robustness criterion (5.2.48) becomes

$$(1 + GK(j\omega))^{-1} > 1 + \varepsilon \tag{5.2.53}$$

EXAMPLE 5.2.2: A SPRING-MASS SYSTEM

Consider the spring-mass-damper system (Figure 5.1.1) again. First, an *lqg* controller, $R = \rho_c I$ and $Q = C^T C$ in Equation 4.8.9, is designed (Matlab Program 5.2.1) to obtain the controller transfer function $K(s)$. Then, the sensitivity function $S_o(s)$, the complementary sensitivity function $T_o(s)$, and $K(s)S_o(s)$ are plotted for two values (0.01 and 0.1) of the control weighting ρ_c in Figure 5.2.5, Figure 5.2.6, and Figure 5.2.7, respectively.

The sensitivity function for $\rho_c = 0.01$ is smaller than that for $\rho_c = 0.1$ near the natural frequencies of the system. Because of the constraint $S_o(s) + T_o(s) = I$, the complementary sensitivity function for $\rho_c = 0.01$ is larger than that for $\rho_c = 0.1$ near the natural frequencies of system. But at all frequencies, the control input or $K(s)S_o(s)$ for $\rho_c = 0.1$ is significantly smaller than that for $\rho_c = 0.01$.

MATLAB PROGRAM 5.2.1: LQG CONTROL OF A SPRING-MASS SYSTEM

```
%
clear all
close all
%
k1=100;
k2=500;
```

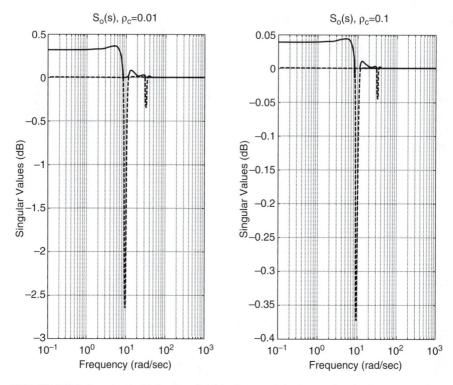

FIGURE 5.2.5 Output sensitivity functions for linear quadratic Gaussian control of a spring-mass-damper system.

```
K=[k1+k2  -k2;-k2  k1+k2];
al=1;
%
A=[zeros(2,2),eye(2);-K  -al*eye(2)];
B=[0  0  1  0;0  0  0  1]';
C=[1  0  0  0;0  1  0  0];
D=zeros(2,2);
%
sysG=ss(A,B,C,D);
%
for iplot=1:2
%LQG Controller
rhoc=0.001*(10)^iplot;
```

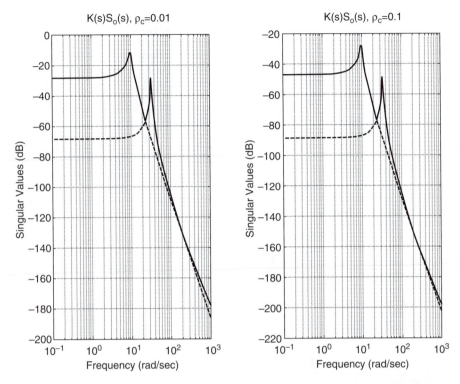

FIGURE 5.2.6 Output complementary sensitivity functions for linear quadratic Gaussian control of a spring-mass-damper system.

```
V=[eye(4) zeros(4,2);zeros(2,4) 0.01*eye(2)];

W=[C'*C zeros(4,2);zeros(2,4) rhoc*eye(2)];

[Af,Bf,Cf,Df]=lqg(A,B,C,D,W,V);

sysK=ss(Af,Bf,Cf,Df);

%

sysId=ss([],[],[],eye(2));

%

sysGK=series(sysK,sysG);

%

%Sensitivity Function

sys_Sens=feedback(sysId,sysGK);

figure(1)

subplot(1,2,iplot)
```

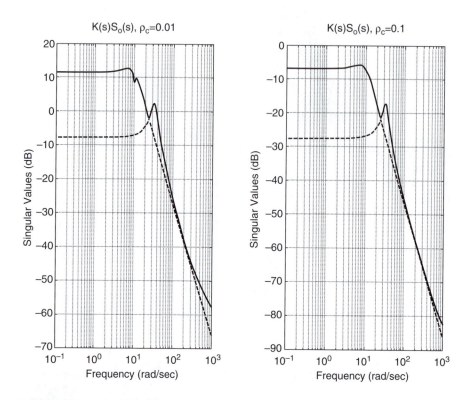

FIGURE 5.2.7 Control inputs or for linear quadratic Gaussian control of a spring-mass-damper system.

```
sigma(sys_Sens)
[sv_Sens,freq]=sigma(sys_Sens);
grid
xlabel('Frequency','Fontsize',12)
ylabel('Singular Values','Fontsize',12)
if(iplot==1)
title('S_o(s), \rho_c=0.01','Fontsize',12)
end
if(iplot==2)
title('S_o(s), \rho_c=0.1','Fontsize',12)
end
%
%
```

```
%Complementary Sensitivity Function
sys_CSens=feedback(sysGK,sysId);
[A_CSens,B_CSens,C_CSens,D_CSens]=branch(sys_CSens);
eig(A_CSens)
figure(2)
subplot(1,2,iplot)
sigma(sys_CSens,freq)
grid
xlabel('Frequency','Fontsize',12)
ylabel('Singular Values','Fontsize',12)
if(iplot==1)
title('T_o(s), \rho_c=0.01','Fontsize',12)
end
if(iplot==2)
title('T_o(s), \rho_c=0.1','Fontsize',12)
end
%
figure(3)
subplot(1,2,iplot)
%K(s)S(s)
sys_KSens=series(sys_Sens,sysK);
sigma(sys_KSens,freq)
grid
xlabel('Frequency','Fontsize',12)
ylabel('Singular Values','Fontsize',12)
if(iplot==1)
title('K(s)S_o(s), \rho_c=0.01','Fontsize',12)
end
if(iplot==2)
title('K(s)S_o(s), \rho_c=0.1','Fontsize',12)
end
end
```

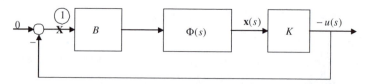

FIGURE 5.3.1 A linear quadratic regular feedback loop.

5.3. ROBUSTNESS OF LQR AND KALMAN FILTER (KF) FEEDBACK LOOPS

5.3.1 LINER QUADRATIC REGULATOR (LQR) FEEDBACK LOOP

Consider the state space model

$$\dot{\mathbf{x}} = A\mathbf{x} + B\mathbf{u}(t) \tag{5.3.1}$$

and the state feedback law

$$\mathbf{u}(t) = -K\mathbf{x}(t) \tag{5.3.2}$$

where K is the optimal state feedback gain matrix (Chapter 3, Section 3.5). Taking Laplace transforms of (5.3.1) and (5.3.2) with zero initial conditions,

$$\mathbf{u}(s) = -K\mathbf{x}(s) \tag{5.3.3}$$

$$\mathbf{x}(s) = \Phi(s)B\mathbf{u}(s) \tag{5.3.4}$$

$$\Phi(s) = (sI - A)^{-1} \tag{5.3.5}$$

Relationships (5.3.3) and (5.3.4) can be represented as an LQR feedback loop shown in Figure 5.3.1. The LQR loop transfer matrix $G_{LQ}(s)$ is given by

$$G_{LQ}(s) = K\Phi(s)B \tag{5.3.6}$$

Breaking the loop at (1) in Figure 5.3.1,

$$\text{Return difference matrix} = I + G_{LQ}(s) \tag{5.3.7}$$

Fact 5.3.1: Return Difference Equality

$$[I + G_{LQ}(-s)]^T R[I + G_{LQ}(s)] = R + [N\Phi(-s)B]^T[N\Phi(s)B] \tag{5.3.8}$$

where

$$Q = N^T N \qquad (5.3.9)$$

Proof

Consider the algebraic Riccati equation (ARE) (Equation 3.5.55 with $N = 0$):

$$-PA - A^T P + PBR^{-1}B^T P = Q \qquad (5.3.10)$$

Equation 5.3.10 can be written as

$$P(sI - A) + (-sI - A^T)P + K^T RK = Q \qquad (5.3.11)$$

because

$$K = R^{-1}B^T P \qquad (5.3.12)$$

Using (5.3.5),

$$P\Phi^{-1}(s) + [\Phi^T(-s)]^{-1}P + K^T RK = Q \qquad (5.3.13)$$

Left-multiplying (5.3.13) by $B^T \Phi^T(-s)$, and right-multiplying (5.3.13) by $\Phi(s)B$,

$$B^T \Phi^T(-s)PB + B^T P\Phi(s)B + B^T \Phi^T(-s)K^T RK \Phi(s)B = B^T \Phi^T(-s)Q\Phi(s)B \quad (5.3.14)$$

From (5.3.12),

$$RK = B^T P \qquad (5.3.15)$$

Substituting (5.3.15) into (5.3.14),

$$B^T \Phi^T(-s)K^T R + RK\Phi(s)B + B^T \Phi^T(-s)K^T RK\Phi(s)B$$
$$= B^T \Phi^T(-s)Q\Phi(s)B \qquad (5.3.16)$$

Adding R to both sides of Equation 5.3.16, using Equation 5.3.6 and Equation 5.3.9,

$$R + G_{LQ}^T(-s)R + RG_{LQ}(s) + G_{LQ}^T(-s)RG_{LQ}(s) = R + B^T \Phi^T(-s)NN^T\Phi(s)B \quad (5.3.17)$$

Equation 5.3.17 can be written as

$$[I + G_{LQ}^T(-s)]R[I + G_{LQ}(s)] = R + [N\Phi(-s)B]^T[N\Phi(s)B] \qquad (5.3.18)$$

This completes the proof.

5.3.2 Gain and Phase Margins of a Single-Input System

For a single-input system, $G_{LQ}(s)$ and R are scalars. Let

$$R = \rho > 0 \qquad (5.3.19)$$

Equation 5.3.18 reduces to

$$[1 + G_{LQ}^T(-s)][1 + G_{LQ}(s)] = 1 + \frac{1}{\rho}[N\Phi(-s)B]^T[N\Phi(s)B] \qquad (5.3.20)$$

Substituting $s = j\omega$ into (5.3.20),

$$\left|1 + G_{LQ}(j\omega)\right|^2 = 1 + \frac{1}{\rho}\left\|N\Phi(j\omega)B\right\|^2 \qquad (5.3.21)$$

Therefore,

$$\left|1 + G_{LQ}(j\omega)\right| \geq 1 \qquad (5.3.22)$$

The Nyquist map of $G_{LQ}(s)$ will be outside the circle of unit radius centered at $-1 + j0$ (Figure 5.3.2).

Let ℓ_1 be the factor by which the magnitude of the input can change. In this case, the loop transfer function will be $\ell_1 G_{LQ}(s)$. Because the Nyquist map of $G_{LQ}(s)$ will be outside the circle of unit radius centered at $-1 + j0$, it will only intersect the negative real axis beyond the point G $(-2, 0)$. Hence, the number of encirclements of $-1 + j0$ will remain unchanged provided

$$\ell_1 > \frac{1}{2} \qquad (5.3.23)$$

Therefore, the gain margin of the single-input LQR feedback loop shown in Figure 5.3.1 is at least between ∞ and 0.5.

Let ϕ_1 be the angle by which the phase of the input can change. In this case, the loop transfer function will be $e^{j\phi_1}G_{LQ}(s)$. Because the Nyquist map of $G_{LQ}(s)$

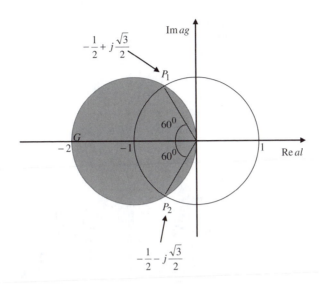

FIGURE 5.3.2 Forbidden unit circle with center at (−1,0).

will be outside the circle of unit radius centered at $-1 + j0$ and unit circles centered

at $(0,0)$ and $(-1,0)$ intersect (Figure 5.3.2) each other at $P_1\left(-\dfrac{1}{2}, \dfrac{\sqrt{3}}{2}\right)$ and

$P_2\left(-\dfrac{1}{2}, -\dfrac{\sqrt{3}}{2}\right)$, the number of encirclements of $(-1, 0)$ will remain unchanged

provided

$$|\phi_1| \le 60° \tag{5.3.24}$$

Therefore, the phase margin of the single-input LQR feedback loop shown in Figure 5.3.1 is at least 60°.

Example 5.3.1: LQ Control of a Simple Mass

Consider the double integrator in Example 4.8. With $\rho_c = 0.05$,

$$G_{LQ}(s) = \frac{2.9907s + 4.4721}{s^2} \tag{5.3.25}$$

Therefore,

$$G_{LQ}(j\omega) = -\frac{4.4721}{\omega^2} - j\frac{2.9907}{\omega} \tag{5.3.26}$$

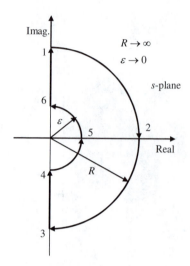

FIGURE 5.3.3A Nyquist path.

and

$$\angle G_{LQ}(j\omega) = \tan^{-1}\frac{-2.9907\omega}{-4.4721} \qquad (5.3.27)$$

Setting $\left|G_{LQ}(j\omega)\right| = 1$,

$$\omega^4 - (2.9907)^2\omega^2 - (4.4721)^2 = 0 \qquad (5.3.28)$$

Solving (5.3.28),

$$\omega = \omega_p = 3.2858 \quad \text{rad/sec} \qquad (5.3.29)$$

and

$$\angle G_{LQ}(j\omega_p) = \tan^{-1}\frac{-2.9907\omega_p}{-4.4721} = 245.53° \qquad (5.3.30)$$

Therefore,

$$\text{Phase Margin} = 245.53 - 180 = 65.53° \qquad (5.3.31)$$

From the Nyquist map (Figures 5.3.3a and 5.3.3b), it is clear that the gain margin is between 0 and ∞.

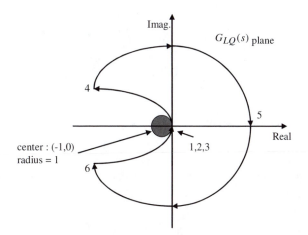

FIGURE 5.3.3B Nyquist map of $G_{LQ}(s)$.

5.3.3 GAIN AND PHASE MARGINS OF A MULTIINPUT SYSTEM

Assume that

$$R = \rho I \tag{5.3.32}$$

Then, from (5.3.8),

$$[I + G_{LQ}(-s)]^T [I + G_{LQ}(s)] = I + \frac{1}{\rho} [N\Phi(-s)B]^T [N\Phi(s)B] \tag{5.3.33}$$

Substituting $s = j\omega$ into (5.3.33),

$$[I + G_{LQ}(-j\omega)]^T [I + G_{LQ}(j\omega)] = I + \frac{1}{\rho} [N\Phi(-j\omega)B]^T [N\Phi(j\omega)B] \tag{5.3.34}$$

Therefore,

$$[I + G_{LQ}(-j\omega)]^T [I + G_{LQ}(j\omega)] \geq I \tag{5.3.35}$$

Fact 5.3.2

Consider the system shown in Figure 5.3.4. If $W^H + W > I$, $I + G_{LQ}(j\omega)W(j\omega)$ is nonsingular (Anderson and Moore, 1990) where W^H is the complex conjugate transpose of W.

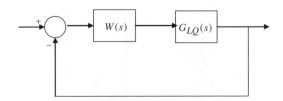

FIGURE 5.3.4 Linear quadratic loop with an uncertainty $W(s)$.

Proof: (by contradiction)

Let $I + G_{LQ}(j\omega)W(j\omega)$ be singular. In this case, there will be a nontrivial vector $\mathbf{q}$ such that

$$(I + G_{LQ}(j\omega)W(j\omega))\mathbf{q} = 0 \tag{5.3.36}$$

or

$$G_{LQ}W\mathbf{q} = -\mathbf{q} \tag{5.3.37}$$

From (5.3.35),

$$G_{LQ}^{H} + G_{LQ} + G_{LQ}^{H}G_{LQ} \geq 0 \tag{5.3.38}$$

where $G_{LQ}^{H}(j\omega)$ is the complex conjugate transpose of $G_{LQ}(j\omega)$. Premultiplying and postmultiplying (5.3.38) by $\mathbf{q}^{H}W^{H}$ and $W\mathbf{q}$, respectively,

$$\mathbf{q}^{H}W^{H}G_{LQ}^{H}W\mathbf{q} + \mathbf{q}^{H}W^{H}G_{LQ}W\mathbf{q} + \mathbf{q}^{H}W^{H}G_{LQ}^{H}G_{LQ}W\mathbf{q} \geq 0 \tag{5.3.39}$$

Using (5.3.37),

$$-\mathbf{q}^{H}W\mathbf{q} - \mathbf{q}^{H}W^{H}\mathbf{q} + \mathbf{q}^{H}\mathbf{q} \geq 0 \tag{5.3.40}$$

or

$$\mathbf{q}^{H}(W + W^{H} - I)\mathbf{q} \leq 0 \tag{5.3.41}$$

Relation (5.3.41) contradicts the assumption $W^{H} + W > I$ in the statement of Fact 5.3.2.

It should be noted that the Nyquist condition for the stability of the system in Figure 5.3.4 is that $I + G_{LQ}(j\omega)W(j\omega)$ is nonsingular. Therefore, the system shown in Figure 5.3.4 is stable when $W^{H} + W > I$.

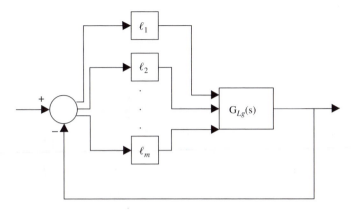

FIGURE 5.3.5 A multiinput/multioutput linear quadratic regular loop with a magnitude uncertainty in each input channel.

Special Case

Let W be a diagonal matrix, with diagonal elements being ℓ_1, ℓ_2, ..., and ℓ_m. Then, the condition $W^H + W > I$ reduces to

$$\ell_i^H + \ell_i > 1 ; \quad i = 1, 2, \ldots, m \tag{5.3.42}$$

Conditions (5.3.42) yield guaranteed levels of gain and phase margins of the LQR feedback loop as follows.

Gain Margin

Assume that ℓ_1, ℓ_2, ..., and ℓ_m are real numbers. Then, conditions (5.3.42) yield

$$\ell_i > \frac{1}{2} ; \quad i = 1, 2, \ldots, m \tag{5.3.43}$$

The LQR feedback loop (Figure 5.3.5) has a guaranteed infinite upward gain margin, and 1/2 downward gain margin in each input channel independently and simultaneously (Athans, 1992).

Phase Margin

Assume that

$$\ell_i = e^{j\phi_i} \tag{5.3.44}$$

Then, conditions (5.3.42) yield

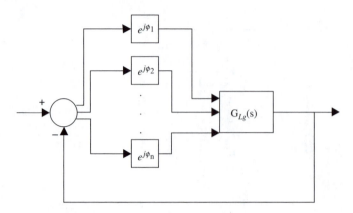

FIGURE 5.3.6 A multiinput/multioutput linear quadratic regular loop with a phase uncertainty in each input channel.

$$e^{-j\phi_i} + e^{j\phi_i} = 2\cos\phi_i > 1 \qquad (5.3.45)$$

Therefore,

$$|\phi_i| < 60° \qquad (5.3.46)$$

The LQR feedback loop (Figure 5.3.6) has a guaranteed $\pm60°$ phase margin in each input channel independently and simultaneously (Athans, 1992).

Note that the stability is not guaranteed if both a phase shift within $\pm60°$ and a gain change in the interval $\left(\dfrac{1}{2}, \infty\right)$ are simultaneously introduced in an input channel.

5.3.2 KALMAN FILTER (KF) LOOP

A Kalman filter (Chapter 4, Section 4.6) is represented by

$$\dot{\hat{x}} = A\hat{x}(t) + Bu(t) + L(y - \hat{y}) \qquad (5.3.47)$$

where

$$\hat{y}(t) = C\hat{x}(t) \qquad (5.3.48)$$

Taking the Laplace transform of (5.3.47) with zero initial conditions,

$$\hat{x}(s) = \Phi(s)Bu(s) + \Phi(s)L(y(s) - \hat{y}(s)) \qquad (5.3.49)$$

where $\Phi(s) = (sI - A)^{-1}$, Equation (5.3.5).

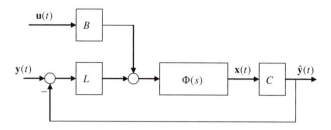

FIGURE 5.3.7 A Kalman filter feedback loop.

The relations (5.3.48) and (5.3.49) are depicted in Figure 5.3.7. The Kalman filter loop transfer matrix $G_{KF}(s)$ is given as

$$G_{KF}(s) = C\Phi(s)L \qquad (5.3.50)$$

Similar to Equation 5.3.8, it can be proved that

$$(I + G_{KF}(s))\Theta(I + G_{KF}(-s))^T = \Theta + (C\Phi(s)N_f)(C\Phi(-s)N_f)^T \qquad (5.3.51)$$

where

$$W = N_f N_f^T \qquad (5.3.52)$$

Assume that

$$\Theta = \chi I ; \quad \chi > 0 \qquad (5.3.53)$$

From (5.3.51) to (5.3.53),

$$(I + G_{KF}(s))(I + G_{KF}(-s))^T = I + \frac{1}{\chi}(C\Phi(s)N_f)(C\Phi(-s)N_f)^T \qquad (5.3.54)$$

Substituting $s = j\omega$ in (5.3.54),

$$(I + G_{KF}(j\omega))(I + G_{KF}(-j\omega))^T = I + \frac{1}{\chi}(C\Phi(j\omega)N_f)(C\Phi(-j\omega)N_f)^T \qquad (5.3.55)$$

As the second term on the right-hand side of (5.3.55) is positive semidefinite,

$$(I + G_{KF}(j\omega)(I + G_{KF}(-j\omega))^T \geq I \qquad (5.3.56)$$

or

$$\underline{\sigma}(I + G_{KF}(j\omega)) \geq 1 \qquad (5.3.57)$$

Because the sensitivity function of the Kalman filter (KF) loop is

$$S_{KF} = (I + G_{KF})^{-1} \qquad (5.3.58)$$

Equation 5.3.57 yields

$$\bar{\sigma}(S_{KF}(j\omega)) \leq 1 ; \quad \forall \omega \qquad (5.3.59)$$

This property indicates that the KF loop will never amplify output disturbances at any frequency.

The complementary sensitivity function, T_{KF}, is given by

$$T_{KF} = I - S_{KF} \qquad (5.3.60)$$

From (5.3.59) and (5.3.60) and properties of singular values (Appendix C),

$$\bar{\sigma}(T_{KF}(j\omega)) \leq 2 ; \quad \forall \omega \qquad (5.3.61)$$

Using (5.2.45), the stability of the KF loop is guaranteed to multiplicative modeling uncertainty provided

$$\bar{\sigma}(\Delta(j\omega)) < 0.5 \qquad (5.3.62)$$

where $\Delta(s)$ relates the actual plant transfer function $G_\Delta(s)$ and the model transfer matrix as follows:

$$G_\Delta(s) = (I + \Delta(s))G(s) \qquad (5.3.63)$$

EXAMPLE 5.3.2: KALMAN FILTER FOR A SIMPLE MASS

Consider the double integrator in Example 4.6. With $\rho = 0.05$,

$$G_{KF}(s) = \frac{4.4721s + 10}{s^2} \qquad (5.3.64)$$

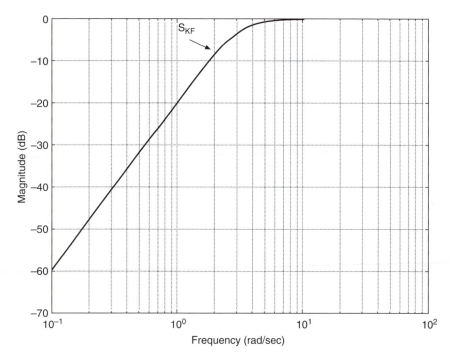

FIGURE 5.3.8 Sensitivity function of the Kalman filter loop (Example 5.3.2).

Therefore,

$$S_{KF} = \frac{s^2}{s^2 + 4.4721s + 10} \quad \text{and} \quad T_{KF} = \frac{4.4721s + 10}{s^2 + 4.4721s + 10} \quad (5.3.65a,b)$$

Both sensitivity and complementary sensitivity functions satisfy (5.3.59) and (5.3.61), respectively (Figure 5.3.8 and Figure 5.3.9).

5.4 LQG/LTR CONTROL

5.4.1 LACK OF GUARANTEED ROBUSTNESS OF LQG CONTROL

Consider the following state space realization (Doyle, 1978):

$$A = \begin{bmatrix} 1 & 1 \\ 0 & 1 \end{bmatrix}; \quad B = \begin{bmatrix} 0 \\ 1 \end{bmatrix} \quad \text{and} \quad C = \begin{bmatrix} 1 & 0 \end{bmatrix} \quad (5.4.1)$$

Let the weighing matrices be

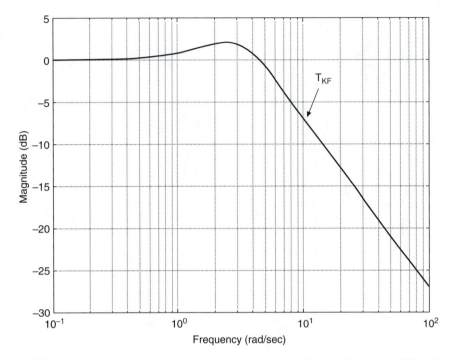

FIGURE 5.3.9 Complementary sensitivity function of the Kalman filter loop (Example 5.3.2).

$$Q = \rho \begin{bmatrix} 1 \\ 1 \end{bmatrix} \begin{bmatrix} 1 & 1 \end{bmatrix}; \quad R = 1 \tag{5.4.2}$$

and

$$W = \sigma \begin{bmatrix} 1 \\ 1 \end{bmatrix} \begin{bmatrix} 1 & 1 \end{bmatrix}; \quad \Theta = 1 \tag{5.4.3}$$

Solving the control algebraic Riccati equation (CARE), Equation 4.8.19,

$$K = \alpha \begin{bmatrix} 1 & 1 \end{bmatrix} \quad \text{where} \quad \alpha = 2 + \sqrt{4 + \rho} \tag{5.4.4}$$

Solving the filter algebraic Riccati equation (FARE), Equation 4.8.16,

$$L = \beta \begin{bmatrix} 1 & 1 \end{bmatrix} \quad \text{where} \quad \beta = 2 + \sqrt{4 + \sigma} \tag{5.4.5}$$

Let the actual plant dynamics be represented by

$$\dot{\mathbf{x}} = A\mathbf{x}(t) + \chi Bu(t) \tag{5.4.6}$$

where $\chi \neq 1$ represents the uncertainty in the plant model. The system matrix for the closed-loop system can be written as

$$A_{sys} = \begin{bmatrix} A & -\chi BK \\ LC & A - BK - LC \end{bmatrix} = \begin{bmatrix} 1 & 1 & 0 & 0 \\ 0 & 1 & -\chi\alpha & -\chi\alpha \\ \beta & 0 & 1-\beta & 1 \\ \beta & 0 & -\alpha-\beta & 1-\alpha \end{bmatrix} \tag{5.4.7}$$

The characteristic equation of A_{sys} can be written as

$$\det(sI - A_{sys}) = s^4 + p_1 s^3 + p_2 s^2 + p_3 s + (1 + (1-\chi)\alpha\beta) = 0 \tag{5.4.8}$$

where coefficients p_1, p_2, and p_3 can be determined. The closed-loop system is unstable if

$$1 + (1-\chi)\alpha\beta < 0 \tag{5.4.9}$$

The condition (5.4.9) leads to

$$\chi > 1 + \frac{1}{\alpha\beta} \tag{5.4.10}$$

With large values of α and β, even a slight increase in the value of χ from its nominal value will render the closed-loop system to be unstable. In other words, the gain margin of the LQG control system can be almost zero. This example clearly shows that the robustness of the LQG control system to modeling errors is not guaranteed.

5.4.2 LOOP TRANSFER RECOVERY (LTR)

LTR stands for loop transfer recovery. The LQ control or KF loop has an excellent robustness property. But the LQG control system (Figure 5.4.1) does not have any guaranteed robustness. Therefore, the LQG/LTR control tries to recover a target loop transfer function $G_{TL}(s)$, which is chosen as the KF loop here:

$$G_{TL}(s) = G_{KF}(s) \tag{5.4.11}$$

It should be noted that the noise vectors in the Kalman filter (Chapter 4, Section 4.6) can be considered "fictitious," and its statistics W and Θ are treated as design parameters to achieve the desired properties of the target loop.

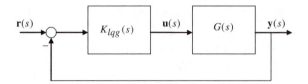

FIGURE 5.4.1 A multiinput/multioutput feedback loop with linear quadratic Gaussian controller.

The LQG controller transfer function matrix is given by

$$K_{lqg}(s) = -K(sI - A + BK + LC)^{-1}L \qquad (5.4.12)$$

where K and L are obtained from (4.8.18) and (4.8.15), respectively.
From (4.8.18), with $R = \rho I$ and $Q = C^T C$,

$$K = \rho^{-1}B^T P \qquad (5.4.13)$$

where

$$PA + A^T P - \rho^{-1}PBB^T P + C^T C = 0 \qquad (5.4.14)$$

Theorem 5.4.1

Assume that $G(s) = C(sI - A)^{-1}B$ is a square matrix and has only minimum phase transmission zeros. If K is obtained by (5.4.13) and L by (4.8.15), pointwise in s,

$$\lim_{\rho \to 0} C(sI - A)^{-1}BK(sI - A + BK + LC)^{-1}L = C(sI - A)^{-1}L \qquad (5.4.15)$$

Equation (5.4.15) states that

$$\lim_{\rho \to 0} G(s)K_{lqg}(s) = G_{TL}(s) \qquad (5.4.16)$$

Proof (Triantafyllou and Hover, 2000)

For a square and minimum phase $G(s)$ (Kwakernaak and Sivan, 1972),

$$\lim_{\rho \to 0} K\sqrt{\rho} = HC \qquad (5.4.17)$$

where H is an orthonormal matrix. Using the matrix inversion lemma (Appendix A),

$$((sI - A + LC) + BK)^{-1} = \Lambda - \Lambda B(I + K\Lambda B)^{-1}K\Lambda \tag{5.4.18}$$

where

$$\Lambda = (sI - A + LC)^{-1} \tag{5.4.19}$$

Therefore,

$$K_{lqg}(s) = K((sI - A + LC) + BK)^{-1}L = [I - K\Lambda B(I + K\Lambda B)^{-1}]K\Lambda L \tag{5.4.20}$$

Using a special case of the matrix inversion lemma (Appendix A),

$$K_{lqg}(s) = (I + K\Lambda B)^{-1}K\Lambda L = (\sqrt{\rho}I + \sqrt{\rho}K\Lambda B)^{-1}\sqrt{\rho}K\Lambda L \tag{5.4.21}$$

Using (5.4.17),

$$\lim_{\rho \to 0} K_{lqg}(s) = (HC\Lambda B)^{-1}HC\Lambda L = (C\Lambda B)^{-1}H^{-1}HC\Lambda L = (C\Lambda B)^{-1}C\Lambda L \tag{5.4.22}$$

Next,

$$C\Lambda = C(sI - A + LC)^{-1} = [I - C\Phi(s)L(I + C\Phi(s)L)^{-1}]C\Phi(s) \tag{5.4.23}$$

where

$$\Phi(s) = (sI - A)^{-1} \tag{5.4.24}$$

Again, using a special case of the matrix inversion lemma (Appendix A),

$$C\Lambda = (I + C\Phi(s)L)^{-1}C\Phi(s) \tag{5.4.25}$$

Substituting (5.4.25) into (5.4.22),

$$\lim_{\rho \to 0} K_{lqg}(s) = [(I + C\Phi(s)L)^{-1}G(s)]^{-1}(I + C\Phi(s)L)^{-1}C\Phi(s)L \tag{5.4.26}$$

where

$$G(s) = C\Phi(s)B \tag{5.4.27}$$

Simplifying (5.4.26),

$$\lim_{\rho \to 0} K_{lqg}(s) = G^{-1}(s)C\Phi(s)L \qquad (5.4.28)$$

From (5.4.28),

$$\lim_{\rho \to 0} G(s)K_{lqg}(s) = C(sI - A)^{-1}L \qquad (5.4.29)$$

This proves the theorem.

In other words, the loop transfer function in the Figure 5.4.1 will converge to the target loop transfer matrix. The LQG/LTR design method was introduced (Doyle and Stein, 1981; Athans, 1986) before the development of the H_∞ method, which is a more general approach and can directly handle many types of modeling uncertainties.

EXAMPLE 5.4.1: A SIMPLE MASS OR A DOUBLE INTEGRATOR

Consider the double integrator plant:

$$A = \begin{bmatrix} 0 & 1 \\ 0 & 0 \end{bmatrix}; \quad B = \begin{bmatrix} 0 \\ 1 \end{bmatrix}; \quad \text{and} \quad C = [1 \quad 0] \qquad (5.4.30)$$

Let

$$K = [k_1 \quad k_2] \quad \text{and} \quad L^T = [\ell_1 \quad \ell_2] \qquad (5.4.31)$$

Then,

$$K(sI - (A - BK - LC))^{-1}L = \frac{(k_1\ell_1 + k_2\ell_2)s + (k_1\ell_2 + k_2\ell_2^2 - k_2\ell_1\ell_2)}{s^2 + (k_2 + \ell_1)s + (k_2\ell_1 + k_1 + \ell_2)} \qquad (5.4.32)$$

From (5.4.13),

$$k_1 = \rho^{-1/2} \quad \text{and} \quad k_2 = \sqrt{2}\rho^{-1/4} \qquad (5.4.33)$$

Substituting (5.4.33) into (5.4.32) and taking limit,

$$\lim_{\rho \to 0} K(sI - (A - BK - LC))^{-1}L = \ell_1 s + \ell_2 \qquad (5.4.34)$$

Therefore,

$$\lim_{\rho \to 0} C(sI - A)^{-1} BK(sI - A + BK + LC)^{-1} L = \frac{\ell_1 s + \ell_2}{s^2} \qquad (5.4.35)$$

Note that

$$C(sI - A)^{-1} L = \frac{\ell_1 s + \ell_2}{s^2} \qquad (5.4.36)$$

Hence, Equation 5.4.16 is verified, or the loop transfer function has been recovered.

5.5 H_2 AND H_∞ NORMS

5.5.1 DEFINITION OF L_2 NORM

Consider a vector signal $\mathbf{x}(t)$, which is a function of time $t \geq 0$. Then, the L_2 norm of this vector signal $\mathbf{x}(t)$ is defined as

$$\left\| \mathbf{x} \right\|_{L_2}^2 = \int_0^\infty \mathbf{x}^T(t)\mathbf{x}(t)dt < \infty \qquad (5.5.1)$$

5.5.2 DEFINITION OF H_2 NORM

Consider a transfer function matrix $G(s)$ with each element being a *strictly proper* and *stable* transfer function. The H_2 norm of the transfer function matrix is given by

$$\left\| G(s) \right\|_{H_2}^2 = \frac{1}{2\pi} Trace \int_{-\infty}^{+\infty} G(j\omega)G^H(j\omega)d\omega \qquad (5.5.2)$$

Using Parseval's theorem (Appendix B),

$$\left\| G(s) \right\|_{H_2}^2 = Trace \int_0^\infty G(t)G^T(t)dt \qquad (5.5.3)$$

where $G(t)$ is the inverse Laplace transform of $G(s)$.

Using the definition of singular values (Appendix C), Equation 5.5.2 can also be written as

$$\left\| G(s) \right\|_{H_2}^2 = \frac{1}{2\pi} \int_{-\infty}^{+\infty} \sum_{i=1}^p [\sigma_i(G(j\omega))]^2 d\omega \qquad (5.5.4)$$

where $\sigma_i(G(j\omega))$ is the ith singular value of $G(j\omega)$, and p is the number of nonzero singular values. It is also clear from the definition of singular value,

$$\sigma_i(G(j\omega)) = \sigma_i(G(-j\omega)) \tag{5.5.5}$$

Furthermore, for the integral in (5.5.4) to be finite, $\sigma_i(G(j\omega)) \to 0$ as $\omega \to \infty$ for each i. This condition is satisfied by a strictly proper transfer function matrix.

5.5.3 Significance of H_2 Norm

I. Connection with Unit Impulse Responses

Consider the MIMO system shown in Figure 5.5.1, for which

$$\mathbf{y}(s) = G(s)\mathbf{u}(s) \tag{5.5.6}$$

Let a unit impulse function, $\delta(t)$, be applied in one input channel at a time. If this input is applied in the ith channel, the corresponding output vector can be written as

$$\mathbf{y}_i(s) = G(s)\mathbf{v}_i \tag{5.5.7}$$

where

$$\mathbf{v}_i = [0 \quad . \quad . \quad 0 \quad 1 \quad 0 \quad . \quad 0]^T \tag{5.5.8}$$

Now,

$$\left\| \mathbf{y}_i(t) \right\|_{L_2}^2 = \int_0^\infty Trace[\mathbf{y}_i(t)\mathbf{y}_i^T(t)]dt \tag{5.5.9}$$

Using Parseval's theorem (Appendix B),

$$\left\| \mathbf{y}_i(t) \right\|_{L_2}^2 = \frac{1}{2\pi} \int_{-\infty}^\infty Trace[\mathbf{y}_i(j\omega)\mathbf{y}_i^H(j\omega)]d\omega \tag{5.5.10}$$

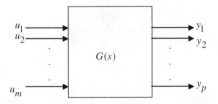

FIGURE 5.5.1 A multiinput/multioutput linear system.

Using (5.5.7),

$$\left\|\mathbf{y}_i(t)\right\|_{L_2}^2 = \frac{1}{2\pi} \int\limits_{-\infty}^{\infty} Trace[G(j\omega)v_i v_i^T G^H(j\omega)]d\omega \qquad (5.5.11)$$

Using (5.5.8),

$$\left\|\mathbf{y}_i(t)\right\|_{L_2}^2 = \frac{1}{2\pi} \int\limits_{-\infty}^{\infty} [|G_{i1}(j\omega)|^2 + \ldots + |G_{ip}(j\omega)|^2]d\omega \qquad (5.5.12)$$

Summing up over all input channels,

$$\sum_{i=1}^{m}\left\|\mathbf{y}_i(t)\right\|_{L_2}^2 = \frac{1}{2\pi} \int\limits_{-\infty}^{\infty} Trace\left[G(j\omega)\left(\sum_{i=1}^{m}v_i v_i^T\right)G^H(j\omega)\right]d\omega \qquad (5.5.13)$$

Using (5.5.8), It can be shown that

$$\sum_{i=1}^{m}\mathbf{v}_i\mathbf{v}_i^T = I_m \qquad (5.5.14)$$

Therefore, the equation (5.5.13) can be written as

$$\sum_{i=1}^{m}\left\|\mathbf{y}_i(t)\right\|_{L_2}^2 = \frac{1}{2\pi} \int\limits_{-\infty}^{\infty} Trace[G(j\omega)G^H(j\omega)]d\omega = \left\|G(s)\right\|_{H_2}^2 \qquad (5.5.15)$$

Therefore, minimizing the H_2 norm of the transfer function matrix is equivalent to minimizing the sum of squares of L_2 norm of outputs due to unit impulse input in each channel.

II.   Connection with the Root Mean Square (RMS) Response

In time domain,

$$\mathbf{y}(t) = \int\limits_{0}^{t} G(t-\tau)\mathbf{u}(\tau)d\tau \qquad (5.5.16)$$

Let each input be an independent zero-mean white noise with unit intensity; i.e.,

$$E(\mathbf{u}(t)\mathbf{u}^T(\tau)) = I_m \delta(t - \tau) \tag{5.5.17}$$

The mean square response can be written (Stein, 1988) as

$$E(\mathbf{y}^T(t)\mathbf{y}(t)) = E(Trace[\mathbf{y}(t)\mathbf{y}^T(t)]) \tag{5.5.18}$$

or

$$E(\mathbf{y}^T(t)\mathbf{y}(t)) = E\left(Trace\left[\int_0^{t_f} G(t-\tau)\mathbf{u}(\tau)d\tau \int_0^{t_f} \mathbf{u}^T(\lambda)G^T(t-\lambda)d\lambda\right]\right) \tag{5.5.19}$$

$$E(\mathbf{y}^T(t)\mathbf{y}(t)) = Trace\left(\int_0^{t_f}\int_0^{t_f} G(t-\tau)E(\mathbf{u}(\tau)\mathbf{u}^T(\lambda))G^T(t-\lambda)d\tau d\lambda\right) \tag{5.5.20}$$

$$E(\mathbf{y}^T(t)\mathbf{y}(t)) = Trace\left(\int_0^{t_f}\int_0^{t_f} G(t-\tau)G^T(t-\lambda)\delta(\tau-\lambda)d\tau d\lambda\right) \tag{5.5.21}$$

$$E(\mathbf{y}^T(t)\mathbf{y}(t)) = Trace\left(\int_0^{t_f} G(t-\lambda)G^T(t-\lambda)d\lambda\right) \tag{5.5.22}$$

Introducing

$$v = t - \lambda \tag{5.5.23}$$

$$E(\mathbf{y}^T(t)\mathbf{y}(t)) = \int_0^t Trace[G(v)G^T(v)]dv \tag{5.5.24}$$

$$\lim_{t\to\infty} E(\mathbf{y}^T(t)\mathbf{y}(t)) = \int_0^\infty Trace[G(v)G^T(v)]dv \tag{5.5.25}$$

Using Parseval's theorem,

$$\lim_{t \to \infty} E(\mathbf{y}^T(t)\mathbf{y}(t)) = \frac{1}{2\pi} \int_{-\infty}^{\infty} Trace[G(j\omega)G^H(j\omega)]d\omega = \|G(s)\|_{H_2}^2 \qquad (5.5.26)$$

Therefore, minimizing the H_2 norm of the transfer function matrix is equivalent to minimizing the RMS of outputs in the statistical steady state due to independent zero-mean white noise inputs.

5.5.4 Definition of H_∞ Norm

Consider a transfer function matrix $G(s)$ with each element being a *proper* and *stable* transfer function. The H_∞ norm of the transfer function matrix is given by

$$\|G(s)\|_{H_\infty} = {}^{sup}_{\omega} \bar{\sigma}(G(j\omega)) \qquad (5.5.27)$$

where supremum (sup) is defined in Appendix C.

Bounded Real Lemma

The condition

$$\|G(s)\|_{H_\infty} < \gamma \qquad (5.5.28)$$

is satisfied if and only if there exists a matrix $P = P^T \geq 0$ that meets the following criteria:

1. It is a solution of $PA + A^T P + C^T C + \dfrac{1}{\gamma^2} PBB^T P = 0$. $\qquad (5.5.29)$

2. The matrix $A + \dfrac{1}{\gamma^2} BB^T P$ is stable. $\qquad (5.5.30)$

5.5.5 Significance of H_∞ Norm

I. Connection with the Worst-Case Frequency Response

Let

$$\mathbf{u}(t) = \mathbf{a}_u e^{j\omega t} \qquad (5.5.31)$$

where

$$\mathbf{a}_u = \begin{bmatrix} a_{u1} & a_{u2} & . & . & a_{um} \end{bmatrix}^T \qquad (5.5.32)$$

Then, in steady state,

$$\mathbf{y}(t) = \mathbf{a}_y e^{j\omega t} \qquad (5.5.33)$$

where

$$\mathbf{a}_y = \begin{bmatrix} a_{y1} & a_{y2} & . & . & a_{yp} \end{bmatrix}^T \qquad (5.5.34)$$

and $\mathbf{a}_u$ and $\mathbf{a}_y$ are in general complex vectors. Then,

$$\mathbf{a}_y = G(j\omega)\mathbf{a}_u \qquad (5.5.35)$$

Utilizing the definition of the maximum singular value,

$$\frac{\|\mathbf{a}_y\|_2}{\|\mathbf{a}_u\|_2} \leq \|G(s)\|_{H_\infty} \qquad (5.5.36)$$

and

$$\sup_{\mathbf{a}_u} \frac{\|\mathbf{a}_y\|_2}{\|\mathbf{a}_u\|_2} = \bar{\sigma}(G(j\omega)) \leq \|G(s)\|_{H_\infty} \qquad (5.5.37)$$

II.   Connection with the Worst-Case Output or Input

Fact 5.5.1 (Vidyasagar, 1985; Stein, 1988)

$$\|G(s)\|_{H_\infty} = \sup_{\mathbf{u}} \frac{\|\mathbf{y}(t)\|_{L_2}}{\|\mathbf{u}\|_{L_2}} \qquad (5.5.38)$$

Proof

$$\|\mathbf{y}(t)\|_{L_2}^2 = \frac{1}{2\pi}\int_{-\infty}^{\infty} \mathbf{y}^H(j\omega)\mathbf{y}(j\omega)d\omega = \frac{1}{2\pi}\int_{-\infty}^{\infty} \|\mathbf{y}(j\omega)\|_2^2 d\omega \qquad (5.5.39)$$

Because $\mathbf{y}(j\omega) = G(j\omega)\mathbf{u}(j\omega)$,

$$\left\|\mathbf{y}(j\omega)\right\|_2 \le \bar{\sigma}(G(j\omega))\left\|\mathbf{u}(j\omega)\right\|_2 \le \left\|G(s)\right\|_{H_\infty}\left\|\mathbf{u}(j\omega)\right\|_2 \qquad (5.5.40)$$

Substituting (5.5.40) into (5.5.39),

$$\left\|\mathbf{y}(t)\right\|_{L_2}^2 \le \frac{1}{2\pi}\int_{-\infty}^{\infty}\left\|G(s)\right\|_{H_\infty}^2\left\|\mathbf{u}(j\omega)\right\|_2^2 d\omega = \left\|G(s)\right\|_{H_\infty}^2 \frac{1}{2\pi}\int_{-\infty}^{\infty}\left\|\mathbf{u}(j\omega)\right\|_2^2 d\omega \quad (5.5.41)$$

Using the Parseval's theorem again,

$$\left\|\mathbf{y}(t)\right\|_{L_2} \le \left\|G(s)\right\|_{H_\infty}\left\|\mathbf{u}(t)\right\|_{L_2} \qquad (5.5.42)$$

In other words, if

$$\left\|G(s)\right\|_{H_\infty} < \gamma \qquad (5.5.43)$$

then

$$\frac{\left\|\mathbf{y}(t)\right\|_{L_2}}{\left\|\mathbf{u}(t)\right\|_{L_2}} < \gamma \qquad (5.5.44)$$

This relation is often used as

$$\left\|\mathbf{y}(t)\right\|_{L_2}^2 - \gamma^2\left\|\mathbf{u}(t)\right\|_{L_2}^2 < 0 \qquad \text{for all } \mathbf{u}(t) \in L_2 \qquad (5.5.45)$$

It should be noted that the relationship (5.5.45) has been derived for zero initial conditions of inputs and outputs.

III. Connection of the Bounded Real Lemma (5.5.28 to 5.5.30) with a Special Linear Quadratic Maximization Problem

Consider the following system:

$$\dot{\mathbf{x}} = A\mathbf{x} + B\mathbf{d} \qquad (5.5.46)$$

and

$$\mathbf{e} = C\mathbf{x} \tag{5.5.47}$$

Note that

$$T_{ed}(s) = C(sI - A)^{-1}B = G(s) \tag{5.5.48}$$

The following optimization problem will be solved:

$$\underset{\mathbf{d}}{\sup}\, J(\mathbf{d}) < \infty \tag{5.5.49}$$

where

$$J(\mathbf{d}) = \int_0^\infty (\mathbf{e}^T\mathbf{e} - \gamma^2\mathbf{d}^T\mathbf{d})dt \tag{5.5.50}$$

and

$$\mathbf{d}(t) = K_d\mathbf{x} \tag{5.5.51}$$

Substituting (5.5.47) and (5.5.51) into (5.5.50),

$$J(K_d) = \int_0^\infty \mathbf{x}^T(C^TC - \gamma^2 K_d^T K_d)\mathbf{x}dt \tag{5.5.52}$$

Substituting (5.5.51) into (5.5.46),

$$\dot{\mathbf{x}} = (A + BK_d)\mathbf{x} \tag{5.5.53}$$

Consider

$$d(\mathbf{x}^T P\mathbf{x}) = (\dot{\mathbf{x}}^T P\mathbf{x} + \mathbf{x}^T P\dot{\mathbf{x}})dt \tag{5.5.54}$$

where P is a symmetric matrix. Substituting (5.5.53) into (5.5.54),

$$d(\mathbf{x}^T P\mathbf{x}) = \mathbf{x}^T((A + BK_d)^T P + P(A + BK_d))\mathbf{x}dt \tag{5.5.55}$$

Integrating both sides from 0 to ∞,

$$\int_0^\infty \mathbf{x}^T((A + BK_d)^T P + P(A + BK_d))\mathbf{x}dt = \mathbf{x}^T(\infty)P\mathbf{x}(\infty) - \mathbf{x}^T(0)P\mathbf{x}(0) \quad (5.5.56)$$

Assuming that $A + BK_d$ is stable, $\mathbf{x}(\infty) = 0$. Therefore,

$$\int_0^\infty \mathbf{x}^T((A + BK_d)^T P + P(A + BK_d))\mathbf{x}dt = -\mathbf{x}^T(0)P\mathbf{x}(0) \quad (5.5.57)$$

Setting

$$(A + BK_d)^T P + P(A + BK_d) + C^T C - \gamma^2 K_d^T K_d = 0 \quad (5.5.58)$$

it can be seen that

$$J = \mathbf{x}^T(0)P\mathbf{x}(0) \quad (5.5.59)$$

The condition for maximization of J, Equation 5.5.59, with respect to K_d is

$$\nabla_{K_d} P = 0 \quad (5.5.60)$$

The gradient matrix $\nabla_{K_d} P$ is defined as follows:

$$(\nabla_{K_d} P)_{ij} = \frac{\partial P}{\partial K_{d_{ij}}} \quad (5.5.61)$$

From (5.5.58),

$$\nabla_{K_d} P(A + BK_d) + B^T P + B^T P + (A + BK_d)^T \nabla_{K_d} P - 2\gamma^2 K_d = 0 \quad (5.5.62)$$

Hence, with condition (5.5.60),

$$K_d = \frac{1}{\gamma^2} B^T P \quad (5.5.63)$$

Substituting (5.5.63) into (5.5.58),

$$PA + A^T P + C^T C + \frac{1}{\gamma^2} PBB^T P = 0 \tag{5.5.64}$$

Hence, the solution of the optimization problem (5.5.49 to 5.5.51) leads to relationships identical to those for the bounded real lemma, Equation 5.5.29 (Green and Limebeer, 1995).

5.5.6 COMPUTATION OF H_2 NORM

Let

$$\mathbf{y}(s) = G(s)\mathbf{u}(s) \tag{5.5.65}$$

where

$$G(s) = C(sI - A)^{-1}B \tag{5.5.66}$$

Taking the inverse Laplace transform,

$$G(t) = Ce^{At}B \tag{5.5.67}$$

The H_2 norm of the transfer matrix is given by

$$\left\| G(s) \right\|_{H_2}^2 = \frac{1}{2\pi} Trace \int_{-\infty}^{+\infty} G(j\omega)G^H(j\omega)d\omega \tag{5.5.68}$$

Using Parseval's theorem (Appendix B),

$$\left\| G(s) \right\|_{H_2}^2 = Trace \int_{0}^{\infty} G(t)G^T(t)dt \tag{5.5.69}$$

Method I

Using (5.5.67),

$$\left\| G(s) \right\|_{H_2}^2 = Trace(CPC^T) \tag{5.5.70}$$

where

$$P = \int_0^\infty e^{At} BB^T e^{A^T t} dt \tag{5.5.71}$$

Fact 5.5.1

P satisfies the Lyapunov equation

$$AP + PA^T + BB^T = 0 \tag{5.5.72}$$

Proof

It can be easily seen that

$$\frac{d}{dt}(e^{At} BB^T e^{A^T t}) = Ae^{At} BB^T e^{A^T t} + e^{At} BB^T e^{A^T t} A^T \tag{5.5.73}$$

Therefore,

$$AP + PA^T = \int_0^\infty \frac{d}{dt}(e^{At} BB^T e^{A^T t}) dt$$

$$= e^{At} BB^T e^{A^T t}\Big|_{t=\infty} - e^{At} BB^T e^{A^T t}\Big|_{t=0} \tag{5.5.74}$$

Because the matrix A is stable,

$$e^{At} \to 0 \quad \text{as} \quad t \to \infty \tag{5.5.75}$$

Therefore, from Equation 5.5.74,

$$AP + PA^T = -BB^T \tag{5.5.76}$$

Method II

Alternatively, from (5.5.69),

$$\|G(s)\|_{H_2}^2 = Trace \int_0^\infty G^T(t)G(t)dt \tag{5.5.77}$$

because

$$Trace(G^T(t)G(t)) = Trace(G(t)G^T(t)) \tag{5.5.78}$$

Using (5.5.67) and (5.5.77),

$$\left\| G(s) \right\|_{H_2}^2 = Trace(B^T \bar{P} B) \tag{5.5.79}$$

where

$$\bar{P} = \int_0^\infty e^{A^T t} C^T C e^{At} dt \tag{5.5.80}$$

Fact 5.5.3

$\bar{P}$ satisfies the following Lyapunov equation:

$$A^T \bar{P} + \bar{P}A + C^T C = 0 \tag{5.5.81}$$

Proof

It can be easily seen that

$$\frac{d}{dt}(e^{A^T t} C^T C e^{At}) = A^T e^{A^T t} C^T C e^{At} + e^{A^T t} C^T C e^{At} A \tag{5.5.82}$$

Therefore,

$$A^T \bar{P} + \bar{P}A = \int_0^\infty \frac{d}{dt}(e^{A^T t} C^T C e^{At}) dt$$

$$= e^{A^T t} C^T C e^{At} \Big|_{t=\infty} - e^{A^T t} C^T C e^{At} \Big|_{t=0} \tag{5.5.83}$$

Because the matrix A is stable,

$$e^{At} \to 0 \quad \text{as} \quad t \to \infty \tag{5.5.84}$$

Therefore, from Equation 5.5.83,

$$A^T \bar{P} + \bar{P}A = -C^T C \tag{5.5.85}$$

EXAMPLE 5.5.1: SPRING-MASS-DAMPER SYSTEM (FIGURE 5.1.1)

Consider Example 5.1.1. Using the Matlab command "lyap,"

$$
P = \begin{bmatrix}
0.0027 & 0.0023 & 0.0000 & 0.0000 \\
0.0023 & 0.0027 & 0.0000 & 0.0000 \\
0.0000 & 0.0000 & 0.5000 & 0.0000 \\
0.0000 & 0.0000 & 0.0000 & 0.5000
\end{bmatrix}
\tag{5.5.86}
$$

and

$$
\bar{P} = \begin{bmatrix}
0.5027 & 0.0023 & 0.0027 & 0.0023 \\
0.0023 & 0.5027 & 0.0023 & 0.0027 \\
0.0027 & 0.0023 & 0.0027 & 0.0023 \\
0.0023 & 0.0027 & 0.0023 & 0.0027
\end{bmatrix}
\tag{5.5.87}
$$

As expected, both equations (5.5.70) and (5.5.79) yield

$$
\left\| G(s) \right\|_{H_2} = 0.0739
\tag{5.5.88}
$$

The Matlab command "normh2" also yields the identical result.

5.5.7 COMPUTATION OF H_∞ NORM

The H_∞ norm of a transfer function matrix is given by

$$
\left\| G(s) \right\|_{H_\infty} = \sup_\omega \bar{\sigma}(G(j\omega))
\tag{5.5.89}
$$

where the supremum (sup) is defined in Appendix C. An important theorem is described as follows.

Theorem 5.5.1

$$
\left\| G(s) \right\|_{H_\infty} < \gamma
\tag{5.5.90}
$$

if and only if the Hamiltonian matrix

$$H = \begin{bmatrix} A & \dfrac{1}{\gamma^2} BB^T \\ -C^T C & -A^T \end{bmatrix} \qquad (5.5.91)$$

has no eigenvalues on the $j\omega$ axis (Zhou et al., 1996).

Using Theorem 5.5.1, an iterative procedure to calculate the H_∞ norm is developed as follows:

1. Arbitrarily select $\gamma = \gamma_1$ and find the eigenvalues of the matrix H.

2. If there are no eigenvalues on the imaginary axis, $\|G(s)\|_{H_\infty} < \gamma_1$ and select $\gamma = \dfrac{\gamma_1}{2}$, and find the eigenvalues of the matrix H. Repeat this process till $\|G(s)\|_{H_\infty} > \gamma$. Set

$$\gamma_\ell = \gamma \quad \text{and} \quad \gamma_u = \gamma_1 .$$

3. If there is at least one eigenvalue on the imaginary axis, $\|G(s)\|_{H_\infty} \geq \gamma_1$ and select $\gamma = 2\gamma_1$, and find the eigenvalues of the matrix H. Repeat this process till $\|G(s)\|_{H_\infty} < \gamma$. Set

$$\gamma_\ell = \gamma_1 \quad \text{and} \quad \gamma_u = \gamma$$

At this stage, it is known that

$$\gamma_\ell \leq \|G(s)\|_{H_\infty} \leq \gamma_u$$

4. Let ε be the required precision. If

$$\gamma_u - \gamma_\ell \leq \varepsilon$$

the computation process is stopped. Otherwise, set

$$\gamma = \frac{\gamma_\ell + \gamma_u}{2}$$

and find the eigenvalues of the matrix H.

5. If there are no eigenvalues on the imaginary axis, $\|G(s)\|_{H_\infty} < \gamma$. Set

$$\gamma_u = \gamma$$

If there is at least one eigenvalue on the imaginary axis, $\left\|G(s)\right\|_{H_\infty} \geq \gamma$. Set

$$\gamma_\ell = \gamma$$

6. Go to Step 4.

EXAMPLE 5.5.2: SPRING-MASS-DAMPER SYSTEM (FIGURE 5.1.1)

Consider Example 5.1.1. The Matlab command "normhinf" yields

$$\left\|G(s)\right\|_{H_\infty} = 0.1002 \tag{5.5.92}$$

Directly from Figure 5.1.2,

$$\left\|G(s)\right\|_{H_\infty}^{P} = 0.1001 \tag{5.5.93}$$

which is quite close to the Matlab result, (5.5.92). However, the result from the plot is dependent on the size of the frequency interval used in making the plot. Therefore, if the frequency grid is not fine enough, the result from the plot can be significantly less than that from the Matlab or the algorithm presented in the preceding text.

5.6 H_2 CONTROL

5.6.1 FULL STATE FEEDBACK H_2 CONTROL (FIGURE 5.6.1)

Consider Figure 5.6.1 and assume that

$$M = \begin{bmatrix} A & B_1 & B_2 \\ C_1 & 0 & D_{12} \\ I & 0 & 0 \end{bmatrix} \tag{5.6.1}$$

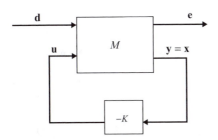

FIGURE 5.6.1 A full state feedback system.

The following assumptions are made:

1. (A, B_1) and (A, B_2) are stabilizable.
2. (C_1, A) is detectable.

From (5.6.1),

$$\dot{\mathbf{x}} = A\mathbf{x} + B_1\mathbf{d}(t) + B_2\mathbf{u}(t) \qquad (5.6.2)$$

$$\mathbf{e}(t) = C_1\mathbf{x}(t) + D_{12}\mathbf{u}(t) \qquad (5.6.3)$$

$$\mathbf{y}(t) = \mathbf{x}(t) \qquad (5.6.4)$$

Assuming that $\mathbf{d}(t)$ is the white noise vector with unit intensity similar to Equation 5.5.17,

$$\left\| T_{ed}(s) \right\|_{H_2}^2 = E(\mathbf{e}^T(t)\mathbf{e}(t)) \qquad (5.6.5)$$

where

$$\mathbf{e}^T\mathbf{e} = \mathbf{x}^T C_1^T C_1 \mathbf{x} + 2\mathbf{x}^T C_1^T D_{12}\mathbf{u} + \mathbf{u}^T D_{12}^T D_{12}\mathbf{u} \qquad (5.6.6)$$

With (5.6.2) and (5.6.5), the minimization of $\left\| T_{ed}(s) \right\|_{H_2}$ is equivalent to the solution of the stochastic regulator problem. Setting

$$Q_f = C_1^T C_1, \quad N_f = C_1^T D_{12}, \quad \text{and} \quad R_f = D_{12}^T D_{12} \qquad (5.6.7)$$

the optimal state feedback law is given by

$$\mathbf{u} = -K\mathbf{x} \qquad (5.6.8)$$

where

$$K = R_f^{-1}(PB_2 + N_f)^T \qquad (5.6.9)$$

and

$$P(A - B_2 R_f^{-1}N_f^T) + (A - B_2 R_f^{-1}N_f^T)^T P - PB_2 R_f^{-1}B_a^T P + Q_f - N_f R_f^{-1}N_f^T = 0 \quad (5.6.10)$$

It should be noted that the gain K is independent of the matrix B_1.

EXAMPLE 5.6.1: H_2 CONTROL OF A DOUBLE INTEGRATOR SYSTEM

Consider the double integrator system:

$$\ddot{x} = u(t) + d(t) \tag{5.6.11}$$

The corresponding elements in (5.6.1) are

$$A = \begin{bmatrix} 0 & 1 \\ 0 & 0 \end{bmatrix}, \quad B_1 = B_2 = \begin{bmatrix} 0 \\ 1 \end{bmatrix}, \quad C_1 = \begin{bmatrix} 1 & 0 \\ 0 & 0 \end{bmatrix}, \quad \text{and} \quad D_{12} = \begin{bmatrix} 0 \\ 1 \end{bmatrix} \tag{5.6.12}$$

Then, from (5.6.7),

$$Q_f = \begin{bmatrix} 1 & 0 \\ 0 & 0 \end{bmatrix}, \quad N_f = 0, \quad \text{and} \quad R_f = 1 \tag{5.6.13}$$

Then, from (5.6.10),

$$1 - p_{12}^2 = 0 \tag{5.6.14}$$

$$p_{11} - p_{12} p_{22} = 0 \tag{5.6.15}$$

$$2 p_{12} - p_{22}^2 = 0 \tag{5.6.16}$$

The solution of (5.6.14) to (5.6.16) yields

$$p_{11} = \sqrt{2}, \quad p_{12} = 1, \quad \text{and} \quad p_{22} = \sqrt{2} \tag{5.6.17}$$

Then, from (5.6.9),

$$K = [p_{12} \quad p_{22}] = [1 \quad \sqrt{2}] \tag{5.6.18}$$

EXAMPLE 5.6.2: ACTIVE SUSPENSION SYSTEM

Consider the optimal suspension system design again, Example 3.11 and Example 4.7. For (5.6.1), define

$$B_2 = B , \quad C_1 = \begin{bmatrix} r_{11} & r_{12} & 0 & 0 \\ r_{21} & r_{22} & 0 & 0 \\ 0 & 0 & 0 & 0 \end{bmatrix} , \quad \text{and} \quad D_{12} = \begin{bmatrix} 0 \\ 0 \\ \sqrt{\rho} \end{bmatrix} \qquad (5.6.19)$$

where

$$\begin{bmatrix} q_1 + q_2 & -q_2 \\ -q_2 & q_2 \end{bmatrix} = \begin{bmatrix} r_{11} & r_{12} \\ r_{21} & r_{22} \end{bmatrix}^T \begin{bmatrix} r_{11} & r_{12} \\ r_{21} & r_{22} \end{bmatrix} \qquad (5.6.20)$$

Because of (5.6.7),

$$C_1^T C_1 = Q , \quad N_f = 0 , \quad \text{and} \quad R_f = \rho \qquad (5.6.21)$$

Therefore, the optimal control law (5.6.8) is identical to the control law (4.7.31).

5.6.2 OUTPUT FEEDBACK H_2 CONTROL (FIGURE 5.6.2)

The matrix M in Figure 5.6.2 is

$$M = \begin{bmatrix} A & B_1 & B_2 \\ C_1 & 0 & D_{12} \\ C_2 & D_{21} & 0 \end{bmatrix} \qquad (5.6.22)$$

Therefore,

$$\dot{\mathbf{x}} = A\mathbf{x} + B_1\mathbf{d}(t) + B_2\mathbf{u}(t) \qquad (5.6.23)$$

$$\mathbf{e}(t) = C_1\mathbf{x}(t) + D_{12}\mathbf{u}(t) \qquad (5.6.24)$$

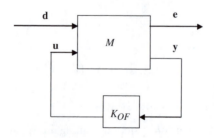

FIGURE 5.6.2 An output feedback system.

$$y(t) = C_2 x(t) + D_{21} d(t) \tag{5.6.25}$$

The following assumptions are made:

1. (A, B_1) and (A, B_2) are stabilizable.
2. (C_1, A) and (C_2, A) are detectable.

Let

$$\mathbf{d}(t) = \begin{bmatrix} \xi(t) \\ \eta(t) \end{bmatrix}, \quad B_1 = [B_{1\xi} \quad \mathbf{0}], \quad \text{and} \quad D_{21} = [\mathbf{0} \quad D_{21\eta}] \tag{5.6.26}$$

where $\xi(t)$ and $\eta(t)$ are vectors with elements as independent white noises with unit intensities. From (5.6.23), (5.6.25), and (5.6.26),

$$\dot{\mathbf{x}} = A\mathbf{x} + B_{1\xi}\xi(t) + B_2\mathbf{u}(t) \tag{5.6.27}$$

$$y(t) = C_2 x(t) + D_{21\eta}\eta(t) \tag{5.6.28}$$

The minimization of the H_2 norm of $T_{ed}(s)$ is equivalent to minimizing

$$E(\mathbf{x}^T Q_f \mathbf{x} + 2\mathbf{x}^T N_f \mathbf{u} + \mathbf{u}^T R_f \mathbf{u}) \tag{5.6.29}$$

for stochastic systems (5.6.27) and (5.6.28), which is exactly the LQG control (Chapter 4, Section 4.8).

EXAMPLE 5.6.3: SIMPLE MASS OR A DOUBLE INTEGRATOR SYSTEM

Consider Example 4.6 again. Define

$$C_2 = [1 \quad 0], \quad C_1 = \begin{bmatrix} 1 & 0 \\ 0 & 0 \\ 0 & 0 \end{bmatrix}, \quad B_{1\xi} = \begin{bmatrix} 0 \\ 1 \end{bmatrix} \tag{5.6.30}$$

$$D_{21\eta} = \sqrt{\rho}, \quad D_{12} = \begin{bmatrix} 0 \\ 0 \\ \sqrt{\rho_c} \end{bmatrix} \tag{5.6.31}$$

Now, considering $\mathbf{w}(t) = B_{1\xi}\xi(t)$ and $\theta(t) = D_{21\eta}\eta(t)$, the output feedback H_2 control is exactly same as the LQG control described in Example 4.9. It should be noted that

$$E(\xi(t)\xi(t+\tau)) = \delta(t-\tau) \tag{5.6.32}$$

$$E(\eta(t)\eta(t+\tau)) = \delta(t-\tau) \tag{5.6.33}$$

5.7 WELL-POSEDNESS, INTERNAL STABILITY, AND SMALL GAIN THEOREM

5.7.1 WELL-POSEDNESS AND INTERNAL STABILITY OF A GENERAL FEEDBACK SYSTEM

Consider the system in Figure 5.7.1 (Zhou et al., 1996, 1998). With respect to the system in Figure 5.2.1, an additional disturbance signal is added to the input to the plant. Breaking the loop at the plant input and ignoring $\mathbf{d}$, $\mathbf{n}$, and $\mathbf{r}$, the difference between the "returned" variable $\mathbf{u}$ and the "injected" variable $\mathbf{u}_p$ (Kwakernaak and Sivan, 1972) is

$$\mathbf{u}_p - \mathbf{u} = (I + L_i)\mathbf{u}_p \tag{5.7.1}$$

where $(I + L_i)$ is called the *input return difference matrix* and

$$L_i = KG \tag{5.7.2}$$

is called the *input loop transfer matrix*. Similarly, breaking the loop at the plant output, the *output loop transfer matrix* L_o is obtained as

$$L_o = GK \tag{5.7.3}$$

and $(I + L_o)$ is called the *output return difference matrix*.

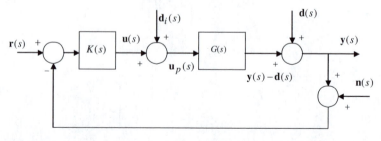

FIGURE 5.7.1 A general multiinput/multioutput system with disturbances.

The following relationships can be easily derived (Zhou et al., 1996, 1998):

$$\mathbf{y} = T_o(\mathbf{r} - \mathbf{n}) + S_o G \mathbf{d}_i + S_o \mathbf{d} \tag{5.7.4}$$

$$\mathbf{r} - \mathbf{y} = S_o(\mathbf{r} - \mathbf{d}) + T_o \mathbf{n} - S_o G \mathbf{d}_i \tag{5.7.5}$$

$$\mathbf{u} = K S_o(\mathbf{r} - \mathbf{n}) - K S_o \mathbf{d} - T_i \mathbf{d}_i \tag{5.7.6}$$

$$\mathbf{u}_p = K S_o(\mathbf{r} - \mathbf{n}) - K S_o \mathbf{d} - S_i \mathbf{d}_i \tag{5.7.7}$$

where

$$S_o = (I + L_o)^{-1} \tag{5.7.8}$$

$$T_o = I - S_o = L_o(I + L_o)^{-1} = (I + L_o)^{-1} L_o \tag{5.7.9}$$

$$S_i = (I + L_i)^{-1} \tag{5.7.10}$$

$$T_i = I - S_i = L_i(I + L_i)^{-1} = (I + L_i)^{-1} L_i \tag{5.7.11}$$

Definition 5.7.1

1. S_o and S_i are *output and input sensitivity functions*, respectively.
2. T_o and T_i are *output and input complementary sensitivity functions*, respectively.

Derivation of (5.7.4) to (5.7.7) follows the same procedure as used in Section 5.2. However, the following relationship is required.

Fact 5.7.1

$$S_o(s)G(s) = G(s)S_i(s) \tag{5.7.12}$$

Proof

Using the matrix inversion lemma (Appendix A),

$$S_o = (I + GK)^{-1} = I - G(KG + I)^{-1}K \tag{5.7.13}$$

Then, postmultiplying (5.7.13) by $G(s)$,

$$S_o G = G - G(KG + I)^{-1} KG = G[I - (I + KG)^{-1} KG] \tag{5.7.14}$$

Therefore,

$$S_o G = G[I - T_i] = GS_i \tag{5.7.15}$$

This completes the proof.

Well-Posedness of Feedback System

A feedback system is said to be *well-posed* if and only if all closed-loop transfer matrices are well defined and proper (Zhou et al., 1996, 1998).

Fact 5.7.2

The feedback system in Figure 5.7.2 is well-posed if and only if the transfer matrix from $[\mathbf{d}_i^T \quad \mathbf{d}^T]^T$ to $\mathbf{u}_p$ exists and is proper.

Fact 5.7.3

The feedback system in Figure 5.7.2 is well-posed if and only if the matrix

$$I - \hat{K}(\infty) G(\infty) \tag{5.7.16}$$

is invertible.

For a strictly proper plant transfer function $G(s)$, $G(\infty) = 0$. Hence, for a strictly proper $G(s)$, the well-posedness of the system in Figure 5.7.2 is guaranteed.

Internal Stability of the System

The concept of internal stability is defined for the system shown in Figure 5.7.2, which is a generic representation of a feedback system; i.e., the plant $G(s)$ with the controller $\hat{K}(s)$.

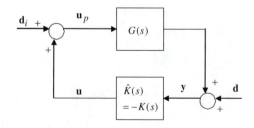

FIGURE 5.7.2 A multiinput/multioutput feedback system with two external inputs.

State Space Description

First, assume that realization for $G(s)$ and $\hat{K}(s)$,

$$G(s)=\left[\begin{array}{c|c} A & B \\ \hline C & D \end{array}\right] \qquad \hat{K}(s)=\left[\begin{array}{c|c} A_{\hat{K}} & B_{\hat{K}} \\ \hline C_{\hat{K}} & D_{\hat{K}} \end{array}\right] \qquad (5.7.17)$$

is stabilizable and detectable. Second, because external inputs do not influence the stability of a linear system, $\mathbf{d}$ and $\mathbf{d}_i$ are set to zero.

For the system $G(s)$, a state space realization is

$$\dot{\mathbf{x}} = A\mathbf{x} + B\mathbf{u} \qquad (5.7.18)$$

$$\mathbf{y} = C\mathbf{x} + D\mathbf{u} \qquad (5.7.19)$$

and for the system $\hat{K}(s)$, a state space realization is

$$\dot{\boldsymbol{\xi}} = A_{\hat{K}}\boldsymbol{\xi} + B_{\hat{K}}\mathbf{y} \qquad (5.7.20)$$

$$\mathbf{u} = C_{\hat{K}}\boldsymbol{\xi} + D_{\hat{K}}\mathbf{y} \qquad (5.7.21)$$

Combining (5.7.18) to (5.7.21),

$$\left[\begin{array}{c} \dot{\mathbf{x}} \\ \dot{\boldsymbol{\xi}} \end{array}\right] = A_{cl}\left[\begin{array}{c} \mathbf{x} \\ \boldsymbol{\xi} \end{array}\right] \qquad (5.7.22)$$

where

$$A_{cl} = \left[\begin{array}{cc} A & 0 \\ 0 & A_{\hat{K}} \end{array}\right] - \left[\begin{array}{cc} B & 0 \\ 0 & B_{\hat{K}} \end{array}\right]\left[\begin{array}{cc} I & -D_{\hat{K}} \\ D_{\hat{K}} & I \end{array}\right]^{-1}\left[\begin{array}{cc} 0 & C_{\hat{K}} \\ C & 0 \end{array}\right] \qquad (5.7.23)$$

The system is called *internally stable* provided the origin of the state space model (5.7.22) is asymptotically stable, which will require that all the eigenvalues of A_{cl} are in the left half of the complex plane.

Transfer Function Description

For the system in Figure 5.7.2,

$$\left[\begin{array}{c} \mathbf{u}_p(s) \\ \mathbf{y}(s) \end{array}\right] = G_{cl}(s)\left[\begin{array}{c} \mathbf{d}_i(s) \\ \mathbf{d}(s) \end{array}\right] \qquad (5.7.24)$$

where

$$G_{cl}(s) = \begin{bmatrix} I & -\hat{K}(s) \\ -G(s) & I \end{bmatrix}^{-1} \tag{5.7.25}$$

The system shown in Figure 5.7.2 is called *internally stable* if and only if each element of the transfer function matrix $G_{cl}(s)$ is proper and stable. To see the elements of the transfer function matrix $G_{cl}(s)$, it can be shown (Zhou et al., 1996, 1998) using (5.7.4) and (5.7.7) that

$$\begin{bmatrix} I & -\hat{K} \\ -G & I \end{bmatrix}^{-1} = \begin{bmatrix} S_i & \hat{K}S_o \\ S_oG & S_o \end{bmatrix} \tag{5.7.26}$$

5.7.2 SMALL GAIN THEOREM

Let $M(s)$ be a stable and proper $p \times q$ transfer function matrix, and let $\gamma > 0$. Then, the interconnected system, Figure 5.7.3, is well-posed and internally stable for a stable and proper transfer matrix $\Delta(s)$ when either of the following two conditions is satisfied:

1. $\|\Delta\|_\infty \leq \dfrac{1}{\gamma}$ if and only if $\|M(s)\|_\infty < \gamma$. $\qquad\qquad$ (5.7.27a)

2. $\|\Delta\|_\infty < \dfrac{1}{\gamma}$ if and only if $\|M(s)\|_\infty \leq \gamma$. $\qquad\qquad$ (5.7.27b)

For either of the two conditions (5.7.27a) and (5.7.27b),

$$\|M(s)\|_\infty \|\Delta(s)\|_\infty < 1 \tag{5.7.28}$$

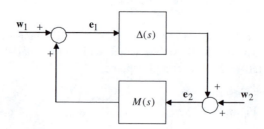

FIGURE 5.7.3 An interconnected system.

Because of (5.7.28),

$$\left\| M(s)\Delta(s) \right\|_\infty < 1 \tag{5.7.29}$$

Because external inputs do not affect the stability of a linear system, set $\mathbf{w}_1 = \mathbf{w}_2 = 0$. In this case, $\mathbf{e}_1$ and $\mathbf{e}_2$ are also outputs of $M(s)$ and $\Delta(s)$, respectively.

Consider a case of nonzero $\mathbf{e}_1$, which is equivalent to nonzero initial conditions. Because of (5.7.29), the norm of the output of $M(s)$ will be less than that of the signal entering $\Delta(s)$, or the the norm of the previous output of $M(s)$. This process will continue in a cyclic way, and the norm of the signal will asymptotically go to zero.

Without any loss of generality, let $\gamma = 1$, and relation (5.7.27a) becomes

$$\left\| \Delta \right\|_\infty \le 1 \text{ if and only if } \left\| M(s) \right\|_\infty < 1 \tag{5.7.30}$$

Proof (Zhou et al., 1996, 1998)

Part I: Sufficiency. For the stability of the closed-loop system, the Nyquist stability criterion (Section 5.2.3) states that

$$\underline{\sigma}(I - M(j\omega)\Delta(j\omega)) > 0 \tag{5.7.31}$$

$$\underline{\sigma}(I - M(j\omega)\Delta(j\omega)) \ge \overline{\sigma}(I) - \overline{\sigma}(M(j\omega)\Delta(j\omega)) = 1 - \overline{\sigma}(M(j\omega)\Delta(j\omega)) \tag{5.7.32}$$

and

$$1 - \overline{\sigma}(M(j\omega)\Delta(j\omega)) \ge 1 - \sup_{\omega} \overline{\sigma}(M(j\omega)\Delta(j\omega)) = 1 - \left\| M(s)\Delta(s) \right\|_\infty \tag{5.7.33}$$

$$1 - \left\| M(s)\Delta(s) \right\|_\infty \ge 1 - \left\| M(s) \right\|_\infty \left\| \Delta(s) \right\|_\infty \ge 1 - \left\| M(s) \right\|_\infty \tag{5.7.34}$$

Combining (5.7.32) to (5.7.34),

$$\underline{\sigma}(I - M(j\omega)\Delta(j\omega)) \ge 1 - \left\| M(s) \right\|_\infty \tag{5.7.35}$$

Therefore, if $\left\| M(s) \right\|_\infty < 1$, the condition (5.7.31) is satisfied, and the system is stable.

Part II: Necessity. Assume that $\left\| M(s) \right\|_\infty \ge 1$, and it will be shown that a $\Delta(s)$ can be found with $\left\| \Delta \right\|_\infty \le 1$ such that the interconnected system in Figure 5.7.3 is unstable.

First, a frequency ω_o is found such that

$$\bar{\sigma}(M(j\omega_o)) \geq 1 \tag{5.7.36}$$

Next, the singular value decomposition of $M(j\omega_o)$ is obtained:

$$M(j\omega_o) = U(j\omega_o)\Sigma(j\omega_o)V(j\omega_o) \tag{5.7.37}$$

where

$$U(j\omega_o) = [\mathbf{u}_1 \quad \mathbf{u}_2 \quad . \quad . \quad \mathbf{u}_p] \tag{5.7.38}$$

$$V(j\omega_o) = [\mathbf{v}_1 \quad \mathbf{v}_2 \quad . \quad . \quad \mathbf{v}_p] \tag{5.7.39}$$

$$\Sigma(j\omega_o) = diag[\sigma_1 \quad \sigma_2 \quad \sigma_3 \quad . \quad .] \tag{5.7.40}$$

and

$$\sigma_1 > \sigma_2 > \sigma_3 > ... \tag{5.7.41}$$

Because of (5.7.36),

$$\sigma_1 \geq 1 \tag{5.7.42}$$

Now, a $\Delta(s)$ can be constructed as

$$\Delta(j\omega_o) = \frac{1}{\sigma_1}\mathbf{v}_1\mathbf{u}_1^* \tag{5.7.43}$$

Using (5.7.37) and (5.7.43),

$$\det(I - M(j\omega_o)\Delta(j\omega_o)) = \det\left(I - \frac{1}{\sigma_1}U\Sigma V\mathbf{v}_1\mathbf{u}_1^*\right) \tag{5.7.44}$$

Using the identity $\det(I_n - PQ) = \det(I_m - QP)$ where P and Q are $n \times m$ and $m \times n$ matrices respectively,

$$\det\left(I - \frac{1}{\sigma_1}U\Sigma V\mathbf{v}_1\mathbf{u}_1^*\right) = 1 - \frac{1}{\sigma_1}\mathbf{u}_1^*U\Sigma V\mathbf{v}_1 = 0 \tag{5.7.45}$$

From (5.7.44) and (5.7.45),

$$\det(I - M(j\omega_o)\Delta(j\omega_o)) = 0 \qquad (5.7.46)$$

Equation 5.7.46 indicates that the interconnected system is unstable when $\bar{\sigma}(M(j\omega_o)) \geq 1$ and a $\Delta(s)$ satisfying (5.7.43). Next, a proper and stable transfer matrix $\Delta(s)$ satisfying $\|\Delta\|_\infty \leq 1$ and Equation 5.7.43 will be constructed.

Case I

$$\omega_o = 0 \quad \text{or} \quad \infty$$

In this case, U and V are real matrices and $\Delta(s)$ can be chosen as

$$\Delta(s) = \frac{1}{\sigma_1} \mathbf{v}_1 \mathbf{u}_1^* \qquad (5.7.47)$$

Case II

$$0 < \omega_o < \infty$$

In this case, U and V are not real matrices. First, $\mathbf{u}_1$ and $\mathbf{v}_1$ are written in the following forms:

$$\mathbf{u}_1^* = [u_{11}e^{j\theta_1} \quad u_{12}e^{j\theta_2} \quad . \quad . \quad u_{1p}e^{j\theta_p}] \qquad (5.7.48)$$

and

$$\mathbf{v}_1 = \begin{bmatrix} v_{11}e^{j\phi_1} \\ v_{12}e^{j\phi_2} \\ . \\ . \\ v_{1q}e^{j\phi_q} \end{bmatrix} \qquad (5.7.49)$$

where u_{1i} and v_{1i} are real and positive numbers. Next, $a_i \geq 0$ and $b_i \geq 0$ are found such that

$$\frac{j\omega_o - a_i}{j\omega_o + a_i} = e^{j\phi_i}; \quad i = 1, 2, ..., q \qquad (5.7.50)$$

and

$$\frac{j\omega_o - b_i}{j\omega_o + b_i} = e^{j\theta_i} \; ; \quad i = 1, 2, \ldots, p \tag{5.7.51}$$

Lastly, a proper and stable transfer matrix $\Delta(s)$ satisfying $\|\Delta\|_\infty \leq 1$ is given by

$$\Delta(s) = \frac{1}{\sigma_1} \boldsymbol{\alpha}(s)\boldsymbol{\beta}(s) \tag{5.7.52}$$

where

$$\boldsymbol{\alpha}(s) = \frac{1}{\sigma_1}
\begin{bmatrix}
v_{11} \dfrac{s - a_1}{s + a_1} \\[2mm]
v_{12} \dfrac{s - a_2}{s + a_2} \\[2mm]
\cdot \\
\cdot \\
v_{1q} \dfrac{s - a_q}{s + a_q}
\end{bmatrix} \tag{5.7.53}$$

$$\boldsymbol{\beta}(s) = \frac{1}{\sigma_1}
\begin{bmatrix}
u_{11} \dfrac{s - b_1}{s + b_1} & u_{12} \dfrac{s - b_2}{s + b_2} & \cdot & \cdot & u_{1p} \dfrac{s - b_p}{s + b_p}
\end{bmatrix} \tag{5.7.54}$$

To calculate the singular value of $\Delta(j\omega)$,

$$\Delta(j\omega)\Delta^*(j\omega) = \frac{1}{\sigma_1^2} \boldsymbol{\alpha}(j\omega)\boldsymbol{\beta}(j\omega)(j\omega)\boldsymbol{\alpha}^*(j\omega) \tag{5.7.55}$$

Utilizing the unitary property of U,

$$\boldsymbol{\beta}(j\omega)\boldsymbol{\beta}^*(j\omega) = 1 \tag{5.7.56}$$

Therefore,

$$\det(\lambda I - \Delta(j\omega)\Delta^*(j\omega)) = \det\left(\lambda I - \frac{1}{\sigma_1^2}\boldsymbol{\alpha}(j\omega)\boldsymbol{\alpha}^*(j\omega)\right) \tag{5.7.57}$$

Using the identity $\det(I_n - PQ) = \det(I_m - QP)$ where P and Q are $n \times m$ and $m \times n$ matrices, respectively,

$$\det\left(\lambda I - \frac{1}{\sigma_1^2}\boldsymbol{\alpha}(j\omega)\boldsymbol{\alpha}^*(j\omega)\right) = \lambda - \frac{1}{\sigma_1^2}\boldsymbol{\alpha}^*(j\omega)\boldsymbol{\alpha}(j\omega) \qquad (5.7.58)$$

Utilizing the unitary property of V,

$$\boldsymbol{\alpha}^*(j\omega)\boldsymbol{\alpha}(j\omega) = 1 \qquad (5.7.59)$$

From (5.7.57) to (5.7.59),

$$\det(\lambda I - \Delta(j\omega)\Delta^*(j\omega)) = \lambda - \frac{1}{\sigma_1^2} \qquad (5.7.60)$$

Equation (5.7.60) yields the only nonzero singular value of $\Delta(j\omega)$. Note that the rank of the matrix $\Delta(j\omega)$ is 1. Therefore,

$$\left\|\Delta(s)\right\|_\infty = \frac{1}{\sigma_1} \qquad (5.7.61)$$

Using (5.7.42) and (5.7.61),

$$\left\|\Delta(s)\right\|_\infty \leq 1 \qquad (5.7.62)$$

Note that Equation 5.7.52 yields an algorithm to find a destabilizing perturbation when the condition for stability from the small gain theorem is violated.

5.7.3 Analysis for Application of Small Gain Theorem

Consider a two-port system $G_{2p}(s)$ in Figure 5.7.4a, where $K(s)$ and $\Delta(s)$ can be viewed as controller and uncertainty transfer function matrices, respectively. To apply the small gain theorem, the system shown in Figure 5.7.4a is converted to that in Figure 5.7.4b as follows. Let the state space model for the system $G_{2p}(s)$ be

$$\begin{bmatrix} \dot{\mathbf{x}} \\ \mathbf{e}(t) \\ \mathbf{y}(t) \end{bmatrix} = \begin{bmatrix} A & B_1 & B_2 \\ C_1 & D_{11} & D_{12} \\ C_2 & D_{21} & D_{22} \end{bmatrix} \begin{bmatrix} \mathbf{x}(t) \\ \mathbf{d}(t) \\ \mathbf{u}(t) \end{bmatrix} \qquad (5.7.63)$$

or, equivalently,

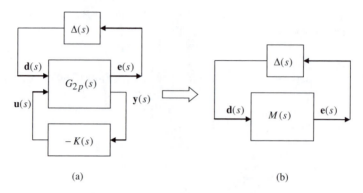

(a)   (b)

FIGURE 5.7.4 Conversion of a two-port system with controller and uncertainty transfer matrices to an M–Δ diagram.

$$\dot{\mathbf{x}} = A\mathbf{x}(t) + B_1\mathbf{d}(t) + B_2\mathbf{u}(t) \tag{5.7.64}$$

$$\mathbf{e}(t) = C_1\mathbf{x}(t) + D_{11}\mathbf{d}(t) + D_{12}\mathbf{u}(t) \tag{5.7.65}$$

$$\mathbf{y}(t) = C_2\mathbf{x}(t) + D_{21}\mathbf{d}(t) + D_{22}\mathbf{u}(t) \tag{5.7.66}$$

and the state space model for the system $-K(s)$ is given as

$$\begin{bmatrix} \dot{\hat{\mathbf{x}}} \\ \mathbf{u}(t) \end{bmatrix} = \begin{bmatrix} A_k & B_k \\ C_k & D_k \end{bmatrix} \begin{bmatrix} \hat{\mathbf{x}}(t) \\ \mathbf{y}(t) \end{bmatrix} \tag{5.7.67}$$

or, equivalently,

$$\dot{\hat{\mathbf{x}}} = A_k\hat{\mathbf{x}}(t) + B_k\mathbf{y}(t) \tag{5.7.68}$$

$$\mathbf{u}(t) = C_k\hat{\mathbf{x}}(t) + D_k\mathbf{y}(t) \tag{5.7.69}$$

Substituting (5.7.69) into (5.7.66), and solving for $\mathbf{y}(t)$,

$$\mathbf{y} = \Lambda C_2\mathbf{x} + \Lambda D_{21}\mathbf{d} + \Lambda D_{22}C_k\hat{\mathbf{x}} \tag{5.7.70}$$

where

$$\Lambda = (I - D_{22}D_k)^{-1} \tag{5.7.71}$$

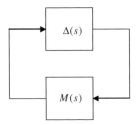

FIGURE 5.8.1 M–Δ system.

Substituting (5.7.70) into (5.7.69),

$$\mathbf{u} = D_k \Lambda C_2 \mathbf{x} + (C_k + D_k \Lambda D_{22} C_k)\hat{\mathbf{x}} + D_k \Lambda D_{21} \mathbf{d} \qquad (5.7.72)$$

Substituting (5.7.72) into (5.7.64) and (5.7.65), and substituting (5.7.70) into (5.7.68),

$$\begin{bmatrix} \dot{\mathbf{x}} \\ \dot{\hat{\mathbf{x}}} \\ \mathbf{e} \end{bmatrix} =$$

$$\begin{bmatrix} A + B_2 D_k \Lambda C_2 & B_2(C_k + D_k \Lambda D_{22} C_k) & B_1 + B_2 D_k \Lambda D_{21} \\ B_k \Lambda C_2 & A_k + B_k \Lambda D_{22} C_k & B_k \Lambda D_{21} \\ C_1 + D_{12} D_k \Lambda C_2 & D_{12}(C_k + D_k \Lambda D_{22} C_k) & D_{11} + D_{12} D_K \Lambda D_{21} \end{bmatrix} \begin{bmatrix} \mathbf{x} \\ \hat{\mathbf{x}} \\ \mathbf{d} \end{bmatrix} \qquad (5.7.73)$$

Equation 5.7.73 represents the state space model for $M(s)$ in Figure 5.7.4b. The Matlab routine *lftf* converts the system in Figure 5.7.4a to Figure 5.7.4b.

5.8 FORMULATION OF SOME ROBUST CONTROL PROBLEMS WITH UNSTRUCTURED UNCERTAINTIES

The system shown in Figure 5.7.3 with zero external inputs can also be drawn as that in Figure 5.8.1, where $M(s)$ and $\Delta(s)$ are viewed as the nominal system and an uncertainty in the system, respectively. The condition (5.7.27) yields the size of the uncertainty matrix $\Delta(s)$ beyond which the system is unstable. Note that $\|\Delta\|_\infty$ is larger if $\|M\|_\infty$ is smaller. This leads to a fundamental idea for the design of a robust control: minimize the H_∞ norm of the nominal system as much as possible.

5.8.1 MULTIPLICATIVE UNCERTAINTY

Let the plant transfer matrix be described by

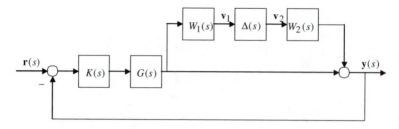

FIGURE 5.8.2 A multiinput/multioutput feedback system with multiplicative uncertainty.

$$G_\Delta(s) = G(s) + W_1(s)\Delta(s)W_2(s)G(s) = (I + W_1(s)\Delta(s)W_2(s))G(s) \quad (5.8.1)$$

where $G(s)$ is the nominal transfer matrix and $W_1(s)\Delta(s)W_2(s)G(s)$ is the uncertainty. The transfer matrices $W_1(s)$ and $W_2(s)$ are weight functions, which are included here for a general treatment. Of course, they can also be identity matrices. It is assumed that the transfer matrices $W_1(s)$, $W_2(s)$, and $\Delta(s)$ are proper and stable.

Consider the feedback system in Figure 5.8.2, which is stable with $\Delta(s) = 0$. In other words, the controller $K(s)$ has been designed to stabilize the feedback system with the nominal plant $G(s)$ alone. Then the question is: What is the size of $\Delta(s)$ beyond which the feedback system in Figure 5.8.2 will be unstable? To answer this question, with the external input $\mathbf{r}(s) = 0$, loop is broken at $\mathbf{v}_1$ and $\mathbf{v}_2$, and it is easily derived that

$$-GK\mathbf{y} + W_2\mathbf{v}_2 = \mathbf{y} \quad (5.8.2)$$

$$\mathbf{v}_1 = -W_1 GK\mathbf{y} \quad (5.8.3)$$

From (5.8.2) and (5.8.3),

$$\mathbf{v}_2(s) = M(s)\mathbf{v}_1(s) \quad (5.8.4)$$

where

$$M(s) = -W_1(s)T_o(s)W_2(s) \quad (5.8.5)$$

The feedback system can be converted to the $M-\Delta$ system in Figure 5.8.1. Applying the small gain theorem, the condition for the stability of the feedback system is as follows:

$$\|\Delta\|_\infty \le \frac{1}{\gamma} \text{ if and only if } \|W_1(s)T_o(s)W_2(s)\|_\infty < \gamma \quad (5.8.6)$$

Without any loss of generality, γ is often set to 1.

Note: Multiplicative uncertainty can be on the input side of G(s); i.e.,

$$G_\Delta(s) = G(s)(I + W_1(s)\Delta(s)W_2(s))$$

In this case, the condition for the stability of the feedback system is as follows:

$$\left\|\Delta\right\|_\infty \le \frac{1}{\gamma} \text{ if and only if } \left\|W_1(s)T_i(s)W_2(s)\right\|_\infty < \gamma \tag{5.8.7}$$

5.8.2 ADDITIVE UNCERTAINTY

Let the plant transfer matrix be described by

$$G_\Delta(s) = G(s) + W_1(s)\Delta(s)W_2(s) \tag{5.8.8}$$

where $G(s)$ is the nominal transfer matrix and $W_1(s)\Delta(s)W_2(s)G(s)$ is the uncertainty. The transfer matrices $W_1(s)$ and $W_2(s)$ are weight functions, which are included here for a general treatment. Of course, they can also be identity matrices. It is assumed that the transfer matrices $W_1(s)$, $W_2(s)$, and $\Delta(s)$ are proper and stable.

Consider the feedback system in Figure 5.8.3, which is stable with $\Delta(s) = 0$. In other words, the controller $K(s)$ has been designed to stabilize the feedback system with the nominal plant $G(s)$ alone. Then the question is: What is the size of $\Delta(s)$ beyond which the feedback system in Figure 5.8.3 will be unstable? To answer this question, with the external input $\mathbf{r}(s) = 0$, loop is broken at $\mathbf{v}_1$ and $\mathbf{v}_2$, and it is easily derived that

$$-G K \mathbf{y} + W_1 \mathbf{v}_2 = \mathbf{y} \tag{5.8.9}$$

$$\mathbf{v}_1 = -W_2 K \mathbf{y} \tag{5.8.10}$$

From (5.8.9) to (5.8.10),

$$\mathbf{v}_2(s) = M(s)\mathbf{v}_1(s) \tag{5.8.11}$$

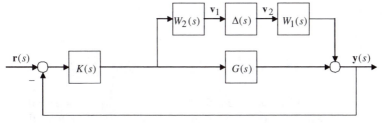

FIGURE 5.8.3 A multiinput/multioutput feedback system with additive uncertainty.

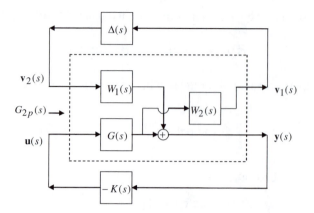

FIGURE 5.8.4 Alternate representation of Figure 5.8.3.

where

$$M(s) = -W_2(s)K(s)S_o(s)W_1(s) \qquad (5.8.12)$$

The feedback system can be converted to the $M-\Delta$ system in Figure 5.8.1. Applying the small gain theorem, the condition for the stability of the feedback system is as follows:

$$\left\|\Delta\right\|_\infty \leq \frac{1}{\gamma} \text{ if and only if } \left\|W_2(s)K(s)S_o(s)W_1(s)\right\|_\infty < \gamma \qquad (5.8.13)$$

Without any loss of generality, γ is often set to 1.

EXAMPLE 5.8.1

The system in Figure 5.8.3 can be redrawn to be as shown in Figure 5.8.4, which has the form of Figure 5.7.4a. Then, the state space representation of $M(s)$ can be obtained via Equation 5.7.73 or the Matlab routine: lftf.

EXAMPLE 5.8.2: RESIDUAL VIBRATORY MODE AS MULTIPLICATIVE UNCERTAINTY

The transfer function for the system shown in Figure 5.8.5 is

$$\frac{y(s)}{u(s)} = \frac{(\alpha s + k)}{ms^2(ms^2 + 2\alpha s + 2k)} \qquad (5.8.14)$$

The controller will be designed on the basis of the rigid mode only, and the vibratory mode will be considered as the residual mode. Therefore, the transfer function (5.8.14) can be written in the form of a multiplicative uncertainty as

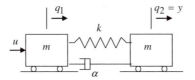

FIGURE 5.8.5 A two-mass system with damping.

$$\frac{y(s)}{u(s)} = G(s)[1 + \Delta(s)] \tag{5.8.15}$$

where

$$G(s) = \frac{1}{2ms^2} \quad \text{and} \quad \Delta(s) = -\frac{ms^2}{ms^2 + 2\alpha s + 2k} \tag{5.8.16a,b}$$

Then, corresponding to Figure 5.7.4, $G_{2p}(s)$ can be defined as

$$\begin{bmatrix} \dot{x}_1 \\ \dot{x}_2 \\ v_1 \\ y \end{bmatrix} = \begin{bmatrix} 0 & 1 & 0 & 0 \\ 0 & 0 & 0 & 0.5m^{-1} \\ 1 & 0 & 0 & 0 \\ 1 & 0 & 1 & 0 \end{bmatrix} \begin{bmatrix} x_1 \\ x_2 \\ v_2 \\ u \end{bmatrix} \tag{5.8.17}$$

and the estimated state feedback controller $K(s)$ can be written as

$$\begin{bmatrix} \dot{\hat{x}}_1 \\ \dot{\hat{x}}_2 \\ u \end{bmatrix} = \begin{bmatrix} -\ell_1 & 1 & \ell_1 \\ -\ell_2 - \dfrac{0.5k_1}{m} & -\dfrac{0.5k_2}{m} & \ell_2 \\ -k_1 & -k_2 & 0 \end{bmatrix} \begin{bmatrix} \hat{x}_1 \\ \hat{x}_2 \\ y \end{bmatrix} \tag{5.8.18}$$

where ℓ_1 and ℓ_2 are observer gains. To place controller poles at $-1 \pm 1j$, $k_1 = k_2 = 4$, and to place observer poles at $-4 \pm 4j$, $\ell_1 = 8$ and $\ell_2 = 32$. Bode magnitude plots of $M_K(s)$ and $\Delta(s)$ are shown in Figure 5.8.6.

It is found that

$$\left\| M_K(s) \right\|_\infty = 1.4603 \quad \text{and} \quad \left\| \Delta(s) \right\|_\infty = 8.8552 \tag{5.8.19}$$

Even though $\left\| M_K(s) \right\|_\infty \left\| \Delta(s) \right\|_\infty > 1$, the eigenvalues of the system (4.9.24) indicate that the closed-loop system is stable in the presence of residual modes. This can be understood by realizing that the small gain theorem only states that there is at least

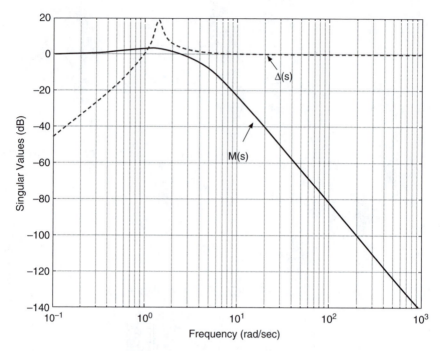

FIGURE 5.8.6 Singular values of $M_K(s)$.

one $\Delta(s)$ that will destabilize the system. It does not rule out the existence of a $\Delta(s)$ satisfying $\left\|M_K(s)\right\|_\infty \left\|\Delta(s)\right\|_\infty > 1$, for which the closed-loop system is stable.

5.9 FORMULATION OF ROBUST CONTROL PROBLEMS WITH STRUCTURED UNCERTAINTIES

5.9.1 LINEAR FRACTIONAL TRANSFORMATION (LFT)

A linear fractional transformation is the mapping of the form (Zhou et al., 1996, 1998)

$$F(s) = \frac{a + bs}{c + ds} \tag{5.9.1}$$

where a, b, c, and d are in general complex numbers. If $c \neq 0$, $F(s)$ can also be written as

$$F(s) = \alpha + \frac{\beta s}{1 - \gamma s} = \alpha + \beta s(1 - \gamma s)^{-1} \tag{5.9.2}$$

where α, β, and γ are complex numbers. Next, partition a complex $(p_1 + p_2) \times (q_1 + q_2)$ matrix M as follows:

$$M = \begin{bmatrix} M_{11} & M_{12} \\ M_{21} & M_{22} \end{bmatrix} \qquad (5.9.3)$$

where M_{11}, M_{12}, M_{21}, and M_{22} are $p_1 \times q_1$, $p_1 \times q_2$, $p_2 \times q_1$, and $p_2 \times q_2$ matrices, respectively.

Lower LFT

Consider a complex $q_2 \times p_2$ matrix Δ_ℓ. Then, a lower LFT with respect to Δ_ℓ is defined as

$$F_\ell(M, \Delta_\ell) = M_{11} + M_{12}\Delta_\ell(I - M_{22}\Delta_\ell)^{-1}M_{21} \qquad (5.9.4)$$

From Figure 5.9.1,

$$\begin{bmatrix} \mathbf{z}_1 \\ \mathbf{y}_1 \end{bmatrix} = M \begin{bmatrix} \mathbf{w}_1 \\ \mathbf{u}_1 \end{bmatrix} = \begin{bmatrix} M_{11} & M_{12} \\ M_{21} & M_{22} \end{bmatrix} \begin{bmatrix} \mathbf{w}_1 \\ \mathbf{u}_1 \end{bmatrix} \qquad (5.9.5)$$

and

$$\mathbf{u}_1 = \Delta_\ell \mathbf{y}_1 \qquad (5.9.6)$$

Dimensions of vectors $\mathbf{w}_1$, $\mathbf{u}_1$, $\mathbf{z}_1$, and $\mathbf{y}_1$ are $\mathbf{q}_1$, $\mathbf{q}_2$, $\mathbf{p}_1$, and $\mathbf{p}_2$, respectively. From (5.9.5),

$$\mathbf{z}_1 = M_{11}\mathbf{w}_1 + M_{12}\mathbf{u}_1 \qquad (5.9.7)$$

$$\mathbf{y}_1 = M_{21}\mathbf{w}_1 + M_{22}\mathbf{u}_1 \qquad (5.9.8)$$

Substituting (5.9.6) into (5.9.8),

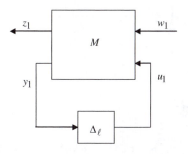

FIGURE 5.9.1 Visualization of lower linear fractional transformation.

$$\mathbf{y}_1 = (I - M_{22}\Delta_\ell)^{-1} M_{21}\mathbf{w}_1 \tag{5.9.9}$$

From (5.9.6) and (5.9.9),

$$\mathbf{u}_1 = \Delta_\ell (I - M_{22}\Delta_\ell)^{-1} M_{21}\mathbf{w}_1 \tag{5.9.10}$$

Substituting (5.9.10) into (5.9.7),

$$\mathbf{z}_1 = T_{z_1 w_1}\mathbf{w}_1 \tag{5.9.11}$$

where

$$T_{z_1 w_1} = M_{11} + M_{12}\Delta_\ell (I - M_{22}\Delta_\ell)^{-1} M_{21} \tag{5.9.12}$$

is the transfer function from $\mathbf{w}_1$ to $\mathbf{z}_1$ when the signal $\mathbf{y}_1$ is fed back according to (5.9.6). Note that $T_{z_1 w_1}$ is the lower LFT $F_\ell(M, \Delta_\ell)$ defined by Equation 5.9.4.

Upper LFT

Consider a complex $q_1 \times p_1$ matrix Δ_u. Then, an upper LFT with respect to Δ_u is defined as

$$F_u(M, \Delta_u) = M_{22} + M_{21}\Delta_u (I - M_{11}\Delta_u)^{-1} M_{12} \tag{5.9.13}$$

From Figure 5.9.2,

$$\begin{bmatrix} y_2 \\ z_2 \end{bmatrix} = M \begin{bmatrix} \mathbf{u}_2 \\ \mathbf{w}_2 \end{bmatrix} = \begin{bmatrix} M_{11} & M_{12} \\ M_{21} & M_{22} \end{bmatrix} \begin{bmatrix} \mathbf{u}_2 \\ \mathbf{w}_2 \end{bmatrix} \tag{5.9.14}$$

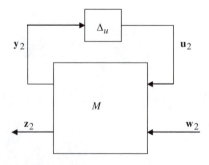

FIGURE 5.9.2 Visualization of upper linear fractional transformation.

and

$$\mathbf{u}_2 = \Delta_u \mathbf{y}_2 \qquad (5.9.15)$$

Dimensions of vectors $\mathbf{u}_2$, $\mathbf{w}_2$, $\mathbf{y}_2$, and $\mathbf{z}_2$ are $\mathbf{q}_1$, $\mathbf{q}_2$, $\mathbf{p}_1$, and $\mathbf{p}_2$, respectively. From (5.9.14),

$$\mathbf{y}_2 = M_{11}\mathbf{u}_2 + M_{12}\mathbf{w}_2 \qquad (5.9.16)$$

$$\mathbf{z}_2 = M_{21}\mathbf{u}_2 + M_{22}\mathbf{w}_2 \qquad (5.9.17)$$

Substituting (5.9.15) into (5.9.16),

$$\mathbf{y}_2 = (I - M_{11}\Delta_u)^{-1} M_{12}\mathbf{w}_2 \qquad (5.9.18)$$

From (5.9.15) and (5.9.18),

$$\mathbf{u}_2 = \Delta_u (I - M_{11}\Delta_u)^{-1} M_{12}\mathbf{w}_1 \qquad (5.9.19)$$

Substituting (5.9.19) into (5.9.17),

$$\mathbf{z}_2 = T_{z_2 w_2} \mathbf{w}_2 \qquad (5.9.20)$$

where

$$T_{z_2 w_2} = M_{22} + M_{21}\Delta_u (I - M_{11}\Delta_u)^{-1} M_{12} \qquad (5.9.21)$$

is the transfer function from $\mathbf{w}_2$ to $\mathbf{z}_2$ when the signal $\mathbf{y}_2$ is fed back according to (5.9.15). Note that $T_{z_2 w_2}$ is the upper LFT $F_u(M, \Delta_u)$ defined by Equation 5.9.13.

EXAMPLE 5.9.1: REPRESENTATION OF A STATE SPACE MODEL (FIGURE 5.9.3)

Comparing transfer function matrix

$$G(s) = D + C(sI - A)^{-1} B \qquad (5.9.22)$$

to Equation 5.9.12,

$$M_{22} = D, \ M_{21} = C, \ M_{11} = A, \ M_{12} = B, \text{ and } \ \Delta_u = \frac{1}{s} I \qquad (5.9.23)$$

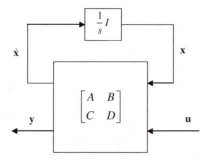

FIGURE 5.9.3 An upper linear fractional transformation representation of a state space model.

Hence,

$$G(s) = F_u\left(M, \frac{1}{s}I\right) \tag{5.9.24}$$

where

$$M = \begin{bmatrix} A & B \\ C & D \end{bmatrix} \tag{5.9.25}$$

5.9.2 STRUCTURED PARAMETRIC UNCERTAINTIES

Consider a linear system (A_δ, B_δ, C_δ, and D_δ) with k uncertain parameters δ_1, δ_2, ..., δ_k (Zhou et al., 1996):

$$A_\delta = A + \sum_{i=1}^{k} \delta_i \hat{A}_i ; \quad B_\delta = B + \sum_{i=1}^{k} \delta_i \hat{B}_i \tag{5.9.26a, b}$$

$$C_\delta = C + \sum_{i=1}^{k} \delta_i \hat{C}_i ; \quad D_\delta = D + \sum_{i=1}^{k} \delta_i \hat{D}_i \tag{5.9.27c, d}$$

where (A, B, C, D) is the model of the plant. Matrices $\hat{A}_i$, $\hat{B}_i$, $\hat{C}_i$, and $\hat{D}_i$ are known, and they describe how uncertain parameters δ_1, δ_2, ..., δ_k enter into the model. It will be assumed that there are n states, m inputs, and p outputs. Following Equation 5.9.13,

$$G_\delta(s) = D_\delta + C_\delta(sI - A_\delta)^{-1}B_\delta = F_u\left(N_\delta, \frac{1}{s}I\right) \tag{5.9.28}$$

where

$$N_\delta = \begin{bmatrix} A + \sum_{i=1}^{k} \delta_i \hat{A}_i & B + \sum_{i=1}^{k} \delta_i \hat{B}_i \\ C + \sum_{i=1}^{k} \delta_i \hat{C}_i & D + \sum_{i=1}^{k} \delta_i \hat{D}_i \end{bmatrix} \qquad (5.9.29)$$

Equation 5.9.29 can be written as

$$N_\delta = \begin{bmatrix} A & B \\ C & D \end{bmatrix} + \sum_{i=1}^{k} \delta_i P_i \qquad (5.9.30)$$

where

$$P_i = \begin{bmatrix} \hat{A}_i & \hat{B}_i \\ \hat{C}_i & \hat{D}_i \end{bmatrix} \qquad (5.9.31)$$

If the rank of the matrix P_i is q_i (Zhou et al., 1996),

$$P_i = \begin{bmatrix} L_i \\ W_i \end{bmatrix} \begin{bmatrix} R_i^H & Z_i^H \end{bmatrix} \qquad (5.9.32)$$

where L_i, W_i, R_i, and Z_i are $n \times q_i$, $p \times q_i$, $n \times q_i$, and $m \times q_i$ matrices, respectively. Hence,

$$\delta_i P_i = \begin{bmatrix} L_i \\ W_i \end{bmatrix} \delta_i I_{q_i} \begin{bmatrix} R_i^H & Z_i^H \end{bmatrix} \qquad (5.9.33)$$

Substituting (5.9.33) into (5.9.30),

$$N_\delta = \begin{bmatrix} A & B \\ C & D \end{bmatrix} + \begin{bmatrix} L_1 & L_2 & . & . & L_k \\ W_1 & W_2 & . & . & W_k \end{bmatrix} \begin{bmatrix} \delta_1 I_{q_1} & . & 0 \\ . & . & . \\ 0 & . & \delta_k I_{q_k} \end{bmatrix} \begin{bmatrix} R_1^H & Z_1^H \\ R_2^H & Z_2^H \\ . & . \\ . & . \\ R_k^H & Z_k^H \end{bmatrix}$$

$$\qquad (5.9.34)$$

Comparing to Equation 5.9.14,

$$N_\delta = F_\ell(M_\delta, \Delta_p) \tag{5.9.35}$$

where

$$M_\delta = \begin{bmatrix} M_{11} & M_{12} \\ M_{21} & 0 \end{bmatrix} \tag{5.9.36}$$

$$M_{11} = \begin{bmatrix} A & B \\ C & D \end{bmatrix} \quad M_{12} = \begin{bmatrix} L_1 & L_2 & \cdot & \cdot & L_k \\ W_1 & W_2 & \cdot & \cdot & W_k \end{bmatrix} \quad M_{22} = 0 \tag{5.9.37a,b,c}$$

$$M_{21} = \begin{bmatrix} R_1^H & Z_1^H \\ R_2^H & Z_2^H \\ \cdot & \cdot \\ \cdot & \cdot \\ R_k^H & Z_k^H \end{bmatrix} \quad \Delta_p = \begin{bmatrix} \delta_1 I_{q_1} & \cdot & 0 \\ \cdot & \cdot & \cdot \\ 0 & \cdot & \delta_k I_{q_k} \end{bmatrix} \tag{5.9.38a,b}$$

Substituting Equation 5.9.35 into Equation 5.9.28,

$$G_\delta(s) = F_u\left(F_\ell(M_\delta, \Delta_p), \frac{1}{s} I\right) \tag{5.9.39}$$

Rewriting (5.9.36) as

$$M_\delta = \begin{bmatrix} A & B & B_2 \\ C & D & D_{12} \\ C_2 & D_{21} & D_{22} \end{bmatrix} \tag{5.9.40}$$

$$B_2 = \begin{bmatrix} L_1 & L_2 & \cdot & \cdot & L_k \end{bmatrix} \tag{5.9.41}$$

$$D_{12} = \begin{bmatrix} L_1 & L_2 & \cdot & \cdot & L_k \end{bmatrix} \tag{5.9.42}$$

$$C_2^H = \begin{bmatrix} R_1 & R_2 & \cdot & \cdot & R_k \end{bmatrix} \tag{5.9.43}$$

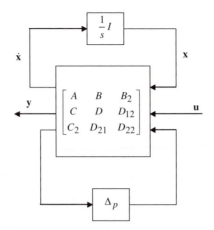

FIGURE 5.9.4 Extracting parametric uncertainties.

$$D_{21}^H = \begin{bmatrix} Z_1 & Z_2 & . & . & Z_k \end{bmatrix} \tag{5.9.44}$$

$$D_{22} = 0 \tag{5.9.45}$$

the relationship (5.9.39) can be visualized as shown in Figure 5.9.4.

EXAMPLE 5.9.2: A TWO-MASS SYSTEM WITH AN UNCERTAIN STIFFNESS

Consider the system shown in Figure 5.9.5 where the stiffness k is the uncertain parameter. Then,

$$A = \begin{bmatrix} 0 & 0 & 1 & 0 \\ 0 & 0 & 0 & 1 \\ -k & k & 0 & 0 \\ k & -k & 0 & 0 \end{bmatrix} = \begin{bmatrix} 0 & 0 & 1 & 0 \\ 0 & 0 & 0 & 1 \\ -k_0 & k_0 & 0 & 0 \\ k_0 & -k_0 & 0 & 0 \end{bmatrix} + \delta k \begin{bmatrix} 0 & 0 & 0 & 0 \\ 0 & 0 & 0 & 0 \\ -1 & 1 & 0 & 0 \\ 1 & -1 & 0 & 0 \end{bmatrix}$$

$$\tag{5.9.46}$$

$$B^T = \begin{bmatrix} 0 & 0 & 1 & 0 \end{bmatrix} \tag{5.9.47}$$

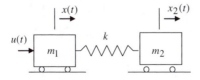

FIGURE 5.9.5 A two-mass system.

$$C = \begin{bmatrix} 0 & 1 & 0 & 0 \end{bmatrix} \tag{5.9.48}$$

$$D = 0 \tag{5.9.49}$$

Therefore,

$$\hat{A}_1 = \begin{bmatrix} 0 & 0 & 0 & 0 \\ 0 & 0 & 0 & 0 \\ -1 & 1 & 0 & 0 \\ 1 & -1 & 0 & 0 \end{bmatrix}; \quad \hat{B}_1^T = \begin{bmatrix} 0 & 0 & 0 & 0 \end{bmatrix} \tag{5.9.50}$$

$$\hat{C}_1 = \begin{bmatrix} 0 & 0 & 0 & 0 \end{bmatrix}; \quad \hat{D}_1 = 0 \tag{5.9.51}$$

Hence,

$$P_1 = \begin{bmatrix} \hat{A}_1 & \hat{B}_1 \\ \hat{C}_1 & \hat{D}_1 \end{bmatrix} = \begin{bmatrix} 0 & 0 & 0 & 0 & 0 \\ 0 & 0 & 0 & 0 & 0 \\ -1 & 1 & 0 & 0 & 0 \\ 1 & -1 & 0 & 0 & 0 \\ 0 & 0 & 0 & 0 & 0 \end{bmatrix} \tag{5.9.52}$$

The rank of P_1 is 1. Therefore,

$$P_1 = \begin{bmatrix} 0 \\ 0 \\ -1 \\ 1 \\ 0 \end{bmatrix} \begin{bmatrix} 1 & -1 & 0 & 0 & 0 \end{bmatrix} \tag{5.9.53}$$

$$M_\delta = \begin{bmatrix} A & B & B_2 \\ C & D & D_{12} \\ C_2 & D_{21} & D_{22} \end{bmatrix} = \begin{bmatrix} 0 & 0 & 1 & 0 & 0 & 0 \\ 0 & 0 & 0 & 1 & 0 & 0 \\ -k_0 & k_0 & 0 & 0 & 1 & 1 \\ k_0 & -k_0 & 0 & 0 & 0 & -1 \\ 0 & 1 & 0 & 0 & 0 & 0 \\ 1 & -1 & 0 & 0 & 0 & 0 \end{bmatrix} \tag{5.9.54}$$

$$\Delta_p = \delta k \tag{5.9.55}$$

EXAMPLE 5.9.3: APPLICATION OF SMALL GAIN THEOREM TO EXAMPLE 5.9.2

Consider Example 5.9.2 with the nominal spring constant $k_0 = 1\, N/m$ and the mass $m = 1$ kg. An LQG controller is designed via the Matlab Program 5.9.1. Then, with respect to Figure 5.9.4 with the feedback controller, or Figure 5.7.4b,

$$\left\| M_\delta(s) \right\|_\infty = 6.2387 \tag{5.9.56}$$

Therefore, the closed loop system will be stable provided

$$\left\| \Delta(s) \right\|_\infty = \left| \delta k \right| < \frac{1}{6.2387} \tag{5.9.57}$$

MATLAB PROGRAM 5.9.1: LQG CONTROL (EXAMPLE 5.9.3)

```
%
Ii=eye(2);
Zi=0*Ii;
Ks=[1 -1;-1 1];
a=[Zi,Ii;-Ks,Zi];
b1=[0 0 1 0]';
c1=[0 1 0 0];
d11=0;
b2=-[0 0 1 -1]';
c2=[1 -1 0 0];
d12=0;
d21=0;
d22=0;
%
W=eye(5);
V=eye(5);
[AF,BF,CF,DF]=lqg(a,b1,c1,d11,W,V);
[Ac,Bc,Cc,Dc]=feedback(a,b1,c1,d11,AF,BF,CF,DF);
```

```
CFF=-CF;
[AA,BB,CC,DD]...
=lftf(a,b2,b1,c2,c1,d22,d21,d12,d11,AF,BF,CFF,DF);
[hinfn]=normhinf(AA,BB,CC,DD)
```

5.10 H_∞ CONTROL

5.10.1 FULL STATE FEEDBACK H_∞ CONTROL (FIGURE 5.10.1)

Assume that

$$M = \begin{bmatrix} A & B_1 & B_2 \\ C_1 & 0 & D_{12} \\ I & 0 & 0 \end{bmatrix} \tag{5.10.1}$$

The following assumptions are made:

1. (A, B_1) and (A, B_2) are stabilizable.
2. (C_1, A_1) is detectable.
3. $C_1^T D_{12} = 0$ and $D_{12}^T D_{12} = I$.

Therefore,

$$\dot{\mathbf{x}} = A\mathbf{x} + B_1\mathbf{d}(t) + B_2\mathbf{u}(t) \tag{5.10.2}$$

$$\mathbf{e}(t) = C_1\mathbf{x}(t) + D_{12}\mathbf{u}(t) \tag{5.10.3}$$

$$\mathbf{y}(t) = \mathbf{x}(t) \tag{5.10.4}$$

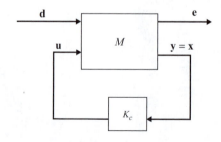

FIGURE 5.10.1 A full state feedback control system.

The condition

$$\left\|T_{ed}(s)\right\|_{H_\infty} < \gamma \tag{5.10.5}$$

implies

$$\underset{\mathbf{u} \quad \mathbf{d}}{\inf \sup} \, J(\mathbf{u}, \mathbf{d}) < \infty \tag{5.10.6}$$

$$J(\mathbf{u}, \mathbf{d}) = \int_0^\infty (\mathbf{e}^T \mathbf{e} - \gamma^2 \mathbf{d}^T \mathbf{d}) dt \tag{5.10.7}$$

where infinimum (inf) and supremum (sup) are defined in Appendix C. Assume that the worst-case disturbance $\mathbf{d}(t)$ and the optimal control $\mathbf{u}(t)$ have the following structure (Stein, 1988):

$$\mathbf{d}(t) = K_d \mathbf{x}(t) \quad \text{and} \quad \mathbf{u}(t) = K_c \mathbf{x}(t) \tag{5.10.8}$$

Then,

$$\mathbf{e}(t) = (C_1 + D_{12} K_c) \mathbf{x}(t) \tag{5.10.9}$$

Using assumption 3,

$$\mathbf{e}^T \mathbf{e} = \mathbf{x}^T (C_1^T C_1 + K_c^T K_c) \mathbf{x} \tag{5.10.10}$$

Therefore,

$$J = \int_0^\infty \mathbf{x}^T (C_1^T C_1 + K_c^T K_c - \gamma^2 K_d^T K_d) \mathbf{x} dt \tag{5.10.11}$$

From Equation 5.10.2 and Equation 5.10.8,

$$\dot{\mathbf{x}} = (A + B_1 K_d + B_2 K_c) \mathbf{x} \tag{5.10.12}$$

Hence, under the assumption that (5.10.12) is stable,

$$J = \mathbf{x}^T(0)P\mathbf{x}(0) \tag{5.10.13}$$

where P satisfies the Lyapunov equation,

$$P(A + B_1 K_d + B_2 K_c) + (A + B_1 K_d + B_2 K_c)^T P + C_1^T C_1 + K_c^T K_c - \gamma^2 K_d^T K_d = 0 \tag{5.10.14}$$

The condition for maximization of J, Equation 5.10.13, with respect to K_d (Stein, 1988) is

$$\nabla_{K_d} P = 0 \tag{5.10.15}$$

The gradient matrix $\nabla_{K_d} P$ is defined as follows:

$$(\nabla_{K_d} P)_{ij} = \frac{\partial P}{\partial K_{d_{ij}}} \tag{5.10.16}$$

From (5.10.14),

$$\nabla_{K_d} P(A + B_1 K_d + B_2 K_c) + B_1^T P + B_1^T P + (A + B_1 K_d + B_2 K_c)^T \nabla_{K_d} P - 2\gamma^2 K_d = 0 \tag{5.10.17}$$

Using (5.10.15),

$$K_d = \frac{1}{\gamma^2} B_1^T P \tag{5.10.18}$$

Similarly, the condition for minimization of J, Equation 5.10.13, with respect to K_c is

$$\nabla_{K_c} P = 0 \tag{5.10.19}$$

Therefore, it can be derived that

$$K_c = -B_2^T P \tag{5.10.20}$$

Substituting (5.10.18) and (5.10.20) into (5.10.14),

$$PA + A^T P + C_1^T C_1 - P\left(B_2 B_2^T - \frac{1}{\gamma^2} B_1 B_1^T\right)P = 0 \qquad (5.10.21)$$

The condition $\left\|T_{ed}(s)\right\|_{H_\infty} < \gamma$ is satisfied provided

1. $\mathbf{u}(t) = K_c \mathbf{x}(t)$ where K_c is given by Equation 5.10.20.   (5.10.22a)

2. $P \geq 0$.   (5.10.22b)

3. The matrix $(A + B_1 K_d + B_2 K_c)$ is stable.   (5.10.22c)

Using the iterative procedure, one can find the minimum value of γ such that

$$\left\|T_{ed}(s)\right\|_{H_\infty} < \gamma_{min} \qquad (5.10.23)$$

Fact 5.10.1

Because K_d represents the worst-case disturbance, the stability of $(A + B_1 K_d + B_2 K_c)$ implies the stability of $A + B_2 K_c$, which is the closed-loop system matrix.

Proof

Choose the following Lyapunov function:

$$V = \mathbf{x}^T(t) P \mathbf{x}(t) \qquad (5.10.24)$$

Therefore,

$$\dot{V} = 2\mathbf{x}^T(t) P (A + B_1 K_d + B_2 K_c)\mathbf{x}(t) < 0 \qquad (5.10.25)$$

Substituting the expression for K_d,

$$2\mathbf{x}^T(t) P (A + \gamma^{-2} B_1 B_1^T P + B_2 K_c)\mathbf{x}(t) < 0 \qquad (5.10.26)$$

Therefore,

$$2\mathbf{x}^T(t) P (A + B_2 K_c)\mathbf{x}(t) < 0 \qquad (5.10.27)$$

Hence, $\dot{V} < 0$ for the solution of $\dot{\mathbf{x}} = (A + B_2 K_c)\mathbf{x}$ also. This completes the proof.

EXAMPLE 5.10.1: A SIMPLE MASS

Consider the system (5.6.12) again, but with

$$B_1 = \begin{bmatrix} 1 \\ 0 \end{bmatrix} \tag{5.10.28}$$

Expanding (5.10.21), the following three equations for the elements of the matrix P are obtained:

$$1 + \frac{p_{11}^2}{\gamma^2} - p_{12}^2 = 0 \tag{5.10.29}$$

$$p_{11} + \frac{p_{11}p_{12}}{\gamma^2} - p_{12}p_{22} = 0 \tag{5.10.30}$$

$$2p_{12} + \frac{p_{12}^2}{\gamma^2} - p_{22}^2 = 0 \tag{5.10.31}$$

Using (5.10.29) and (5.10.31),

$$p_{11}^2 = \gamma^2(p_{12}^2 - 1) \tag{5.10.32}$$

and

$$p_{22}^2 = 2p_{12} + \frac{p_{12}^2}{\gamma^2} \tag{5.10.33}$$

From (5.10.30),

$$p_{11}(1 + \frac{p_{12}}{\gamma^2}) = p_{12}p_{22} \tag{5.10.34}$$

Squaring (5.10.34), and using (5.10.32) and (5.10.33),

$$p_{12}^2 - 2p_{12}\frac{\gamma^2}{\gamma^4 - 1} - \frac{\gamma^4}{\gamma^4 - 1} = 0 \tag{5.10.35a}$$

Solving (5.10.35a),

$$P_{12} = \frac{\gamma^2}{\gamma^4 - 1}(1 \pm \gamma^2) \qquad (5.10.35b)$$

For a positive definite solution for P, a "+" sign should be chosen in (5.10.35b). The solution for P is obtained as follows:

$$P_{12} = \frac{\gamma^2}{\gamma^2 - 1}, \qquad P_{22} = \gamma \frac{\sqrt{2\gamma^2 - 1}}{\gamma^2 - 1} \qquad (5.10.36)$$

EXAMPLE 5.10.2: ACTIVE SUSPENSION SYSTEM

Consider the active suspension system presented in Example 3.11. The state equations are written as

$$\begin{bmatrix} \dot{x}_1 \\ \dot{x}_2 \\ \dot{x}_3 \\ \dot{x}_4 \end{bmatrix} = \begin{bmatrix} 0 & 0 & 1 & 0 \\ 0 & 0 & 0 & 1 \\ -\frac{\lambda_1}{m_1} & 0 & 0 & 0 \\ 0 & 0 & 0 & 0 \end{bmatrix} \begin{bmatrix} x_1 \\ x_2 \\ x_3 \\ x_4 \end{bmatrix} + \begin{bmatrix} 0 \\ 0 \\ -\frac{1}{m_1} \\ \frac{1}{m_2} \end{bmatrix} u(t) + \begin{bmatrix} 0 \\ 0 \\ 1 \\ 0 \end{bmatrix} d(t) \qquad (5.10.37)$$

where the disturbance $d(t)$ is described as

$$d(t) = \frac{\lambda_1}{m_1} x_0(t) \qquad (5.10.38)$$

Consider the quadratic objective function, Equation 3.5.81, for which the weighting matrix on states is rewritten as

$$Q = \begin{bmatrix} q_1 + q_2 & -q_2 & 0 & 0 \\ -q_2 & q_2 & 0 & 0 \\ 0 & 0 & 0 & 0 \\ 0 & 0 & 0 & 0 \end{bmatrix} \qquad (5.10.39)$$

and the scalar input weighting is ρ. Let $\rho^{-1}Q$ be written as

$$\rho^{-1}Q = C_{11}C_{11}^T \qquad (5.10.40)$$

where the order of the matrix C_{11} is 4×2 because the rank of the matrix Q is 2. Therefore, corresponding to the quadratic objective function (3.5.81) (see Chapter 3), matrices C_1 and D_{12} in Equation 5.10.1 are chosen as

$$C_1 = \begin{bmatrix} C_{11}^T \\ 0 \end{bmatrix} \tag{5.10.41}$$

and

$$D_{12} = [0 \quad 0 \quad 1]^T \tag{5.10.42}$$

With the values of $q_1 = 10$, $q_2 = 1$, and $\rho = 10^{-9}$,

$$C_1 = 10^5 \begin{bmatrix} -1.0484 & 0.1038 & 0 & 0 \\ 0.0296 & 0.2987 & 0 & 0 \\ 0 & 0 & 0 & 0 \end{bmatrix} \tag{5.10.43}$$

A full state H_∞ controller gain vector to achieve $\left\| T_{ed}(s) \right\|_\infty < 30$ is

$$K = 10^5 \begin{bmatrix} -1.5809 & 0.4240 & -0.0437 & 0.0797 \end{bmatrix} \tag{5.10.44}$$

The singular values of $T_{ed}(s)$ are plotted for both H_∞ and H_2 control systems in Figure 5.10.2. Although the peak singular value for the H_∞ controller is lower, the bandwidth of $T_{ed}(s)$ is higher than that for the H_2 controller.

5.10.2 FULL STATE FEEDBACK H_∞ CONTROL UNDER DISTURBANCE FEEDFORWARD (FIGURE 5.10.3)

Assume that

$$M_{DF} = \begin{bmatrix} A & B_1 & B_2 \\ C_1 & 0 & D_{12} \\ C_2 & I & 0 \end{bmatrix} \tag{5.10.45}$$

The following assumptions are made:

1. (A, B_1) and (A, B_2) are stabilizable.
2. (C_1, A_1) is detectable.

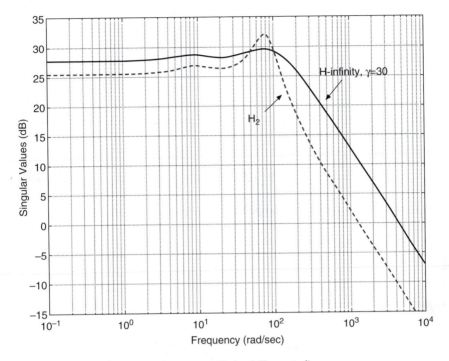

FIGURE 5.10.2 Active suspension system H_2 (and H_∞ control).

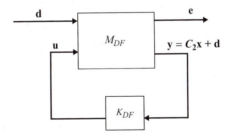

FIGURE 5.10.3 Disturbance feedforward in outputs.

3. $C_1^T D_{12} = 0$ and $D_{12}^T D_{12} = I$.
4. $(A - B_1 C_2)$ is stable.

The state dynamics is given by

$$\dot{\mathbf{x}} = A\mathbf{x} + B_1 \mathbf{d}(t) + B_2 \mathbf{u}(t) \qquad (5.10.46)$$

Consider the following controller:

$$\dot{\mathbf{x}} = A\hat{\mathbf{x}} + B_1(\mathbf{y}(t) - C_2\hat{\mathbf{x}}) + B_2 u(t) \tag{5.10.47}$$

and

$$\mathbf{u}(t) = K_c\hat{\mathbf{x}} \tag{5.10.48}$$

Substituting (5.10.48) into (5.10.47),

$$\dot{\hat{\mathbf{x}}} = (A - B_1C_2 + B_2K_c)\hat{\mathbf{x}} + B_1\mathbf{y} \tag{5.10.49}$$

Hence, the controller can be described in the state space form by

$$\mathbf{K}_{DF}^{ss} = \begin{bmatrix} A - B_1C_2 + B_2K_c & B_1 \\ K_c & 0 \end{bmatrix} \tag{5.10.50}$$

Subtracting (5.10.49) from (5.10.47),

$$\frac{d(\mathbf{x} - \hat{\mathbf{x}})}{dt} = (A - B_1C_2)(\mathbf{x} - \hat{\mathbf{x}}) \tag{5.10.51}$$

Because $(A - B_1C_2)$ has been assumed to be stable,

$$\hat{\mathbf{x}} \to \mathbf{x} \quad \text{as} \quad t \to \infty \tag{5.10.52}$$

Therefore, asymptotically,

$$\mathbf{u}(t) = K_c\mathbf{x} \tag{5.10.53}$$

Note that the definition of H_∞ norm is based on the frequency response of the transfer function matrix. The frequency response is the steady state response or the response of the system as $t \to \infty$. Hence, $\left\|T_{ed}(s)\right\|_{H_\infty}$ for M_{DF} with $\mathbf{u} = K_c\hat{\mathbf{x}}$ will be same as that for M with $\mathbf{u} = K_c\mathbf{x}$; (Figure 5.10.4). Therefore, to achieve

$$\left\|T_{ed}(s)\right\|_{H_\infty} < \gamma \tag{5.10.54}$$

K_c in Equation 5.10.48 should again be computed by Equation 5.10.20.

5.10.3 GUARANTEED H_∞ NORM VIA STATE ESTIMATION (FIGURE 5.10.5)

Assume that

$$M_{SE} = \begin{bmatrix} A & B_1 & B_2 \\ C_1 & 0 & I \\ C_2 & D_{21} & 0 \end{bmatrix} \tag{5.10.55}$$

The following assumptions are made:

1. (A, B_1) is stabilizable.
2. (C_1, A) and (C_2, A) are detectable.
3. $C_1^T D_{12} = 0$ and $D_{12}^T D_{12} = I$.
4. $(A - B_2 C_1)$ is stable.

The state dynamics is given by

$$\dot{\mathbf{x}} = A\mathbf{x} + B_1\mathbf{d}(t) + B_2\mathbf{u}(t) \tag{5.10.56}$$

$$\mathbf{y} = C_2\mathbf{x} + D_{21}\mathbf{d} \tag{5.10.57}$$

Note that M_{SE} and M_{DF} are dual to each other (Equation 5.10.37 and Equation 5.10.55). Hence, the controller, which will be dual to the full state feedback control (5.10.48), will be in the form of a state estimator. The dual of M_{SE} can be written as follows:

$$M_{SE}^{dual} = M_{SE}^T = \begin{bmatrix} A^T & C_1^T & C_2^T \\ B_1^T & 0 & D_{21}^T \\ B_2^T & I & 0 \end{bmatrix} \tag{5.10.58}$$

The form of M_{DF} is the same as that of M_{SE}^{dual}. Comparing (5.10.45) and (5.10.58),

$$A \rightarrow A^T , \quad B_1 \rightarrow C_1^T , \quad B_2 \rightarrow C_2^T \tag{5.10.59}$$

Therefore, from Equation 5.10.50,

$$K_{DF}^{ss,dual} = \begin{bmatrix} A^T - C_1^T B_2^T + C_2^T K_E^T & C_1^T \\ K_E^T & 0 \end{bmatrix} \tag{5.10.60}$$

where

$$K_E^T = -C_2 Q \tag{5.10.61}$$

and

$$QA^T + AQ + B_1 B_1^T - Q\left(C_2^T C_2 - \frac{1}{\gamma^2} C_1^T C_1\right)Q = 0 \tag{5.10.62}$$

Lastly, the controller K_{SE} for (5.10.55) to achieve $\left\|T_{ed}(s)\right\|_{H_\infty} < \gamma$ would be dual to (5.10.60):

$$K_{SE} = \begin{bmatrix} A - B_2 C_1 + K_E C_2 & K_E \\ C_1 & 0 \end{bmatrix} \tag{5.10.63}$$

Similar to conditions (5.10.22), the following conditions must be checked:

1. $Q \geq 0$. $\hspace{6cm}$ (5.10.64)

2. The matrix $\left(A^T + C_2^T K_E^T + C_1^T \dfrac{1}{\gamma^2} C_1 Q\right)$ or, equivalently,

$\left(A + K_E C_2 + \dfrac{1}{\gamma^2} Q C_1^T C_1\right)$ is stable. $\hspace{3cm}$ (5.10.65)

5.10.4 OUTPUT FEEDBACK H_∞ CONTROL

Assume that (Stein, 1988)

$$M_{OF} = \begin{bmatrix} A & B_1 & B_2 \\ C_1 & 0 & D_{12} \\ C_2 & D_{21} & 0 \end{bmatrix} \tag{5.10.66}$$

The following assumptions are made:

1. (A, B_1) and (A, B_2) are stabilizable. $\hspace{3cm}$ (5.10.67)

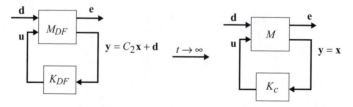

FIGURE 5.10.4 Asymptotic property of disturbance feedforward problem.

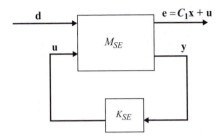

FIGURE 5.10.5 H_∞ state estimation.

2. (C_1, A) and (C_2, A) are detectable.   (5.10.68)

3. $C_1^T D_{12} = 0$ and $D_{12}^T D_{12} = I$.   (5.10.69)

4. $B_1 D_{21}^T = 0$ and $D_{21} D_{21}^T = I$.   (5.10.70)

Define the following set of new variables:

$$\mathbf{r} = \mathbf{d} - K_d \mathbf{x} \tag{5.10.71}$$

$$\mathbf{v} = \mathbf{u} - K_c \mathbf{x} \tag{5.10.72}$$

Based on (5.10.71) and (5.10.72), the system, Figure 5.10.6, yields the system shown in Figure 5.10.7.

Theorem 5.10.1

The system shown in Figure 5.10.6 is stable with $\left\|T_{ed}\right\|_{H_\infty} < \gamma$ if and only if the system shown in Figure 5.10.7 is stable with $\left\|T_{vr}\right\|_{H_\infty} < \gamma$ (Stein, 1988).

Lemma 5.10.1(Performance equivalence of systems shown in Figure 5.10.6 and Figure 5.10.7):

$$\left\|T_{vr}\right\|_{H_\infty} < \gamma \text{ implies } \left\|T_{ed}\right\|_{H_\infty} < \gamma, \text{ and vice versa.}$$

Proof

Choose the following Lyapunov function:

$$V = \mathbf{x}^T(t) P \mathbf{x}(t) \tag{5.10.73}$$

where P satisfies Equation 5.10.21:

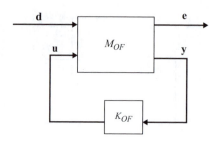

FIGURE 5.10.6 H_∞ output feedback.

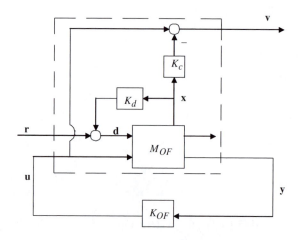

FIGURE 5.10.7 H_∞ output feedback with new variables.

$$PA + A^T P + C_1^T C_1 - P\left(B_2 B_2^T - \frac{1}{\gamma^2} B_1 B_1^T\right)P = 0 \qquad (5.10.74)$$

Differentiating (5.10.73) with respect to time,

$$\dot{V} = \dot{\mathbf{x}}^T P \mathbf{x} + \mathbf{x}^T P \dot{\mathbf{x}} \qquad (5.10.75)$$

State equations are

$$\dot{\mathbf{x}} = A\mathbf{x} + B_1 \mathbf{d} + B_2 \mathbf{u} \qquad (5.10.76)$$

Substituting (5.10.76) into (5.10.75),

$$\dot{V} = (A\mathbf{x} + B_1 \mathbf{d} + B_2 \mathbf{u})^T P \mathbf{x} + \mathbf{x}^T P(A\mathbf{x} + B_1 \mathbf{d} + B_2 \mathbf{u}) \qquad (5.10.77)$$

Rearranging (5.10.77),

$$\dot{V} = \mathbf{x}^T (A^T P + PA)\mathbf{x} + 2\mathbf{d}^T B_1^T P\mathbf{x} + 2\mathbf{u}^T B_2^T P\mathbf{x} \qquad (5.10.78)$$

Substituting (5.10.74) into (5.10.78),

$$\dot{V} = \mathbf{x}^T \left[C_1^T C_1 - P\left(B_2 B_2^T - \frac{1}{\gamma^2} B_1 B_1^T \right)P \right]\mathbf{x} + 2\mathbf{d}^T B_1^T P\mathbf{x} + 2\mathbf{u}^T B_2^T P\mathbf{x} \qquad (5.10.79)$$

Using (5.10.18) and (5.10.20),

$$\dot{V} = \mathbf{x}^T (-C_1^T C_1 + K_c^T K_c - \gamma^2 K_d^T K_d)\mathbf{x} + 2\gamma^2 \mathbf{d}^T K_d \mathbf{x} - 2\mathbf{u}^T K_c \mathbf{x} \qquad (5.10.80)$$

Equation 5.10.80 can be written as

$$\dot{V} = -(C_1\mathbf{x})^T C_1\mathbf{x} + (\mathbf{u} - K_c\mathbf{x})^T (\mathbf{u} - K_c\mathbf{x}) - \gamma^2 (\mathbf{d} - K_d\mathbf{x})^T (\mathbf{d} - K_d\mathbf{x}) - \mathbf{u}^T\mathbf{u} + \gamma^2 \mathbf{d}^T\mathbf{d}$$

$$(5.10.81)$$

Using (5.10.71) and (5.10.72),

$$\dot{V} = -(C_1\mathbf{x})^T C_1\mathbf{x} - \mathbf{u}^T\mathbf{u} + \gamma^2 \mathbf{d}^T\mathbf{d} + \mathbf{v}^T\mathbf{v} - \gamma^2 \mathbf{r}^T\mathbf{r} \qquad (5.10.82)$$

Now,

$$\mathbf{e} = C_1\mathbf{x} + D_{12}\mathbf{u} \qquad (5.10.83)$$

Because $C_1^T D_{12} = 0$ and $D_{12}^T D_{12} = I$, Equation 5.10.82 leads to

$$\dot{V} = -(\mathbf{e}^T\mathbf{e} - \gamma^2 \mathbf{d}^T\mathbf{d}) + (\mathbf{v}^T\mathbf{v} - \gamma^2 \mathbf{r}^T\mathbf{r}) \qquad (5.10.84)$$

Integrating both sides from 0 to ∞,

$$\int_0^\infty \frac{dV}{dt} dt = -[\|\mathbf{e}\|_{L_2}^2 - \gamma^2 \|\mathbf{d}\|_{L_2}^2] + [\|\mathbf{v}\|_{L_2}^2 - \gamma^2 \|\mathbf{r}\|_{L_2}^2] \qquad (5.10.85)$$

or

$$-[\|\mathbf{e}\|_{L_2}^2 - \gamma^2 \|\mathbf{d}\|_{L_2}^2] + [\|\mathbf{v}\|_{L_2}^2 - \gamma^2 \|\mathbf{r}\|_{L_2}^2] = V(\infty) - V(0) = \mathbf{x}^T(\infty)P\mathbf{x}(\infty) - \mathbf{x}^T(0)P\mathbf{x}(0)$$

$$(5.10.86)$$

Because $\mathbf{x} \in L_2$, $\mathbf{x}(\infty) = 0$. Therefore,

$$-[\|\mathbf{e}\|_{L_2}^2 - \gamma^2 \|\mathbf{d}\|_{L_2}^2] + [\|\mathbf{v}\|_{L_2}^2 - \gamma^2 \|\mathbf{r}\|_{L_2}^2] < 0 \qquad (5.10.87)$$

Therefore,

$$[\|\mathbf{e}\|_{L_2}^2 - \gamma^2 \|\mathbf{d}\|_{L_2}^2] < 0 \Leftrightarrow [\|\mathbf{v}\|_{L_2}^2 - \gamma^2 \|\mathbf{r}\|_{L_2}^2] < 0 \qquad (5.10.88)$$

In other words,

$$\|T_{ed}\|_{H_\infty} < \gamma \Leftrightarrow \|T_{vr}\|_{H_\infty} < \gamma \qquad (5.10.89)$$

Equivalence of Stability of Systems Shown in Figure 5.10.6 and Figure 5.10.7

The system, Figure 5.10.7, is not implementable because it requires states $\mathbf{x}$, which are not known. Therefore, the system shown in Figure 5.10.8 is constructed. Note that the system shown in Figure 5.10.8 can be represented as the system in Figure 5.10.9.

Lemma 5.10.2

Response of the system in Figure 5.10.8 or Figure 5.10.9 becomes that of the system shown in Figure 5.10.6 asymptotically. Therefore, if the system shown in Figure 5.10.8 or Figure 5.10.9 is stable, the system shown in Figure 5.10.6 is also stable and vice versa (Stein, 1988).

Proof

In Figure 5.10.8,

$$\mathbf{r} = \mathbf{d} - K_d\mathbf{x} \quad \text{and} \quad \mathbf{v} = \hat{\mathbf{u}} - K_c\hat{\mathbf{x}} \qquad (5.10.90)$$

Then, the dynamics of states $\mathbf{x}(t)$ can be written by using the definition 5.10.66 of M_{OF} in subsystem N:

$$\dot{\mathbf{x}} = A\mathbf{x} + B_1\mathbf{d} + B_2[(\hat{\mathbf{u}} - K_c\hat{\mathbf{x}}) + K_c\mathbf{x}] \qquad (5.10.91)$$

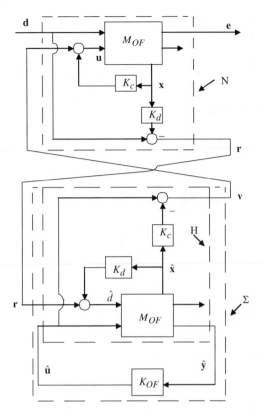

FIGURE 5.10.8 Graphical representation of an important aspect of the output feedback H_∞ control.

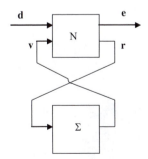

FIGURE 5.10.9 Schematic representation of Figure 5.10.8.

And the dynamics of estimated states $\hat{\mathbf{x}}$ can be written by using the definition 5.10.66 of M_{OF} in subsystem H:

$$\dot{\hat{\mathbf{x}}} = A\hat{\mathbf{x}} + B_1[(\mathbf{d} - K_d\mathbf{x}) + K_d\hat{\mathbf{x}}] + B_2\hat{\mathbf{u}} \qquad (5.10.92)$$

Subtracting (5.10.92) from (5.10.91),

$$\frac{d(\mathbf{x} - \hat{\mathbf{x}})}{dt} = (A + B_1K_d + B_2K_c)(\mathbf{x} - \hat{\mathbf{x}}) \qquad (5.10.93)$$

Because the matrix $(A + B_1K_d + B_2K_c)$ is stable,

$$\mathbf{x} - \hat{\mathbf{x}} \to 0 \quad \text{as} \quad t \to \infty \qquad (5.10.94)$$

In this case,

$$\hat{\mathbf{d}} = \mathbf{r} + K_d\hat{\mathbf{x}} \to \mathbf{d} = \mathbf{r} + K_d\mathbf{x} \quad \text{as} \quad t \to \infty \qquad (5.10.95)$$

$$\hat{\mathbf{y}} = C_2\hat{\mathbf{x}} + D_{21}\hat{\mathbf{d}} \to \mathbf{y} = C_2\mathbf{x} + D_{21}\mathbf{d} \quad \text{as} \quad t \to \infty \qquad (5.10.96)$$

$$\hat{\mathbf{u}}(t) \to \mathbf{u}(t) \quad \text{as} \quad t \to \infty \qquad (5.10.97)$$

and states of systems in Figure 5.10.6 and Figure 5.10.8 become identical asymptotically as they will be both governed by

$$\dot{\mathbf{x}} = A\mathbf{x} + B_1\mathbf{d} + B_2\mathbf{u} \qquad (5.10.98)$$

with an identical input $\mathbf{u}$ given by

$$\mathbf{u}(s) = K_{OF}(s)\mathbf{y}(s) \qquad (5.10.99)$$

Lemma 5.10.3

If Σ is stable with $\left\|T_{vr}\right\|_{H_\infty} = \left\|\Sigma\right\|_{H_\infty} < \gamma$, the system shown in Figure 5.10.8 or Figure 5.10.9 is stable.

Proof

From Figure 5.10.8,

$$\mathbf{r} = \mathbf{d} - K_d\mathbf{x} \quad \text{and} \quad \mathbf{v} = \mathbf{u} - K_c\mathbf{x} \qquad (5.10.100)$$

Using (5.10.66) and (5.10.100), the system N in Figure 5.10.8 can be described as

$$\begin{bmatrix} \mathbf{e} \\ \mathbf{r} \end{bmatrix} = N \begin{bmatrix} \mathbf{d} \\ \mathbf{v} \end{bmatrix} \tag{5.10.101}$$

where

$$N = \begin{bmatrix} A + B_2 K_c & B_1 & B_2 \\ C_1 & 0 & D_{12} \\ -K_d & I & 0 \end{bmatrix} = \begin{bmatrix} N_{11} & N_{12} \\ N_{21} & N_{22} \end{bmatrix} \tag{5.10.102}$$

Therefore,

$$\begin{bmatrix} \mathbf{e} \\ \gamma \mathbf{r} \end{bmatrix} = \begin{bmatrix} I & 0 \\ 0 & \gamma I \end{bmatrix} N \begin{bmatrix} \gamma^{-1} I & 0 \\ 0 & I \end{bmatrix} \begin{bmatrix} \gamma \mathbf{d} \\ \mathbf{v} \end{bmatrix} = \begin{bmatrix} \gamma^{-1} N_{11} & N_{12} \\ N_{21} & \gamma N_{22} \end{bmatrix} \begin{bmatrix} \gamma \mathbf{d} \\ \mathbf{v} \end{bmatrix} \tag{5.10.103}$$

From Equation 5.10.86,

$$\left\| \mathbf{e} \right\|_{L_2}^2 + \gamma^2 \left\| \mathbf{r} \right\|_{L_2}^2 = \gamma^2 \left\| \mathbf{d} \right\|_{L_2}^2] + \left\| \mathbf{v} \right\|_{L_2}^2 - \mathbf{x}^T(0) P \mathbf{x}(0) \tag{5.10.104}$$

Equation 5.10.104 can be written as

$$\left\| \begin{matrix} \mathbf{e} \\ \gamma \mathbf{r} \end{matrix} \right\|_{L_2} \leq \left\| \begin{matrix} \gamma \mathbf{d} \\ \mathbf{v} \end{matrix} \right\|_{L_2} \tag{5.10.105}$$

From (5.10.103),

$$\left\| \begin{bmatrix} \gamma^{-1} N_{11} & N_{12} \\ N_{21} & \gamma N_{22} \end{bmatrix} \right\|_{H_\infty} \leq 1 \tag{5.10.106}$$

Therefore,

$$\left\| \gamma N_{22} \right\|_{H_\infty} \leq 1 \Rightarrow \left\| N_{22} \right\|_{H_\infty} \leq \frac{1}{\gamma} \tag{5.10.107}$$

Using the lower LFT formulae,

$$T_{ed} = N_{11} - N_{12}(I - \Sigma N_{22})^{-1} N_{21} \qquad (5.10.108)$$

Furthermore,

$$\left\|\Sigma N_{22}\right\|_{H_\infty} \leq \left\|\Sigma\right\|_{H_\infty} \left\|N_{22}\right\|_{H_\infty} < \gamma\gamma^{-1} = 1 \qquad (5.10.109)$$

The system shown in Figure 5.10.9 is stable because $\det(I - \Sigma N_{22})$ cannot encircle the origin of the complex plane.

Lemma 5.10.4

If the system shown in Figure 5.10.6 is stable with $\left\|T_{ed}\right\|_{H_\infty} < \gamma$, Σ is stable with $\left\|T_{vr}\right\|_{H_\infty} = \left\|\Sigma\right\|_{H_\infty} < \gamma$ (Stein, 1988).

Proof

If the system shown in Figure 5.10.6 is stable with $\left\|T_{ed}\right\|_{H_\infty} < \gamma$, the system shown in Figure 5.10.7 is stable with $\left\|T_{ed}\right\|_{H_\infty} < \gamma$. In this case, Redheffer (1960) has shown that the system Σ is stable with $\left\|T_{vr}\right\|_{H_\infty} = \left\|\Sigma\right\|_{H_\infty} < \gamma$.

Controller Design

In Figure 5.10.8,

$$\hat{\mathbf{d}} = \mathbf{r} + K_d \hat{\mathbf{x}} \quad \text{and} \quad \mathbf{v} = \hat{\mathbf{u}} - K_c \hat{\mathbf{x}} \qquad (5.10.110)$$

Using (5.10.110), (5.10.66), and the assumption $B_1 D_{21}^T = 0$, the subsystem H in Σ can be described as

$$\begin{bmatrix} \dot{\hat{\mathbf{x}}} \\ \mathbf{v} \\ \hat{\mathbf{y}} \end{bmatrix} = H \begin{bmatrix} \hat{\mathbf{x}} \\ \mathbf{r} \\ \hat{\mathbf{u}} \end{bmatrix} \qquad (5.10.111)$$

where

$$H = \begin{bmatrix} A + B_1 K_d & B_1 & B_2 \\ -K_c & 0 & I \\ C_2 & D_{21} & 0 \end{bmatrix} \qquad (5.10.112)$$

The structure of the system (5.10.112) is identical to that of the system (5.10.55). Therefore, the controller that will stabilize Σ and guarantee $\left\|T_{vr}\right\|_{H_\infty} = \left\|\Sigma\right\|_{H_\infty} < \gamma$ is given by

$$K_{OF} = \begin{bmatrix} A + B_1 K_d + B_2 K_c + K_e C_2 & K_e \\ -K_c & 0 \end{bmatrix} \qquad (5.10.113)$$

where

$$K_e = -P_t C_2^T \qquad (5.10.114)$$

$$P_t (A + B_1 K_d)^T + (A + B_1 K_d) P_t + B_1 B_1^T - P_t (C_2^T C_2 - \gamma^{-2} K_c^T K_c) P_t = 0 \qquad (5.10.115)$$

Note that K_c and K_d are obtained by the solution of (5.10.21), the Riccati-like equation. While solving Equation 5.10.115, following conditions must be checked:

$$P_t \geq 0 \qquad (5.10.116)$$

and

$$A + B_1 K_d + K_e C_2 + \frac{1}{\gamma^2} P_t K_c^T K_c \quad \text{is stable.} \qquad (5.10.117)$$

Equation 5.10.115 can be transformed to have the structure of (5.10.21), the Riccati-like equation, with the following substitution:

$$P_t = (I - \gamma^{-2} P_e P_c)^{-1} P_e \qquad (5.10.118)$$

From (5.10.115) and (5.10.118),

$$(I - \gamma^{-2} P_e P_c)^{-1} P_e (A + B_1 K_d)^T + (A + B_1 K_d)(I - \gamma^{-2} P_e P_c)^{-1} P_e + B_1 B_1^T$$
$$-(I - \gamma^{-2} P_e P_c)^{-1} P_e (C_2^T C_2 - \gamma^{-2} K_c^T K_c)(I - \gamma^{-2} P_e P_c)^{-1} P_e = 0 \qquad (5.10.119)$$

Premultiplying (5.10.119) by $(I - \gamma^{-2} P_e P_c)$,

$$P_e (A + B_1 K_d)^T + (I - \gamma^{-2} P_e P_c)(A + B_1 K_d)(I - \gamma^{-2} P_e P_c)^{-1} P_e$$
$$+(I - \gamma^{-2} P_e P_c) B_1 B_1^T - P_e (C_2^T C_2 - \gamma^{-2} K_c^T K_c)(I - \gamma^{-2} P_e P_c)^{-1} P_e = 0 \qquad (5.10.120)$$

Postmultiplying (5.10.120) by $(I - \gamma^{-2} P_c P_e)$,

$$P_e(A + B_1 K_d)^T (I - \gamma^{-2} P_c P_e)$$

$$+(I - \gamma^{-2} P_e P_c)(A + B_1 K_d)(I - \gamma^{-2} P_e P_c)^{-1} P_e (I - \gamma^{-2} P_c P_e)$$

$$+(I - \gamma^{-2} P_e P_c) B_1 B_1^T (I - \gamma^{-2} P_c P_e)$$ (5.10.121)

$$-P_e(C_2^T C_2 - \gamma^{-2} K_c^T K_c)(I - \gamma^{-2} P_e P_c)^{-1} P_e (I - \gamma^{-2} P_c P_e) = 0$$

Lemma 5.10.4

$$(I - \gamma^{-2} P_e P_c)^{-1} P_e (I - \gamma^{-2} P_c P_e) = P_e$$

Proof

From the matrix inversion lemma,

$$(I - \gamma^{-2} P_e P_c)^{-1} = I + \gamma^{-2} P_e (I - \gamma^{-2} P_c P_e)^{-1} P_c$$

Therefore,

$$(I - \gamma^{-2} P_e P_c)^{-1} P_e (I - \gamma^{-2} P_c P_e) = P_e (I - \gamma^{-2} P_c P_e) +$$

$$P_e (I - \gamma^{-2} P_c P_e)^{-1} \gamma^{-2} P_c P_e (I - \gamma^{-2} P_c P_e)$$

Because $(I - A)^{-1} A = A(I - A)^{-1}$,

$$(I - \gamma^{-2} P_e P_c)^{-1} P_e (I - \gamma^{-2} P_e P_c) = P_e (I - \gamma^{-2} P_c P_e) +$$

$$P_e \gamma^{-2} P_c P_e (I - \gamma^{-2} P_c P_e)^{-1} (I - \gamma^{-2} P_c P_e)$$

$$(I - \gamma^{-2} P_e P_c)^{-1} P_e (I - \gamma^{-2} P_e P_c) = P_e - \gamma^{-2} P_e P_c P_e + \gamma^{-2} P_e P_c P_e = P_e$$

Applying Lemma 5.10.4 to Equation 5.10.121,

$$P_e(A^T + K_d^T B_1^T)(I - \gamma^{-2} P_c P_e) + (I - \gamma^{-2} P_e P_c)(A + B_1 K_d)P_e$$

$$+(I - \gamma^{-2} P_e P_c) B_1 B_1^T (I - \gamma^{-2} P_c P_e) - P_e(C_2^T C_2 - \gamma^{-2} K_c^T K_c)P_e = 0$$ (5.10.122)

Substituting $K_d = \gamma^{-2} B_1^T P_c$ and $K_c = -B_2^T P_c$ into (5.10.122),

$$(P_e A^T + AP_e + B_1 B_1^T - P_e C_2^T C_2 P_e)$$
$$-\gamma^{-2} P_e (A^T P_c + \gamma^{-2} P_c B_1 B_1^T P_c + P_c A - P_c B_2 B_2^T P_c) P_e = 0 \qquad (5.10.123)$$

Using (5.10.21),

$$P_e A^T + AP_e + B_1 B_1^T - P_e (C_2^T C_2 - \gamma^{-2} C_1^T C_1) P_e = 0 \qquad (5.10.124)$$

While solving (5.10.124), the following conditions must be checked:

$$P_e \geq 0 \qquad (5.10.125)$$

$$\rho(P_e P_c) < \gamma^2 \qquad (5.10.126)$$

$$A - P_e (C_2^T C_2 - \gamma^{-2} C_1^T C_1) \quad \text{is stable.} \qquad (5.10.127)$$

Summary

$$K_{OF} = \begin{bmatrix} A + B_1 K_d + B_2 K_c + K_e C_2 & K_e \\ -K_c & 0 \end{bmatrix} \qquad (5.10.128)$$

$$K_e = -(I - \gamma^{-2} P_e P_c)^{-1} P_e C_2^T \qquad (5.10.129)$$

$$K_d = \frac{1}{\gamma^2} B_1^T P_c \qquad (5.10.130)$$

$$K_c = -B_2^T P_c \qquad (5.10.131)$$

General Case

Now, assumptions (5.10.69) and (5.10.70) will be relaxed (Stein, 1988); i.e.,

1. $C_1^T D_{12} \neq 0$ and $D_{12}^T D_{12} = R \neq I$ $\qquad (5.10.132)$

2. $B_1 D_{21}^T \neq 0$ and $D_{21} D_{21}^T = \Theta \neq I$ $\qquad (5.10.133)$

Then, the output feedback H_∞ control is given by

$$K_{OF} = \begin{bmatrix} A + (B_1 - K_e D_{21})K_d + B_2 K_c + K_e C_2 & K_e \\ -K_c & 0 \end{bmatrix} \quad (5.10.134)$$

$$K_d = \frac{1}{\gamma^2} B_1^T P_c \quad (5.10.135)$$

$$K_c = -R^{-1}(B_2^T P_c + D_{12}^T C_1) \quad (5.10.136)$$

$$P_c(A - B_2 R^{-1} D_{12}^T C_1) + (A - B_2 R^{-1} D_{12}^T C_1)^T P_c + \tilde{C}_1^T \tilde{C}_1$$
$$- P_c \left(B_2 R^{-1} B_2^T - \frac{1}{\gamma^2} B_1 B_1^T \right) P_c = 0 \quad (5.10.137)$$

$$\tilde{C}_1 = (I - D_{12} R^{-1} D_{12}^T) C_1 \quad (5.10.138)$$

$$P_e(A - B_1 D_{21}^T \Theta^{-1} C_2)^T + (A - B_1 D_{21}^T \Theta^{-1} C_2) P_e + \tilde{B}_1 \tilde{B}_1^T$$
$$- P_e(C_2^T \Theta^{-1} C_2 - \gamma^{-2} C_1^T C_1) P_e = 0 \quad (5.10.139)$$

$$\tilde{B}_1 = B_1(I - D_{21}^T \Theta^{-1} D_{21}) \quad (5.10.140)$$

$$K_e = -(I - \gamma^{-2} P_e P_c)^{-1} P_e C_2^T \quad (5.10.141)$$

5.10.5 Poles or Zeros on Imaginary Axis

When a plant has poles or zeros on the imaginary axis, numerical algorithms to compute H_∞ optimal control fail (Safanov, 1986; Chiang and Safanov 1992a). This problem is solved by introducing a bilinear transform:

$$\bar{s} = \frac{s + a}{1 + bs} \quad (5.10.142)$$

where a and b are nonnegative real numbers. A special case of bilinear transform is obtained by setting $b = 0$,

$$\bar{s} = s + a \quad (5.10.143)$$

Therefore, the imaginary axis gets shifted to the left by a distance a, and poles or zeros, or both, are no longer on the imaginary axis of the $\bar{s}$ plane. The plant transfer function can be expressed as

$$C(sI - A)^{-1}B + D = C(\bar{s}I - \bar{A})^{-1}B + D \qquad (5.10.144)$$

where

$$\bar{A} = A + aI \qquad (5.10.145)$$

Hence, for the H_∞ controller design, the system matrix is chosen as $\bar{A}$. Other matrices B, C, and D are not affected. Then, the inverse bilinear transform is applied to the controller transfer function:

$$C_k(\bar{s}I - \bar{A}_k)^{-1}B_k + D_k = C_k(sI - A_k)^{-1}B_k + D_k \qquad (5.10.146)$$

where

$$A_k = \bar{A}_k - aI \qquad (5.10.147)$$

An important feature of this bilinear transform is that the following inequality is guaranteed:

$$\left\| T_{ed}(s) \right\|_\infty \leq \left\| \bar{T}_{ed}(\bar{s}) \right\|_\infty < \gamma \qquad (5.10.148)$$

where $\bar{T}_{ed}(\bar{s})$ is the transfer function whose H_∞ norm is made to be less than γ by the controller design in the $\bar{s}$ domain. The result (5.10.148) says that the H_∞ norm of the original transfer function $T_{ed}(s)$ is guaranteed to be less than γ.

5.11 LOOP SHAPING

5.11.1 TRADE-OFF BETWEEN PERFORMANCE AND ROBUSTNESS VIA H_∞ CONTROL

We require the sensitivity function $S(j\omega)$ to be small in the lower frequency range. With a suitable weight function $W_1(s)$, it can be specified as

$$\left| W_1(j\omega) \right| \bar{\sigma}(S(j\omega)) \leq 1 \quad \text{or} \quad \bar{\sigma}(S(j\omega)) \leq \left| W_1^{-1}(j\omega) \right| \qquad (5.11.1)$$

The sensitivity function $S(j\omega)$ is made small by having a large loop transfer matrix $L(j\omega)$. Assuming that $\underline{\sigma}(L(j\omega)) \gg 1$,

$$S(j\omega) = (I + L(j\omega))^{-1} \approx L^{-1}(j\omega) \tag{5.11.2}$$

From (5.11.1) and (5.11.2),

$$\underline{\sigma}(L(j\omega)) >> |W_1(j\omega)| \tag{5.11.3}$$

Similarly, we require the complementary sensitivity function $T(j\omega)$ to be small in the higher frequency range. With a suitable weight function $W_3(s)$, it can be specified as

$$|W_3(j\omega)| \bar{\sigma}(T(j\omega)) \le 1 \quad \text{or} \quad \bar{\sigma}(T(j\omega)) \le |W_3^{-1}(j\omega)| \tag{5.11.4}$$

The complementary sensitivity function $T(j\omega)$ is made small by having a small loop transfer matrix $L(j\omega)$. Assuming that $\bar{\sigma}(L(j\omega)) << 1$,

$$T(j\omega) = L(j\omega)(I + L(j\omega))^{-1} \approx L(j\omega) \tag{5.11.5}$$

From (5.11.4) and (5.11.5),

$$\bar{\sigma}(L(j\omega)) \le |W_3^{-1}(j\omega)| \tag{5.11.6}$$

Both conditions (5.11.3) and (5.11.6) are graphically shown in Figure 5.11.1. Also, conditions (5.11.1) and (5.11.4) can be satisfied by the following mixed sensitivity condition:

$$\left\| \begin{matrix} W_1(s)S(s) \\ W_3(s)T(s) \end{matrix} \right\|_\infty \le 1 \tag{5.11.7}$$

In Matlab (Chiang and Safanov, 1992b), a routine "augss" is available to set up this mixed sensitivity optimization problem. The "augss" creates the M matrix in Equation 5.10.66 on the basis of the system shown in Figure 5.11.2, which is equivalent to Figure 5.7.1, with $\mathbf{r}(s) = 0$, $\mathbf{n}(s) = 0$, and $\mathbf{d}_i(s) = 0$ and addition of weighting matrices $W_1(s)$, $W_2(s)$, and $W_3(s)$ as shown in Figure 5.11.3.

From (5.7.4),

$$\mathbf{y}(s) = S_o(s)\mathbf{d}(s) \tag{5.11. 8}$$

$$\mathbf{y}(s) - \mathbf{d}(s) = -T_o(s)\mathbf{d}(s) \tag{5.11.9}$$

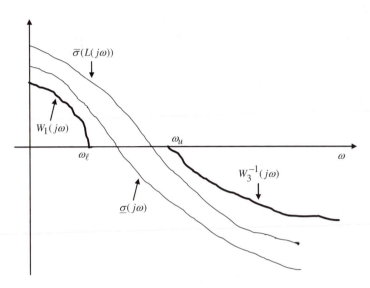

FIGURE 5.11.1 Concept of loop shaping (Chiang and Safanov, 1992b).

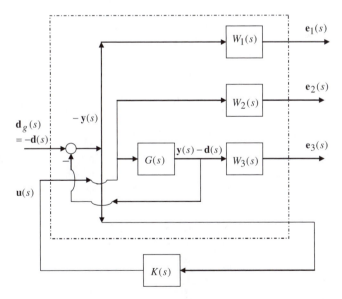

FIGURE 5.11.2 System for Matlab routine "augss" (Chiang and Safanov, 1992b).

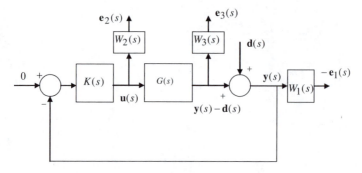

FIGURE 5.11.3 Alternate representation of Figure 5.11.2.

From (5.7.6),

$$\mathbf{u}(s) = -K_o(s)S_o(s)\mathbf{d}(s) \tag{5.11.10}$$

Therefore,

$$\mathbf{e}(s) = \begin{bmatrix} \mathbf{e}_1(s) \\ \mathbf{e}_2(s) \\ \mathbf{e}_3(s) \end{bmatrix} = \begin{bmatrix} -W_1(s)\mathbf{y}(s) \\ W_2(s)\mathbf{u}(s) \\ W_3(s)(\mathbf{y}(s)-\mathbf{d}(s)) \end{bmatrix} = T_{ed}(s)(-\mathbf{d}(s)) \tag{5.11.11}$$

where

$$T_{ed}(s) = \begin{bmatrix} W_1(s)S_o(s) \\ W_2(s)K_o(s)S_o(s) \\ W_3(s)T_o(s) \end{bmatrix} \tag{5.11.12}$$

EXAMPLE 5.11.1: THREE-DISK SYSTEM (*ECP MANUAL*)

Consider the three-disk system shown in Figure 5.11.4, in which J_1, J_2, and J_3 are mass-moments of inertia. The stiffnesses of torsional springs are k_1 and k_2. With the applied torque $T(t)$, the system of differential equations is

$$M\ddot{\boldsymbol{\theta}} + C\dot{\boldsymbol{\theta}} + K\boldsymbol{\theta} = B_f T(t) \tag{5.11.13}$$

where

$$M = \begin{bmatrix} J_1 & 0 & 0 \\ 0 & J_2 & 0 \\ 0 & 0 & J_3 \end{bmatrix}; \quad C = \begin{bmatrix} c_1 & 0 & 0 \\ 0 & c_2 & 0 \\ 0 & 0 & c_3 \end{bmatrix};$$

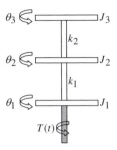

FIGURE 5.11.4 A three-disk system.

TABLE 5.11.1
System Parameters (ECP Manual)

$J_1 = 0.0024$ kg-m², $J_2 = J_3 = 0.0019$ kg-m²
$k_1 = k_2 = 2.8$ N-m/rad
$c_1 = 0.007$ N-m-sec/rad, $c_2 = c_3 = 0.001$ N-m-sec/rad
$k_{hw} = 100/6$ units

$$K = \begin{bmatrix} k_1 & -k_1 & 0 \\ -k_1 & k_1 + k_2 & -k_2 \\ 0 & -k_2 & k_2 \end{bmatrix} \qquad (5.11.14)$$

$$\boldsymbol{\theta} = \begin{bmatrix} \theta_1 \\ \theta_2 \\ \theta_3 \end{bmatrix} \quad \text{and} \quad B_f = \begin{bmatrix} 1 \\ 0 \\ 0 \end{bmatrix} \qquad (5.11.15)$$

The energy dissipation in the system is modeled by equivalent viscous damping coefficients c_1, c_2, and c_3 associated with disks J_1, J_2, and J_3, respectively. The system input $u(t)$ is related to torque $T(t)$ as

$$T(t) = k_{hw}u(t) \qquad (5.11.16)$$

where k_{hw} is the hardware gain. The system parameters are listed in Table 5.11.1.

Natural frequencies and mode shapes of the system are obtained by solving an eigenvalue problem with Matlab command: eig(K,M). Results are shown in Table 5.11.2.

Mode#1 is called a rigid body mode as all the disks move together. The second and third modes are vibratory in nature. A reduced-order model based on the rigid body mode can be developed by defining rigid body modal coordinate q_1 as the following approximation:

TABLE 5.11.2
Natural Frequencies and Mode Shapes

Mode#1: $\omega_1 = 0$; $\boldsymbol{\varphi}_1 = \begin{bmatrix} 12.7 & 12.7 & 12.7 \end{bmatrix}^T$

Mode#2: $\omega_2 = 36.22$ rad/sec; $\boldsymbol{\varphi}_2 = \begin{bmatrix} 14.2975 & -1.7835 & -16.2764 \end{bmatrix}^T$

Mode#3: $\omega_3 = 65.39$ rad/sec; $\boldsymbol{\varphi}_3 = \begin{bmatrix} 7.1385 & -19.0222 & 10.0051 \end{bmatrix}^T$

$$\boldsymbol{\theta}(t) = \boldsymbol{\varphi}_1 q_1 \tag{5.11.17}$$

Substituting (5.11.17) into (5.11.13) and premultiplying by $\boldsymbol{\varphi}_1^T$,

$$\boldsymbol{\varphi}_1^T M \boldsymbol{\varphi}_1 \ddot{q}_1 + \boldsymbol{\varphi}_1^T C \boldsymbol{\varphi}_1 \dot{q}_1 + \boldsymbol{\varphi}_1^T K \boldsymbol{\varphi}_1 q_1 = \boldsymbol{\varphi}_1^T B_f T(t) \tag{5.11.18}$$

Substituting parameter values listed in Table 1 and using (5.11.16),

$$\ddot{q}_1 + 1.4516 \dot{q}_1 = 211.667 u(t) \tag{5.11.19}$$

The transfer function corresponding to (5.11.19) is

$$\frac{q_1(s)}{u(s)} = G(s) = \frac{211.667}{s(s + 1.4516)} \tag{5.11.20}$$

Using Matlab routine hinf, a controller

$$u(s) = -K(s)q_1(s) \tag{5.11.21}$$

has been designed on the basis of the model (5.11.20) to achieve the following objective function:

$$\left\| \begin{matrix} W_1(s)S_o(s) \\ W_3(s)T_o(s) \end{matrix} \right\|_\infty \leq 1 \tag{5.11.22}$$

where

$$S_o(s) = (1 + G(s)K(s))^{-1} \tag{5.11.23}$$

$$T_o(s) = G(s)K(s)(1 + G(s)K(s))^{-1} \tag{5.11.24}$$

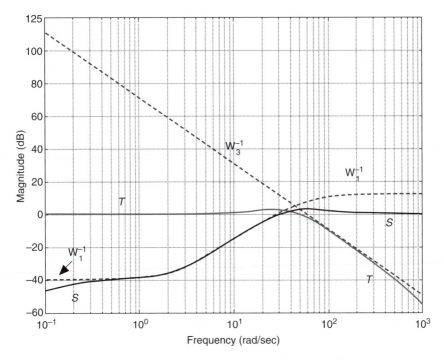

FIGURE 5.11.5 Results of loop shaping via H_∞ control of the three-disk system.

$$W_1^{-1}(s) = \frac{(0.02s + 1)^2}{0.01(0.4s + 1)^2} \tag{5.11.25}$$

$$W_3^{-1}(s) = \frac{3500}{s^2} \tag{5.11.26}$$

The Matlab program 5.11.1 is provided. In Figure 5.11.5, Bode magnitude plots of $W_1^{-1}(s)$, $W_3^{-1}(s)$, $S_o(s)$, and $T_o(s)$ are shown. As expected,

$$\left\|S(s)\right\|_\infty \le \left|W_1^{-1}(j\omega)\right| \quad \text{and} \quad \left\|T(s)\right\|_\infty \le \left|W_3^{-1}(j\omega)\right| \tag{5.11.27a,b}$$

Because of the choice $W_1^{-1}(s)$, the sensitivity function is greater than one beyond 35 rad/sec, which is less than the frequencies of unmodeled second and third modes.

MATLAB PROGRAM 5.11.1: LOOP SHAPING VIA H_∞ CONTROL

```
%1 Mode of Vibration
numg=211.6667;
deng=[1 1.4516 0];
```

```
%
numw1=[4e-4  4.e-2  1];
denw1=0.01*[0.16  0.8  1];
sysw1i=tf(denw1,numw1);
numw3=[1  0  0];
denw3=[0  0  3500];
sysw3i=tf(denw3,numw3)
[ag,bg,cg,dg]=tf2ss(numg,deng);
ag0=ag+0.1*eye(size(ag));
sysP=ss(ag,bg,cg,dg);
ssg=mksys(ag0,bg,cg,dg);
w1=[numw1;denw1];
w2=[];
w3=[numw3;denw3];
%
TSS=augtf(ssg,w1,w2,w3);
[ssf,sscl]=hinf(TSS);
%
[af,bf,cf,df]=branch(ssf);
af=af-0.1*eye(size(af));
[numf,denf]=ss2tf(af,bf,cf,df)
sysK=ss(af,bf,cf,df);
sysPK=series(sysP,sysK);
sys_Id=ss([],[],[],1);
sys_Sens=feedback(sys_Id,sysPK);
sys_cSens=feedback(sysPK,sys_Id);
wrang=logspace(-1,3);
bodemag(sysw1i,sysw3i,sys_Sens,sys_cSens,wrang)
legend('w_1^-^1',  'w_3^-^1','S','T')
```

5.11.2 FUNDAMENTAL CONSTRAINT: BODE'S SENSITIVITY INTEGRALS

SISO System

Let $L(s) = G(s)K(s)$ be the loop transfer function, where $G(s)$ and $K(s)$ are plant and controller transfer functions, respectively. Assuming that $L(s)$ has at least two more poles than zeros (Zhou et al., 1996),

$$\int_0^\infty \ell n \left| S(j\omega) \right| d\omega = \pi \sum_{i=1}^r p_i \geq 0 \qquad (5.11.28)$$

where $S(s)$ is the sensitivity function; i.e.,

$$S(s) = (1 + L(s))^{-1} \qquad (5.11.29)$$

and p_1, p_2, ..., p_r are r poles in the open right half plane; i.e., with real parts greater than zero. For a stable $L(s)$, Equation 5.11.28 yields

$$\int_0^\infty \ell n \left| S(j\omega) \right| d\omega = 0 \qquad (5.11.30)$$

Relationships (5.11.28) and (5.11.30) are known as *Bode's sensitivity integrals*, which represent fundamental constraints or laws that every controller must satisfy. Stein (2003) has described them as *conservation laws*: The integrated value of the log of the magnitude of the sensitivity function is conserved under the action of feedback. This integral is zero for a stable loop transfer function, whereas it is a constant positive number for an unstable loop transfer function.

For all physical systems,

$$L(j\omega) \approx 0 \quad \text{or} \quad \ell n \left| S(j\omega) \right| = 0 \quad \text{for} \quad \omega > \omega_b \qquad (5.11.31)$$

where ω_b is called the *available bandwidth*, which is dependent on the physical hardwares used in the control systems and is independent of the controller. As a result, Equation 5.11.28 can be written as

$$\int_0^{\omega_b} \ell n \left| S(j\omega) \right| d\omega \approx \pi \sum_{i=1}^r p_i \geq 0 \qquad (5.11.32)$$

Then, for a stable $L(s)$, Equation 5.11.32 reduces to

$$\int_{0}^{\omega_b} \ell n \left| S(j\omega) \right| d\omega \approx 0 \tag{5.11.33}$$

Typically, for good performance in the frequency range $\omega \le \omega_d$,

$$\left| S(j\omega) \right| \le 1 \quad \text{or} \quad \ell n \left| S(j\omega) \right| \le 0 ; \quad \omega \le \omega_d \tag{5.11.34}$$

and

$$\int_{0}^{\omega_d} \ell n \left| S(j\omega) \right| d\omega < 0 \tag{5.11.35}$$

Because of constraints (5.11.33),

$$\int_{\omega_d}^{\omega_b} \ell n \left| S(j\omega) \right| d\omega = -\int_{0}^{\omega_d} \ell n \left| S(j\omega) \right| d\omega > 0 ; \quad \omega_d < \omega_b \tag{5.11.36}$$

Therefore, a performance increase in the frequency range $0 \le \omega \le \omega_d$ must be accompanied by an equal performance decrease in the frequency range $\omega_d \le \omega \le \omega_b$ (Figure 5.11.6). Hence, if the integral in (5.11.35) is made more negative by a highly ambitious controller design, a price must be paid in the frequency range $\omega_d \le \omega \le \omega_b$.

For MIMO systems, a constraint similar to (5.11.28) exists in terms of the maximum singular value of the sensitivity function (Zhou et al., 1996).

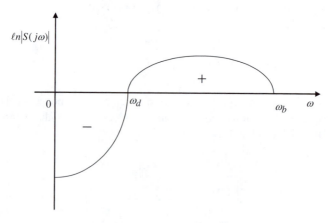

FIGURE 5.11.6 Graphical description of Bode's sensitivity integral.

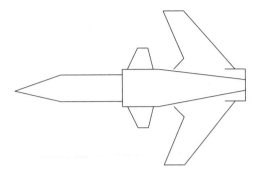

FIGURE 5.11.7 A schematic of X-29 aircraft.

EXAMPLE 5.11.2: X-29 AIRCRAFT

The X-29 aircraft (Stein, 1988; Stein, 2003) is the forward-swept-wing research plane (Figure 5.11.7). It is designed to have static or open-loop instability by locating the center pf pressure (the effective point of action of lift forces) of each wing ahead of its center of gravity. The stability of the flight is achieved by feedback control. The advantages of open-loop instability is reduced weight, maneuverability, and speeds of command response.

The open-loop system, which represents actuators, airframe, sensors, and digital controller, is described by

$$G(s) = \frac{20(s+3)(s-35)}{(s+10p/6)(s-p)(s+20)(s+35)} \; ; \quad p > 0 \qquad (5.11.37)$$

Then, controllers $K(s)$ are designed to minimize the following objective functions:

$$\left\| \begin{matrix} W_1(s)S_o(s) \\ W_2(s)K(s)S_o(s) \end{matrix} \right\|_{H_\infty} \qquad (5.11.38)$$

where

$$W_2(s) = 0.01 \qquad (5.11.39)$$

The Matlab routines "augtf" and "hinfopt" have been used (Matlab Program 5.11.2). Numerical results are generated for two different $W_1(s)$ denoted as

$$W_{1A}(s) = \frac{s+1}{(s+0.001)(1+0.001s)} \qquad (5.11.40)$$

and

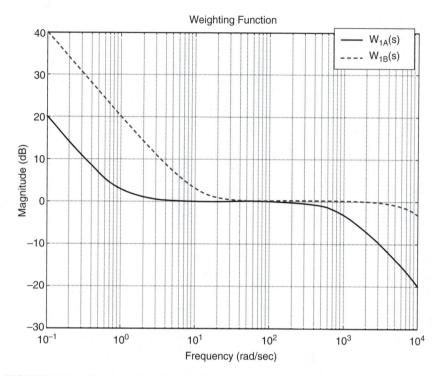

FIGURE 5.11.8 Weighting functions for H_∞ control of X-29 aircraft.

$$W_{1B}(s) = \frac{s+10}{(s+0.01)(1+0.0001s)} \tag{5.11.41}$$

Bode plots for $W_{1A}(s)$ and $W_{1B}(s)$ are shown in Figure 5.11.8. The sensitivity functions are plotted in Figure 5.11.9 for $p = 9$. The sensitivity function in the frequency range of 0.1 to 10 rad/sec is smaller for $W_{1B}(s)$ in comparison to that for $W_{1A}(s)$. However, the sensitivity function in the frequency range of 10 to 60 rad/sec is larger for $W_{1B}(s)$, which is expected from the Bode integral.

MATLAB PROGRAM 5.11.2: X-29 AIRCRAFT

```
%
clear all
close all
%
numw1=[0 1 1];
denw1=conv([1 0.001],[0.001 1]);
%
```

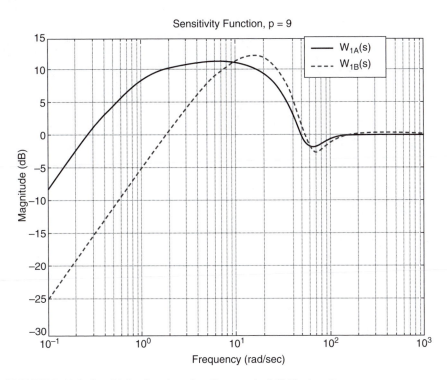

FIGURE 5.11.9 Sensitivity functions for H_∞ control of X-29 aircraft.

```
numw1=[0  1  10];
denw1=conv([1  0.01],[0.0001  1]);
%
numw2=0.01;
denw2=1;
%
w1=[numw1;denw1];
figure(1)
wrang=logspace(-3,3);
bodemag(tf(numw1,denw1),wrang);
grid
w2=[numw2;denw2];
w3=[];
%
```

```
numg=conv([20 60],[1 -35]);
qq=0
for icount=1:3
qq=qq+3;
pq=conv([1 qq*10/6],[1 -qq]);
pqr=conv(pq,[1 20]);
deng=conv(pqr,[1 35]);
[ag,bg,cg,dg]=tf2ss(numg,deng);
sysP=ss(ag,bg,cg,dg);
ssg=mksys(ag,bg,cg,dg);
%
TSS=augtf(ssg,w1,w2,w3);
%[gamopt,ssf,sscl]=hinfopt(TSS);
[ssf,sscl]=h2lqg(TSS);
%
[af,bf,cf,df]=branch(ssf);
[numf,denf]=ss2tf(af,bf,cf,df)
sysK=ss(af,bf,cf,df);
sysPK=series(sysP,sysK);
sys_Id=ss([],[],[],1);
if icount== 1
Sensfn_1B=feedback(sys_Id,sysPK);
Csensfn1B=feedback(sysPK,sys_Id);
end
if icount== 2
Sensfn_2B=feedback(sys_Id,sysPK);
Csensfn2B=feedback(sysPK,sys_Id);
end
if icount== 3
Sensfn_3B=feedback(sys_Id,sysPK);
Csensfn3B=feedback(sysPK,sys_Id);
```

```
end

end

wrang=logspace(-1,3);

figure(2)

bodemag(Sensfn_1B,Sensfn_2B,Sensfn_3B,wrang)

legend('p=3','p=6','p=9')

grid
```

5.12 CONTROLLER BASED ON μ ANALYSIS

The uncertainty matrix, Δ, which represents parametric errors, has a block diagonal structure. It will be shown in Section 5.12.3 that the solution of the robust performance control problem leads to an uncertainty matrix, Δ, which also has a block diagonal structure. But the uncertainty matrix, Δ, in the small gain theorem is a general matrix, and includes the cases of non-block-diagonal structure. As a result, application of the small gain theorem can yield a highly conservative bound of the norm of Δ for robustness. To reduce the conservativeness of the result, structured singular values (μ) are defined (Doyle et al., 2000).

5.12.1 DEFINITIONS AND PROPERTIES OF STRUCTURED SINGULAR VALUES (μ)

Structured singular value $\mu_\Delta(M)$ of a complex $n \times n$ matrix M is defined with respect to the structure $\mathbf{\Delta}$ of a block diagonal complex $n \times n$ matrix. There can be two types of blocks: repeated scalar and full. In general, the structure of a block diagonal complex $n \times n$ matrix can be defined (Zhou et al., 1996) as follows:

$$\mathbf{\Delta} = diag[\delta_1 I_{r_1} \quad . \quad . \quad \delta_S I_{r_S} \quad \Delta_1 \quad . \quad . \quad \Delta_F] \qquad (5.12.1)$$

where δ_i is a complex number and Δ_i is a complex $m_i \times m_i$ matrix. Δ_i is called the ith full block. Note that δ_i is repeated r_i times in the block $\delta_i I_{r_i}$. Therefore, $\delta_i I_{r_i}$ is called a *repeated block*. There are S repeated and F full blocks. Dimensions of these repeated and full blocks must add up to n; i.e.,

$$\sum_{i=1}^{S} r_i + \sum_{j=1}^{F} m_j = n \qquad (5.12.2)$$

Definition 5.12.1

The structured singular value $\mu_\Delta(M)$ of a complex $n \times n$ matrix M is defined as follows:

$$\mu_\Delta(M) = \begin{cases} 0 & if \quad \det(I - M\Delta) \ne 0 \quad for \quad any \quad \Delta \in \mathbf{\Delta} \\ \dfrac{1}{\min\{\overline{\sigma}(\Delta): \quad \Delta \in \mathbf{\Delta}, \quad \det(I - M\Delta) = 0\}} \end{cases} \quad (5.12.3)$$

The following are examples:

1. Let

$$M = \begin{bmatrix} 0 & 0 \\ 1 & 0 \end{bmatrix} \quad and \quad \Delta = \begin{bmatrix} \delta_1 & 0 \\ 0 & \delta_2 \end{bmatrix} \quad (5.12.4)$$

Then,

$$I - M\Delta = \begin{bmatrix} 1 & 0 \\ \delta_1 & 1 \end{bmatrix} \quad and \quad \det(I - M\Delta) = 1 \quad (5.12.5)$$

Because $\det(I - M\Delta) \ne 0$,

$$\mu_\Delta(M) = 0 \quad (5.12.6)$$

2. Let

$$M = \begin{bmatrix} 2 & 1 \\ 2 & 1 \end{bmatrix} \quad and \quad \Delta = \begin{bmatrix} \delta_1 & 0 \\ 0 & \delta_2 \end{bmatrix} \quad (5.12.7)$$

Then,

$$\det(I - M\Delta) = 1 + (\delta_2 - 2\delta_1) \quad (5.12.8)$$

Hence, $\det(I - M\Delta) = 0$ when

$$2\delta_1 - \delta_2 = 1 \quad (5.12.9)$$

The minimum value of $\overline{\sigma}(\Delta)$ equals $1/3$ when $\delta_1 = 1/3$ and $\delta_2 = -1/3$. Therefore,

$$\mu_\Delta(M) = 3 \quad (5.12.10)$$

Properties of μ are as follows:

1. If $\Delta = \delta I_n$,

$$\mu_\Delta(M) = \rho(M) \tag{5.12.11}$$

where $\rho(M)$ is the maximum modulus of eigenvalues of M. Note that $\rho(M)$ is called the *spectral radius* of M, and $S = 1$, $F = 0$, and $r_1 = n$.

2. If Δ is a full matrix,

$$\mu_\Delta(M) = \bar{\sigma}(M) \tag{5.12.12}$$

Note that $S = 0$, $F = 1$, and $r_1 = 0$ in this case.

3. In general, $\rho(M) \le \mu_\Delta(M) \le \bar{\sigma}(M)$ $\tag{5.12.13}$

4. $\displaystyle \max_{U \in \mathbf{U}} \rho(UM) = \mu_\Delta(M) \le \inf_{D \in \mathbf{D}} \bar{\sigma}(DMD^{-1})$ $\tag{5.12.14}$

    where

$$\mathbf{U} = \left\{ U \in \Delta: \quad UU^* = I_n \right\} \tag{5.12.15}$$

$$D = \left\{ diag[D_1 \quad . \quad . \quad D_s \quad d_1 I_{m_1} \quad . \quad . \quad d_{F-1} I_{m_{F-1}} \quad I_{m_F} \right\} \tag{5.12.16}$$

5. $\displaystyle \mu_\Delta(M) = \inf_{D \in \mathbf{D}} \bar{\sigma}(DMD^{-1})$ if $2S + F \le 3$ $\tag{5.12.17}$

Computation of μ: The value of μ can be computed by numerically finding U which maximizes $\rho(UM)$. However, it is difficult to find the global maximum as there can be many local maxima.

The determination of D to minimize $\bar{\sigma}(DMD^{-1})$ can be done numerically. However, this minimum value equals μ only when $2S + F \le 3$. For $2S + F > 3$, it will be an upper bound.

5.12.2 ROBUSTNESS ANALYSIS VIA μ

Basic theorem: Let $M(s)$ and $\Delta(s)$ be stable MIMO $p_1 \times q_1$ and $q_1 \times p_1$ transfer functions. Also, let $\Delta(s)$ be a block diagonal matrix having the structure of Δ defined by (5.12.1) and satisfying

$$\left\| \Delta(s) \right\|_\infty < \beta^{-1}; \quad \beta > 0 \tag{5.12.18}$$

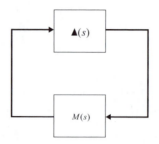

FIGURE 5.12.1 M–Δ system.

The loop shown in Figure 5.12.1 is well-posed and internally stable (Zhou et al., 1996) if and only if

$$\sup_{\omega} \mu_{\Delta}(M(j\omega)) \leq \beta \tag{5.12.19}$$

Loop theorem: Let

$$M = \begin{bmatrix} M_{11} & M_{12} \\ M_{21} & M_{22} \end{bmatrix} \tag{5.12.20}$$

Let Δ_1 and Δ_2 be block structures corresponding to diagonal blocks M_{11} and M_{22}, respectively. Then, the block diagonal structure Δ is defined as

$$\Delta = \begin{bmatrix} \Delta_1 & 0 \\ 0 & \Delta_2 \end{bmatrix}; \quad \Delta_1 \in \Delta_1 \quad \text{and} \quad \Delta_2 \in \Delta_2 \tag{5.12.21}$$

Then, the following results hold:

$$\mu_{\Delta}(M) \leq 1 \iff \mu_{\Delta_2}(M_{22}) \leq 1 \quad \text{and} \quad \sup_{\Delta_2} \mu_{\Delta_1}(F_{\ell}(M, \Delta_2)) \leq 1 \tag{5.12.22}$$

where supremum is over all Δ_2 satisfying $\overline{\sigma}(\Delta_2) \leq 1$ and $F_{\ell}(M, \Delta_2)$ is the lower LFT defined as follows:

$$F_{\ell}(M, \Delta_2) = M_{11} + M_{12}\Delta_2(I - M_{22}\Delta_2)^{-1}M_{21} \tag{5.12.23}$$

5.12.3 ROBUST PERFORMANCE VIA μ ANALYSIS

Consider the system shown in Figure 5.12.2. The nominal model of the uncertain plant P_{Δ} is $P(s)$. Furthermore,

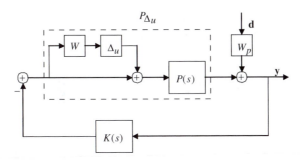

FIGURE 5.12.2 Robust performance problem.

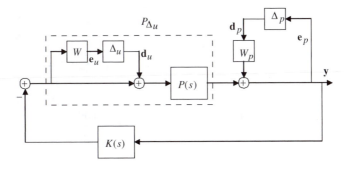

FIGURE 5.12.3 Equivalent robust stability problem.

$$P_{\Delta_u} = P(s)(I + \Delta_u(s)W(s)) \qquad (5.12.24)$$

Also,

$$\mathbf{y}(s) = (I + P_{\Delta_u}K)^{-1}W_p\mathbf{d}(s) \qquad (5.12.25)$$

Hence, the robust performance problem is defined as follows. Find a controller $K(s)$ such that

$$\left\| (I + P_{\Delta_u}K)^{-1}W_p \right\|_\infty \leq 1 \quad \forall \quad \Delta_u(s) \qquad (5.12.26)$$

The robust performance problem can be converted into the robust stability problem (Dahleh, 1992) by introducing a fictitious perturbation $\Delta_p(s)$, Figure 5.12.3, such that

$$\left\| \Delta_p \right\|_\infty < 1 \qquad (5.12.27)$$

Breaking loops at $\mathbf{d}_p$ and $\mathbf{e}_p$,

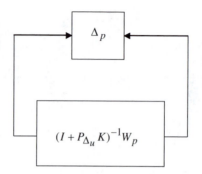

FIGURE 5.12.4 Equivalent to Figure 5.12.3.

$$\mathbf{e}_p(s) = (I + P_{\Delta_u}K)^{-1}W_p\mathbf{d}_p(s) \tag{5.12.28}$$

From the small gain theorem, the system shown in Figure 5.12.3 is stable for all Δ_p if and only if the condition (5.11.26) is satisfied. Hence, the robust performance criterion is equivalent to the robust stability of the system shown in Figure 5.12.4.

To solve this problem, breaking loops at $\mathbf{d}_u$, $\mathbf{e}_u$, $\mathbf{d}_p$, and $\mathbf{e}_p$ in Figure 5.12.3,

$$\begin{bmatrix} \mathbf{e}_p \\ \mathbf{e}_u \end{bmatrix} = M \begin{bmatrix} \mathbf{d}_p \\ \mathbf{d}_u \end{bmatrix} \quad \text{and} \quad \begin{bmatrix} \mathbf{d}_p \\ \mathbf{d}_u \end{bmatrix} = \Delta \begin{bmatrix} \mathbf{e}_p \\ \mathbf{e}_u \end{bmatrix} \tag{5.12.29a,b}$$

where

$$M = \begin{bmatrix} (I + PK)^{-1}W_p & (I + PK)^{-1}P \\ -WK(I + PK)^{-1}W_p & -WK(I + PK)^{-1}P \end{bmatrix} \tag{5.12.30}$$

and

$$\Delta = \begin{bmatrix} \Delta_p & 0 \\ 0 & \Delta_u \end{bmatrix} \tag{5.12.31}$$

Comparing Figure 5.12.4 and Figure 5.12.5,

$$F_\ell(M, \Delta_u) = (I + P_{\Delta_u}K)^{-1}W \tag{5.12.32}$$

where $F_\ell(M, \Delta_u)$ is the lower LFT defined as follows:

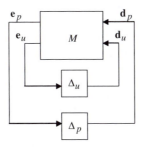

FIGURE 5.12.5 Equivalent to Figure 5.12.4.

$$F_\ell(M, \Delta_u) = M_{11} + M_{12}\Delta_u(I - M_{22}\Delta_u)^{-1}M_{21} \qquad (5.12.33)$$

Applying loop theorem (5.12.22), the robust performance is achieved when

$$\mu_\Delta(M) \leq 1 \qquad (5.12.34)$$

For the structure of Δ, S = 0 and F = 2. Therefore, $2S + F = 2$, and

$$\mu_\Delta(M) = \frac{\inf}{D \in \mathbf{D}} \bar{\sigma}(DMD^{-1}) \qquad (5.12.35)$$

where

$$D = diag[d_1 I_{m_1} \quad I_{m_2}] \qquad (5.12.36)$$

where Δ_p and Δ_u are $m_1 \times m_1$ and $m_2 \times m_2$ matrices, respectively.

EXAMPLE 5.12.1: APPLICATION TO A SISO SYSTEM

In this case,

$$D = diag[d_1 \quad 1] \qquad (5.12.37)$$

and

$$M_D = DMD^{-1} = \begin{bmatrix} M_{11} & d_1 M_{12} \\ \dfrac{1}{d_1} M_{21} & M_{22} \end{bmatrix} \qquad (5.12.38)$$

Now,

$$\bar{\sigma}(M_D) = \bar{\sigma}\begin{bmatrix} M_{11} & d_1 M_{12} \\ \dfrac{1}{d_1} M_{21} & M_{22} \end{bmatrix} \le \bar{\sigma}(N) \qquad (5.12.39)$$

where

$$N = \begin{bmatrix} |M_{11}| & |M_{12}|\alpha \\ \dfrac{1}{\alpha} M_{21} & |M_{22}| \end{bmatrix} \qquad (5.12.40)$$

$$\alpha = |d_1| \qquad (5.12.41)$$

To get the singular values of N,

$$\det(\lambda I - N^T N) = \lambda^2 - (p_1 + p_2)\lambda + (p_1 p_2 - q^2) \qquad (5.12.42)$$

where

$$p_1 = |M_{11}|^2 + \frac{1}{\alpha^2}|M_{21}|^2 \qquad (5.12.43)$$

$$p_2 = \alpha^2 |M_{12}|^2 + |M_{22}|^2 \qquad (5.12.44)$$

$$q = \alpha|M_{11}||M_{12}| + \frac{1}{\alpha}|M_{21}||M_{22}| \qquad (5.12.45)$$

If λ_1 and λ_2 are eigenvalues of $N^T N$,

$$\lambda_1 + \lambda_2 = p_1 + p_2 \qquad (5.12.46)$$

Therefore,

$$\bar{\sigma}(N) \le \sqrt{p_1 + p_2} \qquad (5.12.47)$$

and

$$\inf_{\alpha} \bar{\sigma}(N) \le \sqrt{\inf_{\alpha} (p_1 + p_2)} \tag{5.12.48}$$

To minimize $(p_1 + p_2)$,

$$\frac{d(p_1 + p_2)}{d\alpha} = 2\alpha |M_{12}|^2 - \frac{2}{\alpha^3} |M_{21}|^2 = 0 \tag{5.12.49}$$

or

$$\alpha^2 = \frac{|M_{21}|}{|M_{12}|} \tag{5.12.50}$$

For this solution,

$$\frac{d^2 (p_1 + p_2)}{d\alpha^2} = 2 |M_{12}|^2 + \frac{6}{\alpha^4} |M_{21}|^2 > 0 \tag{5.12.51}$$

$$\inf_{\alpha} (p_1 + p_2) = |M_{11}|^2 + |M_{22}|^2 + 2 |M_{12}||M_{21}| \tag{5.12.52}$$

From (5.12.30),

$$|M_{12}||M_{21}| = |M_{11}||M_{22}| \tag{5.12.53}$$

Substituting (5.12.53) into (5.12.52),

$$\inf_{\alpha} (p_1 + p_2) = (|M_{11}| + |M_{22}|)^2 \tag{5.12.54}$$

From (5.12.48),

$$\inf_{\alpha} \bar{\sigma}(N) \le |M_{11}| + |M_{22}| \tag{5.12.55}$$

From (5.12.35), (5.12.39), and (5.12.55),

$$\mu_\Delta(M) \le |M_{11}| + |M_{22}| \tag{5.12.56}$$

To satisfy (5.12.34),

$$|M_{11}| + |M_{22}| \le 1 \tag{5.12.57}$$

The condition (5.12.57) yields

$$\left|\frac{WL}{1+L}(j\omega)\right| + \left|\frac{W_p}{1+L}(j\omega)\right| \le 1 \quad \forall \omega \tag{5.12.58}$$

where

$$L(s) = P(s)K(s) \tag{5.12.59}$$

Graphical Interpretation (Dahleh, 1992)

The condition (5.12.58) will be satisfied if

$$\frac{|WL|}{|1+L|} + \frac{|W_p|}{|1+L|} \le 1 \tag{5.12.60}$$

or

$$|W(j\omega)L(j\omega)| + |W_p(j\omega)| \le |1 + L(j\omega)| \quad \forall \omega \tag{5.12.61}$$

This condition implies that the circle of radius $|W_p(j\omega)|$ with center at $(-1,0)$ must not intersect the circle of radius $|WL(j\omega)|$ with center at $|L(j\omega)|$ (Figure 5.12.6).

EXAMPLE 5.12.2: BENCHMARK ROBUST CONTROL PROBLEM (WIE AND BERNSTEIN, 1992)

Consider the benchmark problem shown in Figure 5.12.7 with uncertain parameters (m_1, m_2, and k), the disturbance $d(t)$, and the sensor noise $v(t)$. Uncertain parameters are as follows (Braatz and Morari, 1992):

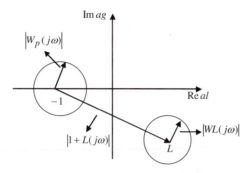

FIGURE 5.12.6 Graphical interpretation of robust performance criterion for a SISO system.

$$y(t) = x_2(t) + v(t)$$

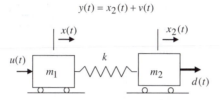

FIGURE 5.12.7 Benchmark robust control problem.

$$k = \bar{k} + w_k \delta_k \; ; \quad m_1 = \bar{m}_1 + w_1 \delta_1 \; ; \quad \text{and} \quad m_2 = \bar{m}_2 + w_2 \delta_2 \quad (5.12.62)$$

where $\bar{k}$, $\bar{m}_1$, and $\bar{m}_2$ are nominal values; w_k, w_1, and w_2 are weights to scale the corresponding uncertainties δ_k, δ_1, and δ_2, respectively. Additional weighting functions are defined for the sensor noise $v(t)$, the disturbance $d(t)$, the control input $u(t)$, and the performance variable $z(t)$ as follows:

$$v = w_v v' \; ; \quad d = w_d d' \; ; \quad u' = w_u u \; ; \quad \text{and} \quad z' = w_z z \quad (5.12.63)$$

Breaking loops at (x) in Figure 5.12.8, Equation 5.12.65 is obtained with the following relationships:

$$q_2 = \delta_1 q_1 \; ; \quad q_4 = \delta_k q_3 \; ; \quad q_6 = \delta_2 q_2 \; ;$$
$$v' = \delta_v z' \; ; \quad \text{and} \quad d' = \delta_d u' \quad (5.12.64)$$

Describing the system (5.12.65) by M, Equation 5.12.63 and Equation 5.12.64 are pictorially represented by Figure 5.12.9 where $K(s)$ is the controller transfer function. The numerical values of parameters and weights are provided in Table 5.12.1.

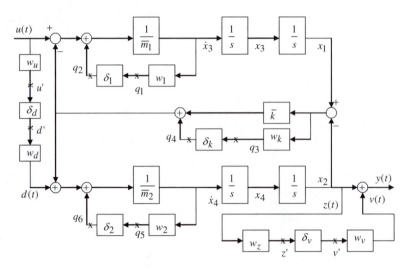

FIGURE 5.12.8 Simulation diagram of a two-mass system with uncertain parameters.

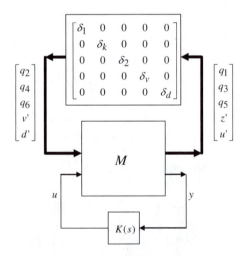

FIGURE 5.12.9 M–Δ–K representation of Figure 5.12.8.

TABLE 5.12.1
Numerical Values

$\bar{k} = 1$, $\bar{m}_1 = 1$, and $\bar{m}_2 = 1$

$w_1 = 0.2$, $w_k = 0.2$, and $w_2 = 0.2$

$w_v = 0.01$, $w_d = 1$, $w_u = 1$, and $w_z = 0.12$

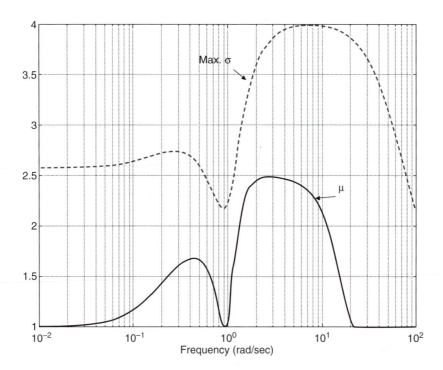

FIGURE 5.12.10 Maximum singular and μ values.

$$
\begin{bmatrix}
\dot{x}_1 \\
\dot{x}_2 \\
\dot{x}_3 \\
\dot{x}_4 \\
- \\
\overline{m}_1 q_1 \\
q_3 \\
\overline{m}_2 q_5 \\
- \\
z' \\
u' \\
- \\
y
\end{bmatrix}
=
\left[
\begin{array}{cccc|ccc|cc|c}
0 & 0 & 1 & 0 & 0 & 0 & 0 & 0 & 0 & 0 \\
0 & 0 & 0 & 1 & 0 & 0 & 0 & 0 & 0 & 0 \\
-\overline{k}/\overline{m}_1 & \overline{k}/\overline{m}_1 & 0 & 0 & 1/\overline{m}_1 & -1/\overline{m}_1 & 0 & 0 & 0 & 1/\overline{m}_1 \\
\overline{k}/\overline{m}_2 & -\overline{k}/\overline{m}_2 & 0 & 0 & 0 & 1/\overline{m}_2 & 1/\overline{m}_2 & 0 & w_d/\overline{m}_2 & 0 \\
\hline
-w_1\overline{k} & w_1\overline{k} & 0 & 0 & w_1 & -w_1 & 0 & 0 & 0 & w_1 \\
w_k & -w_k & 0 & 0 & 0 & 1 & 0 & 0 & 0 & 0 \\
w_2\overline{k} & -w_2\overline{k} & 0 & 0 & 0 & w_2 & w_2 & 0 & w_2 w_d & 0 \\
\hline
0 & w_z & 0 & 0 & 0 & 0 & 0 & 0 & 0 & 0 \\
0 & 0 & 0 & 0 & 0 & 0 & 0 & 0 & 0 & w_u \\
\hline
0 & 1 & 0 & 0 & 0 & 0 & 0 & w_v & 0 & 0
\end{array}
\right]
\begin{bmatrix}
x_1 \\
x_2 \\
x_3 \\
x_4 \\
- \\
q_2 \\
q_4 \\
q_6 \\
- \\
v' \\
d' \\
- \\
u
\end{bmatrix}
$$

$$(5.12.65)$$

Using Matlab routine "hinfopt," a controller $K(s)$ is obtained. Matlab Program 5.12.1 is provided. It should be noted that poles have been shifted to avoid having them on the imaginary axis. Resulting maximum singular and μ values are plotted in Figure 5.12.10. The small gain theorem indicates that the closed-loop system is stable for all uncertainties satisfying

$$\left\|\Delta\right\|_{\infty} < \frac{1}{3.988} = 0.2508 \tag{5.12.66}$$

where

$$\Delta = \begin{bmatrix} \delta_1 & 0 & 0 & 0 & 0 \\ 0 & \delta_k & 0 & 0 & 0 \\ 0 & 0 & \delta_2 & 0 & 0 \\ 0 & 0 & 0 & \delta_v & 0 \\ 0 & 0 & 0 & 0 & \delta_d \end{bmatrix} \tag{5.12.67}$$

and the μ analysis suggests that

$$\left\|\Delta\right\|_{\infty} < \frac{1}{2.4849} = 0.4024 \tag{5.12.68}$$

As expected, the μ analysis is less conservative.

MATLAB PROGRAM 5.12.1: μ ANALYSIS

```
%
k=1;
m1=1;
m2=1;
%
wd=1;
w1=0.2;
w2=0.2;
wk=0.2;
wu=1;
wv=1/100;
wz=0.12;
%
M=[0 0 1 0 0 0 0 0 0 0;0 0 0 1 0 0 0 0 0 0;
   -k/m1 k/m1 0 0 1/m1 -1/m1 0 0 0 1/m1;
   k/m2 -k/m2 0 0 0 1/m2 1/m2 0 wd/m2 0;
```

```
    -w1*k  w1*k  0  0  w1  -w1  0  0  0  w1;

    wk  -wk  0  0  0  1  0  0  0  0;w2*k  -w2*k  0  0  0  w2  w2  0
        w2*wd  0;

    0  wz  0  0  0  0  0  0  0  0;0  0  0  0  0  0  0  0  0  0  wu;0  1
0  0  0  0  0  wv  0  0];
%
A=M(1:4,1:4);
B1=M(1:4,5:9);
B2=M(1:4,10);
C1=M(5:9,1:4);
C2=M(10,1:4);
D11=M(5:9,5:9);
D12=M(5:9,10);
D21=M(10,5:9);
D22=M(10,10);
blksz=[1  1;  1  1;  1  1;1  1;1  1];
%
w=logspace(-2,2,3000);
%
Ag=A+0.1*eye(4);
[gamopt,acpg,bcp,ccp,dcp,acl,bcl,ccl,dcl]=hinfopt
    (Ag,B1,B2,C1,C2,D11,D12,D21,D22);
acp=acpg-0.1*eye(size(acpg))
%
%Merge M and K
%
[aclr,bclr,cclr,dclr]=lftf(A,B1,B2,C1,C2,D11,D12,
    D21,D22,acp,bcp,ccp,dcp);
SV=sigma(aclr,bclr,cclr,dclr,w);
[muf_f,logdf]=ssv(aclr,bclr,cclr,dclr,w,blksz);
semilogx(w,muf_f,w,max(SV),'--')
```

```
xlabel('Frequency (rad./sec.)','Fontsize',12)
grid
```

5.12.4 μ Synthesis: *D–K* Iteration

Consider the case of structured Δ without any scalar block; i.e., $S = 0$. In this case,

$$D(s) = diag[d_1(s)I_{m_1} \quad . \quad . \quad d_{F-1}(s)I_{m_{F-1}} \quad I_{m_F}] \qquad (5.12.69)$$

Therefore,

$$D(s)\Delta(s)D^{-1}(s) = \Delta(s) \qquad (5.12.70)$$

Because

$$\mu_\Delta(M) \le \inf_D \bar{\sigma}(DMD^{-1}) \qquad (5.12.71)$$

the M–Δ system shown in Figure 5.12.11 can be converted to the system shown in Figure 5.12.12. Therefore, to maximize the robustness of the system, $K(s)$ and $D(s)$ are sought to minimize the following function (Chiang and Safanov, 1992b; Zhou et al., 1996, 1998):

$$\left\| DMD^{-1} \right\|_\infty \qquad (5.12.72)$$

where

$$M = F_\ell(P(s), K(s)) \qquad (5.12.73)$$

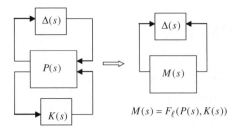

FIGURE 5.12.11 Deriving the M–Δ system.

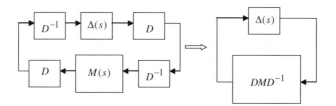

FIGURE 5.12.12 M–Δ System with the diagonal scaling matrix D.

This optimization problem is solved via D–K iteration:

1. Take $D(s) = D_0$ pointwise across frequency.
2. With this D(s), find a $K(s)$ to minimize $\left\| DMD^{-1} \right\|_\infty$. The Matlab routine "hinfopt" may be used for this purpose:

   If $\left\| DMD^{-1} \right\|_\infty \leq 1$, Stop

3. With this K(s), find a diagonal $D(s)$ to minimize $\left\| DMD^{-1} \right\|_\infty$. This process is equivalent to finding $\mu_\Delta(M)$. The Matlab routine "ssv" may be used for this purpose. The output of this routine will be the log Bode plot of the optimal diagonal scaling $D_{new}(j\omega)$.
4. Using a curve fitting method, find a low-order transfer function approximation to the optimal diagonal scaling $D_{new}(j\omega)$ data. The Matlab routine "fitd" may be used for this purpose.

EXERCISE PROBLEMS

P.5.1   Consider the following system:

$$\frac{dx}{dt} = u(t)$$

The feedback law is to be obtained such that the following objective function is minimized:

$$I = \int_0^\infty x^2 + u^2 \, dt$$

a   Find the optimal state feedback gain.
b. Draw the LQR loop and determine its gain and phase margins.

P5.2 Consider Problem P3.17 (Chapter 3). Draw the LQR loop and determine its gain and phase margins.

P5.3 Consider Problem P3.18 (Chapter 3). Draw the LQR loop and determine its gain and phase margins for all three values of ρ .

P5.4 Consider the benchmark robust control problem. Assume that there is no parametric uncertainty in the system. Also, $m_1 = 1$ kg, $m_2 = 1$ kg, and $k = 1$ N/m.

Design an LQG controller on the basis of first mode (rigid body mode) such that

$$J = \lim_{t_f \to \infty} E\left\{ \frac{1}{t_f} \int_0^{t_f} [y^2(t) + \rho u^2(t) dt] \right\}$$

is minimized.

Consider three values of ρ: 0.01, 0.1, and 1. Find gain and phase margins in each case by drawing the Nyquist diagram.

P5.5 Consider the benchmark robust control problem. Assume that there is no parametric uncertainty in the system. Also, $m_1 = 1$ kg, $m_3 = 1$ kg, and $k = 1$ N/m.

a. Design an observer-based controller on the basis of first mode (rigid body mode) only. Locate controller and observer poles on the negative real axis.

b. Evaluate the following integral

$$\int_0^\infty \ln|S(j\omega)| d\omega$$

where $S(j\omega)$ is the output sensitivity function as the controller and observer poles are pushed farther into the left half of the s-plane.

P5.6 Consider the benchmark robust control problem. Parameters m_1 and m_2 are exactly known to be 1 kg. The parameter k is uncertain but is known to be between 0.5 and 2 N/m. The estimated value of the stiffness is 1 N/m.

a. Design an LQG controller for the nominal plant ($k = 1$) with $W = I$ and $V = I$.

b. Using the small gain theorem, determine if the closed-loop system will be stable in the presence of uncertain k as defined earlier.

P5.7 Let a loop transfer matrix $H(s)$ be such that

$$\underline{\sigma}(I + H(j\omega)) \geq \alpha \qquad \text{where } \alpha < 1$$

Prove the following results (Dorato et al., 1995) for gain and phase margin:

$$\frac{1}{1+\alpha} < \ell_i < \frac{1}{1-\alpha}$$

$$-\cos^{-1}\left(1-\frac{\alpha^2}{2}\right) < \phi_i < \cos^{-1}\left(1-\frac{\alpha^2}{2}\right)$$

Hint: $I + HL = ((L^{-1} - I)(I + H)^{-1} + I)(I + H)L$.

P5.8 Consider the system shown in Figure P5.8 where the plant transfer function is

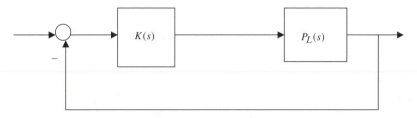

FIGURE P5.8 An uncertain closed loop system.

$$P_L(s) = (1 + \Delta(s))P(s)$$

where $P(s) = \dfrac{1}{s-1}$

Suppose that the controller is the pure gain $K(s) = k$, and the perturbation $\Delta(s)$ is stable and satisfies $\|\Delta(s)\|_\infty \leq 2$. Determine the range of k for which the closed loop is stable for all such perturbations.

P5.9 Consider the system shown in Figure P5.9. The plant transfer function is

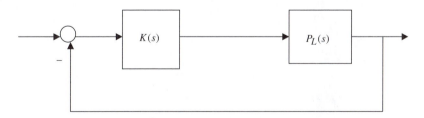

FIGURE P5.9 An uncertain closed loop system.

$$P_L(s) = (1 + \Delta(s)W_2(s))P(s)$$

where

$$P(s) = \frac{1}{s-1} \quad \text{and} \quad W_2(s) = \frac{2}{s+10}$$

Suppose that the controller is the pure gain $K(s) = k$, and the perturbation $\Delta(s)$ is stable and satisfies $\|\Delta(s)\|_\infty \leq 2$.

a. Convert the system shown in Figure P5.9 to M-Δ form.
b. Using the small gain theorem, find the condition for the stability of the system shown in Figure P5.9.
c. Determine the range of k for the stability of the system shown in Figure P5.9.

P5.10　Consider the following system:

$$\frac{dx}{dt} = -0.1x(t) + u(t) + d(t)$$

$$y(t) = x(t)$$

where $u(t)$ is the control input and $d(t)$ is the external disturbance.

Let the the control law be $u(t) = -kx(t)$. Find the value of k such that the H∞ norm of the closed-loop transfer function relating the vector $[y(t)\ u(t)]^T$ and the disturbance $d(t)$ is minimized.

P5.11　Consider the following state space model:

$$\frac{dx_1}{dt} = x_3$$

$$\frac{dx_2}{dt} = x_4$$

$$\frac{dx_3}{dt} = \alpha_{11}x_1 + \alpha_{12}x_2 + \beta_1 u_1$$

$$\frac{dx_4}{dt} = \alpha_{21}x_1 + \alpha_{22}x_2 + \beta_2 u_2$$

Outputs: $y_1 = x_1$ and $y_2 = x_2$.

The parameters α_{12} and β_2 are uncertain. Pull out the uncertainty channels into the $M - \Delta$ form and find M.

P5.12　Consider the bending and torsional vibration control of a flexible plate structure (Kar et al., 2000). The matrices A, $\mathbf{b}$, $\mathbf{b}_d$, and C in Figure P5.12 are as follows.

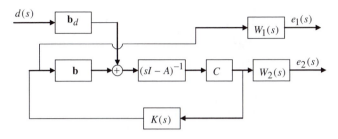

FIGURE P5.12 Vibration control of a flexible plate.

$$A = \begin{bmatrix} A_{11} & A_{12} \\ A_{21} & A_{22} \end{bmatrix}$$

$$A_{11} = \begin{bmatrix} 0 & 0 & 0 & 0 \\ 0 & 0 & 0 & 1.5440 \\ 0 & 0 & 0 & 0 \\ 0 & 0 & 0 & -16.3446 \end{bmatrix} \text{ rad/sec}; \quad A_{21} = I_4 \text{ and } A_{22} = 0$$

$$A_{12} = \begin{bmatrix} -5.5710e+003 & 7.3234e+003 & 2.4439e+003 & 0 \\ 7.3234e+003 & -1.5589e+004 & 1.2548e+004 & 38.8561 \\ 6.8211e+003 & 1.4907e+004 & -1.8008e+004 & 0 \\ -7.3234e+003 & 1.5589e+004 & -1.2548e+004 & -411.3142 \end{bmatrix} \text{ (rad/sec}$$

$$\mathbf{b} = \begin{bmatrix} 0 & 0.6713 & 0 & -7.1063 & 0 & 0 & 0 & 0 \end{bmatrix}^T \text{ (rad/sec)}^2$$

$$\mathbf{b}_d = \begin{bmatrix} 0.2857 & 0.6713 & 0.7975 & 0 & 0 & 0 & 0 & 0 \end{bmatrix}^T \text{ (kg)}^{-1}$$

$$C = \begin{bmatrix} 1 & 0 & 0 & 0 & 4 & 0 & 0 & 0 \\ 0 & 1 & 0 & 0 & 0 & 4 & 0 & 0 \\ 0 & 0 & 1 & 0 & 0 & 0 & 4 & 0 \end{bmatrix}$$

    a. Design the controller $K(s)$ such that $\left\| T_{ed} \right\|_\infty$ is minimized.
    b. Design the controller $K(s)$ such that $\left\| T_{ed} \right\|_{H_2}$ is minimized.
P5.13 Consider the system shown in Figure 5.11.4 in which the torque $T(t)$ is the control input.
    a. Consider the model based on the first two modes of vibration. Select suitable weight functions $W_1(s)$ and $W_3(s)$ and design full state feedback controllers such that

$$\left\| \begin{matrix} W_1 S_o(s) \\ W_3 T_o(s) \end{matrix} \right\|_{H_\infty} \quad \text{and} \quad \left\| \begin{matrix} W_1 S_o(s) \\ W_3 T_o(s) \end{matrix} \right\|_{H_2}$$

are minimized.

b. Discuss your results with respect to the choice of weight functions $W_1(s)$ and $W_3(s)$ and norms used for controller design.

Parameters

$$J_1 = 0.0024 \text{ kg} - m^2, \quad J_2 = 0.0019 \text{ kg} - m^2, \quad J_3 = 0.0019 \text{ kg} - m^2$$

$$k_1 = k_2 = 2.8 \ N - m / \text{rad}$$

$$c_1 = 0.007 \ N - m / \text{rad} / \sec, \quad c_2 = c_3 = 0.001 \ N - m / \text{rad} / \sec$$

$$k_{hw} = \frac{100}{6} \text{units}$$

P5.14   Consider the flexible tetrahedral truss structure (Appendix F) with collocated sensors and actuators. The controller is to be designed on the basis of the first four modes of vibration.

   a. Select suitable weight functions $W_1(s)$ and $W_3(s)$ and design full state and output feedback controllers such that

$$\left\| \begin{matrix} W_1 S_o(s) \\ W_3 T_o(s) \end{matrix} \right\|_{H_\infty} \quad \text{and} \quad \left\| \begin{matrix} W_1 S_o(s) \\ W_3 T_o(s) \end{matrix} \right\|_{H_2}$$

are minimized.

   b. Discuss your results with respect to the choice of weight functions $W_1(s)$ and $W_3(s)$ and norms used for controller design.

   c. Find the controller transfer matrix and determine its stability.

P5.15   Consider the control of molten steel level in a continuous caster (Kitada et al., 1998). Many factors such as upward steel flow in the mold, immersion nozzle clogging, etc. complicate the design of a control system (Figure P5.15A). The block diagram of the control system is shown in Figure P5.15B.

Eddy current sensor transfer function: $G_s(s) = \dfrac{1 + 0.0035s}{1 + 0.35s}$

Slide gate transfer function: $G_g(s) = \dfrac{1}{1 + 0.05s}$

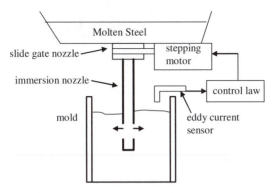

FIGURE P5.15A Control of molten steel level.

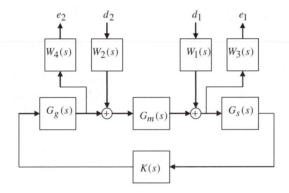

FIGURE P5.15B Block diagram for molten steel level control.

Transfer function for linearized dynamics of the molten steel level:

$$G_m(s) = -0.5274 \frac{1+0.266s}{0.378s} \frac{1-0.6s}{1+0.6s}$$

Weighting functions:

$$W_1(s) = 0.065 \; ; \quad W_2(s) = 0.065 \frac{4[0.25\pi s + (0.25\pi)^2]}{s^2 + 0.25\pi s + (0.25\pi)^2} \; ; \quad W_3(s) = \frac{s+0.05}{s} \; ;$$

$$W_4(s) = 0.9 \frac{s^2 + 8\pi s + (0.8\pi)^2}{s^2 + 0.8\pi s + (0.8\pi)^2} (0.5s + 1)$$

a. Find the controller $K(s)$ such that

$$\left\| \begin{matrix} W_3 S W_1 & W_3 S G_m W_2 \\ -\dfrac{W_4 T W_1}{G_m} & -W_4 T W_2 \end{matrix} \right\|_\infty < 1$$

where

$$S(s) = \frac{1}{1 + G_s G_m G_g K} \quad \text{and} \quad T(s) = \frac{G_s G_m G_g K}{1 + G_s G_m G_g K}$$

b. Examine the stability of the system for the following multiplicative uncertainty:

$$G_{m\Delta} = G_m (1 + \Delta(s)) \quad \text{where} \quad \Delta(s) = \frac{0.9}{(0.24\pi)^{-1} s + 1}$$

P5.16a Design a controller $k_{H_\infty}(s)$ around the LQG controller for a single-link flexible manipulator in P4.8 to minimize the following objective function:

$$\left\| \begin{matrix} W_1(s)(I + G_l(s)k_{H_\infty}(s))^{-1} \\ W_3(s)G_l(s)k_{H_\infty}(s)(I + G_l(s)k_{H_\infty}(s))^{-1} \end{matrix} \right\|_{H_\infty}$$

is minimized. The weighting functions are as follows:

$$W_1(s) = \frac{10}{(15s + 1)} \quad \text{and} \quad W_3(s) = \frac{s}{50}$$

b. Plot the unit step response of the system shown in Figure P5.16.

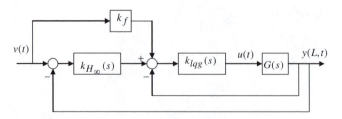

FIGURE P5.16 Control system for a single link flexible manipulator.

    c. Determine the stability of the system in the presence of unmodeled dynamics.

P5.17 Consider the benchmark robust control problem (Figure 5.12.7). Parameters m_1 and m_2 are exactly known to be 1 kg. The parameter k is uncertain but is known to be between 0.5 and 2 N/m. The estimated value of the stiffness is 1 N/m.

    a. Design an LQG controller for the nominal plant ($k = 1$) with $W = V = I$.

    b. Using μ analysis, determine if the closed-loop system will be stable in the presence of uncertain k as defined earlier.

    c. Using μ analysis, determine if the magnitude of the transfer function relating the disturbance w and x_2 will be less than 1 for all possible values of k.

P5.18 Consider the electromagnetic system with coil inductance (Fujita et al., 1995). The nominal plant model is

$$P(s) = \frac{-36.27}{(s + 66.94)(s - 66.94)(s + 45.69)}$$

The uncertainty in the plant is described as

$$P_\Delta(s) = P(s) + W(s)\Delta(s) ; \quad \left\| \Delta(s) \right\|_\infty \le 1$$

where

$$W(s) = \frac{1.4x10^{-5}(1 + s/8)(1 + s/170)(1 + s/420)}{(1 + s/30)(1 + s/35)(1 + s/38)}$$

Design an output feedback control system (Figure 5.12.2) to achieve the following objective:

$$\left\| (I + P_\Delta K)^{-1} Wp \right\|_\infty \le 1 \quad \forall \quad \Delta$$

P5.19 Consider the space shuttle flight control system during reentry (Appendix G). Convert Figure G.2 to Figure P.5.19.

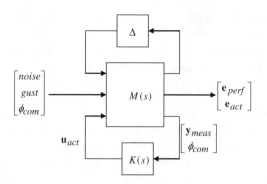

FIGURE P5.19 $M–\Delta–K$ system for the space shuttle.

a. Design an H_∞ control system to minimize $\left\| T_{ev}(s) \right\|_\infty$.

b. Design an H_∞ control system to minimize $\left\| T_{ev}(s) \right\|_\infty$ via $D–K$ iteration.

6 Robust Control: Sliding Mode Methods

First, basic concepts of sliding modes are presented along with the sliding mode control of a linear system with full state feedback. Then, it is shown how H_∞ and sliding mode theories can be blended to control an uncertain linear system with full state feedback. Next, the sliding mode control of a deterministic linear system is developed with the feedback of estimated states. Lastly, the optimal sliding Gaussian (OSG) control theory is presented for a stochastic system.

6.1 BASIC CONCEPTS OF SLIDING MODES

Consider the following single-input second-order linear system:

$$\begin{bmatrix} \dot{x}_1 \\ \dot{x}_2 \end{bmatrix} = \begin{bmatrix} 0 & 1 \\ \alpha & \beta \end{bmatrix} \begin{bmatrix} x_1 \\ x_2 \end{bmatrix} + \begin{bmatrix} 0 \\ 1 \end{bmatrix} u \tag{6.1.1}$$

This system represents a spring-mass-damper system. Define a line in the state space, Figure 6.1, as follows:

$$s(t) = x_2(t) + \lambda x_1(t) ; \quad \lambda > 0 \tag{6.1.2}$$

When $s = 0$, the line passes through the origin of the state space, and this line will be called the *sliding line*. The first objective is to ensure that the system will reach this line from any initial conditions in a finite time and will remain on this line after reaching it. The condition to achieve this objective is called *reaching* or *attractive condition* (Utkin, 1977; Asada and Slotine, 1986) and is described as

$$\dot{s} = -\delta \, \text{sgn}(s) ; \quad \delta > 0 \tag{6.1.3}$$

where the *sign* function *sgn(s)* is defined as follows:

$$\text{sgn}(s) = \begin{cases} +1 & \text{if} \quad s > 0 \\ -1 & \text{if} \quad s < 0 \end{cases} \tag{6.1.4}$$

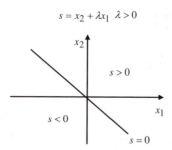

FIGURE 6.1 A sliding line ($s = 0$).

When s > 0, $\dot{s}$ is a negative constant δ, and s will decrease linearly to s = 0 in a finite time. Similarly, when s < 0, $\dot{s}$ is a positive constant δ, and s will increase linearly to s = 0 in a finite time. The condition (6.1.3) is also written as

$$s\dot{s} < 0 \tag{6.1.5}$$

When the system remains on the sliding line, $\dot{s} = 0$. The corresponding input u is called the *equivalent control*, u_{eq}, and is obtained by differentiating (6.1.2) and using state equations (6.1.1):

$$\dot{s} = \alpha x_1 + (\beta + \lambda)x_2 + u \tag{6.1.6}$$

For $\dot{s} = 0$,

$$u_{eq}(t) = -\alpha x_1(t) - (\beta + \lambda)x_2 \tag{6.1.7}$$

Therefore, to satisfy the reaching condition (6.1.3), the control law is

$$u(t) = u_{eq}(t) - \delta \, \text{sgn}(s) \tag{6.1.8}$$

On the sliding line, $u(t) = u_{eq}(t)$ and state equations (6.1.1) become

$$\begin{bmatrix} \dot{x}_1 \\ \dot{x}_2 \end{bmatrix} = \begin{bmatrix} 0 & 1 \\ 0 & -\lambda \end{bmatrix} \begin{bmatrix} x_1 \\ x_2 \end{bmatrix} \tag{6.1.9}$$

The eigenvalues of this system are 0 and $-\lambda$. The zero eigenvalue reflects the fact that the system is constrained to remain on the s = 0 line. Hence, the effective dynamics on the sliding line is represented by $-\lambda$, which is in the left half of the complex plane when $\lambda > 0$. Here, the effective dynamics can also be derived from (6.1.2) by using the first state equation and s = 0:

$$\dot{x}_1 + \lambda x_1 = 0 \tag{6.1.10}$$

In summary, the sliding mode control input has two parts: linear and nonlinear. The nonlinear part, *sgn* function, ensures the reaching condition. After the system has reached the sliding line, the control law ($u_{eq}(t)$) is linear full state feedback (Equation 6.1.7).

6.2 SLIDING MODE CONTROL OF A LINEAR SYSTEM WITH FULL STATE FEEDBACK

Consider the following uncertain linear system:

$$\dot{\mathbf{x}} = A\mathbf{x} + B\mathbf{u} \tag{6.2.1}$$

For m inputs, m hyperplanes in the state space are defined as

$$s_i(t) = \mathbf{g}_i^T \mathbf{x}(t) \ ; \ i = 1, 2, \ldots, m \tag{6.2.2}$$

These hyperplanes pass through the origin of the state space for $s_i = 0$. From (6.2.1) and (6.2.2),

$$\dot{s}_i(t) = \mathbf{g}_i^T \dot{\mathbf{x}}(t) = \mathbf{g}_i^T A x + \mathbf{g}_i^T B \mathbf{u} \ ; \ i = 1, 2, \ldots, m \tag{6.2.3}$$

Equation 6.2.3 can be written in the matrix form as

$$\dot{\mathbf{s}}(t) = GA\mathbf{x} + GB\mathbf{u} \tag{6.2.4}$$

where

$$\mathbf{s} = \begin{bmatrix} s_1 & s_2 & \cdot & \cdot & s_m \end{bmatrix}^T \tag{6.2.5}$$

and

$$G = \begin{bmatrix} \mathbf{g}_1 & \mathbf{g}_2 & \cdot & \cdot & \mathbf{g}_m \end{bmatrix}^T \tag{6.2.6}$$

It should also be noted that

$$\mathbf{s}(t) = G\mathbf{x}(t) \tag{6.2.7}$$

and the vector $\mathbf{s}(t) = 0$ represents the intersection of all the m sliding hyperplanes passing through the origin of the state space.

The equivalent control input, $\mathbf{u}_{eq}(t)$, corresponds to $\dot{\mathbf{s}} = 0$. Therefore, from (6.2.4),

$$\mathbf{u}_{eq}(t) = -(GB)^{-1}GA\mathbf{x} \tag{6.2.8}$$

Note that the matrix GB has been assumed to be nonsingular. The reaching condition for each hyperplane is

$$s_i \dot{s}_i < 0 \tag{6.2.9}$$

Conditions (6.2.9) are satisfied by the following control law:

$$\mathbf{u}(t) = \mathbf{u}_{eq}(t) - (GB)^{-1} diag(\boldsymbol{\eta}) \mathrm{sgn}(\mathbf{s}(t)) \tag{6.2.10}$$

where $diag(\boldsymbol{\eta})$ is a diagonal matrix with the ith diagonal element equal to a positive number η_i.

6.2.1 COMPUTATION OF SLIDING HYPERPLANE MATRIX G

Define a similarity transformation

$$\mathbf{q}(t) = H\mathbf{x}(t) \tag{6.2.11}$$

where

$$H = \begin{bmatrix} N & B \end{bmatrix}^T \tag{6.2.12}$$

and the columns of the $nx(n-m)$ matrix N are composed of basis vectors of the null space (Utkin and Yang, 1978; Strang, 1988) of B^T. From (6.2.1) and (6.2.11),

$$\dot{\mathbf{q}} = \bar{A}\mathbf{q}(t) + \bar{B}\mathbf{u}(t) \tag{6.2.13}$$

where

$$\bar{A} = HAH^{-1} \tag{6.2.14}$$

and

$$\bar{B} = HB \qquad (6.2.15)$$

Because of the special structure of the matrix H, the first $(n-m)$ rows of $\bar{B}$ turn out to be zeros. Hence, the vector $\mathbf{q}$ is decomposed as follows:

$$\mathbf{q} = \begin{bmatrix} \mathbf{q}_1 \\ \mathbf{q}_2 \end{bmatrix} \qquad (6.2.16)$$

where $\mathbf{q}_1$ and $\mathbf{q}_2$ are $(n-m)$- and m-dimensional vectors, respectively. Partitioning (6.2.13),

$$\begin{bmatrix} \dot{\mathbf{q}}_1 \\ \dot{\mathbf{q}}_2 \end{bmatrix} = \begin{bmatrix} \bar{A}_{11} & \bar{A}_{12} \\ \bar{A}_{21} & \bar{A}_{22} \end{bmatrix} \begin{bmatrix} \mathbf{q}_1 \\ \mathbf{q}_2 \end{bmatrix} + \begin{bmatrix} 0 \\ \bar{B}_r \end{bmatrix} \mathbf{u} \qquad (6.2.17)$$

and sliding hyperplanes can be described as

$$\mathbf{s}(t) = \mathbf{q}_2(t) + K\mathbf{q}_1(t) \qquad (6.2.18)$$

From (6.2.7) and (6.2.18),

$$G = \begin{bmatrix} K & I_m \end{bmatrix} H \qquad (6.2.19)$$

Equation 6.2.19 indicates that the matrix G can be determined via the matrix K. The first objective behind the choice of the matrix G or the matrix K is to ensure the system stability on the intersection of all hyperplanes. When $\mathbf{s}(t) = 0$,

$$\mathbf{q}_2(t) = -K\mathbf{q}_1(t) \qquad (6.2.20)$$

Equation 6.2.17 yields

$$\dot{\mathbf{q}}_1 = \bar{A}_{11}\mathbf{q}_1 + \bar{A}_{12}\mathbf{q}_2 \qquad (6.2.21)$$

and

$$\dot{\mathbf{q}}_2 = \bar{A}_{21}\mathbf{q}_1 + \bar{A}_{22}\mathbf{q}_2 + \bar{B}_r\mathbf{u} \qquad (6.2.22)$$

For the system (6.2.21), $\mathbf{q}_2$ (t) can be viewed as the input, and Equation 6.2.20 as the state feedback law. Utkin and Yang (1978) have shown that $(\bar{A}_{11}, \bar{A}_{12})$ is controllable if (A, B) is controllable. Therefore, the matrix K can be selected to arbitrarily place the eigenvalues of $(\bar{A}_{11} - \bar{A}_{12}K)$ in the left half of the complex plane.

From (6.2.1) and (6.2.8), the system dynamics on the intersection of sliding hyperplanes is given by

$$\dot{\mathbf{x}} = [A - B(GB)^{-1}GA]\mathbf{x} \tag{6.2.23}$$

When G is chosen according to Equation 6.2.19, $(n\text{-}m)$ eigenvalues of $[A - B(GB)^{-1}GA]$ are those of $(\bar{A}_{11} - \bar{A}_{12}K)$ and remaining m eigenvalues are 0, which is a reflection of the fact that m degrees of freedom of the system is lost when $\mathbf{s}(t) = 0$.

EXAMPLE 6.1: A HELICOPTER

Consider the state equations for a CH-47 tandem rotor helicopter (Doyle and Stein, 1981):

$$A = \begin{bmatrix} -0.02 & 0.005 & 2.4 & -32 \\ -0.14 & 0.44 & -1.3 & -30 \\ 0 & 0.018 & -1.6 & 1.2 \\ 0 & 0 & 1 & 0 \end{bmatrix} ; \quad B = \begin{bmatrix} 0.14 & -0.12 \\ 0.36 & -8.6 \\ 0.35 & 0.009 \\ 0 & 0 \end{bmatrix} \tag{6.2.24a,b}$$

Independent vectors in the null space of B^T are found by using the Matlab command "null." From Equation 6.2.12,

$$H = \begin{bmatrix} -0.9331 & 0.0134 & 0.3594 & 0 \\ 0 & 0 & 0 & 1 \\ 0.14 & 0.36 & 0.35 & 0 \\ -0.12 & -8.6 & 0.009 & 0 \end{bmatrix} \tag{6.2.25}$$

Therefore,

$$\bar{A}_{11} = \begin{bmatrix} -1.0335 & 29.8877 \\ 0.3594 & 0 \end{bmatrix} \quad \text{and} \quad \bar{A}_{12} = \begin{bmatrix} -7.0221 & -0.2965 \\ 2.4853 & 0.1046 \end{bmatrix} \tag{6.2.26a,b}$$

To locate the eigenvalues of the reduced-order system (6.2.21) at −1 and −4, the state feedback gain matrix is

$$K = \begin{bmatrix} 45 & 1880 \\ -1064 & -44623 \end{bmatrix} \tag{6.2.27}$$

And from Equation 6.2.19, the sliding hyperplane matrix is defined by

$$G = \begin{bmatrix} -42 & 1 & 16 & 1880 \\ 993 & -23 & -382 & -44623 \end{bmatrix} \tag{6.2.28}$$

Lastly, it is obvious that the matrix K can also be obtained by minimizing a quadratic objective function involving the state vector $\mathbf{q}_1$ and the effective input $\mathbf{q}_2$, that is, by solving a liner quadratic (LQ) problem. This brings us to a new type of controller that is called the OS (Optimal Sliding Mode) controller.

6.2.2 OPTIMAL SLIDING MODE (OS) CONTROLLER

For the linear system

$$\dot{\mathbf{x}} = A\mathbf{x}(t) + B\mathbf{u}(t) \tag{6.2.29}$$

the objective is to minimize

$$J = \int_0^\infty \mathbf{x}^T Q \mathbf{x} dt \tag{6.2.30}$$

subject to constraints

$$\mathbf{s}(t) = G\mathbf{x}(t) = 0 \tag{6.2.31}$$

For the LQ control, it is necessary to have the input term in the quadratic objective function (Chapter 3, Section 3.5). Here, the input term is not present in the objective function (6.2.30), and the constraints are that the system is on the intersection on m sliding hyperplanes.

Furthermore, the matrix G is not specified *a priori* and will come out as a solution to the problem.

Using the similarity transformation (6.2.11),

$$J = \int_0^\infty \mathbf{q}^T Q_q \mathbf{q} dt \tag{6.2.32}$$

where

$$Q_q = (H^{-1})^T Q H^{-1} \tag{6.2.33}$$

If $Q = Q^T \geq 0$, then $Q_q = Q_q^T \geq 0$ because the signs of eigenvalues are preserved under congruence transformation (Strang, 1988). Partition Q_q to conform to the partition of $\mathbf{q}$ in (6.2.17) as follows:

$$Q_q = \begin{bmatrix} Q_r & N \\ N^T & R \end{bmatrix}$$

(6.2.34)

Then, from (6.2.32),

$$J = \int_0^\infty \mathbf{q}_1^T Q_r \mathbf{q}_1 + 2\mathbf{q}_1^T N \mathbf{q}_2 + \mathbf{q}_2^T R \mathbf{q}_2 \, dt$$

(6.2.35)

When $s(t) = 0$, the $(n-m)$ dimensional dynamics is represented by

$$\dot{\mathbf{q}}_1 = \bar{A}_{11} \mathbf{q}_1 + \bar{A}_{12} \mathbf{q}_2$$

(6.2.36)

Equation 6.2.35 and Equation 6.2.36 constitute a standard LQ problem provided $R > 0$. If Q is chosen to be positive definite, R is guaranteed to be positive definite. In general, R is not guaranteed to be positive definite if Q is positive semidefinite. If R does not turn out to be positive definite, it has to be arbitrarily chosen to be a positive definite matrix (Sinha and Miller, 1995). In this case, a new Q will be defined according to (6.2.34).

The gain matrix K for the minimum value of J is

$$K = R^{-1}(\bar{A}_{12}^T P + N^T)$$

(6.2.37)

where

$$P(\bar{A}_{11} - \bar{A}_{12} R^{-1} N^T) + (\bar{A}_{11}^T - NR^{-1}\bar{A}_{12}^T)P - P\bar{A}_{12}R^{-1}\bar{A}_{12}^T P +$$
$$Q_q - NR^{-1}N^T = 0$$

(6.2.38)

Finally, the optimal sliding hyperplane matrix G is obtained by substituting K from (6.2.37) into (6.2.19).

EXAMPLE 6.2: A DOUBLE INTEGRATOR SYSTEM

Consider the following linear system:

$$\begin{bmatrix} \dot{x}_1 \\ \dot{x}_2 \end{bmatrix} = \begin{bmatrix} 0 & 1 \\ 0 & 0 \end{bmatrix} \begin{bmatrix} x_1 \\ x_2 \end{bmatrix} + \begin{bmatrix} 0 \\ 1 \end{bmatrix} u(t)$$

(6.2.39)

The objective is to minimize

$$I = \frac{1}{2} \int_0^\infty x_1^2 + x_2^2 \, dt \qquad (6.2.40)$$

subject to the constraint

$$x_1 + \alpha x_2 = 0 \qquad (6.2.41)$$

Equation 6.2.39 is already in the form of Equation 6.2.17. Hence, the state space equation for minimization of the objective function (6.2.40) is

$$\dot{x}_1 = x_2 \qquad (6.2.42)$$

Treating x_2 as the input, Equation 6.2.42 and Equation 6.2.40 form a standard LQ problem. The ARE (6.2.38) is

$$1 - P^2 = 0 \qquad (6.2.43)$$

and the optimal gain (6.2.37) is

$$K = P = +1 \qquad (6.2.44)$$

Therefore, from Equation 6.2.20,

$$x_2 = -x_1 \qquad (6.2.45)$$

This relationship describes the sliding line; i.e., $\alpha = 1$ in Equation 6.2.41. From the second state equation,

$$\dot{x}_2 = u(t) \qquad (6.2.46)$$

Using (6.2.42) and (6.2.45),

$$u(t) = -x_2 \qquad (6.2.47)$$

In other words, the state feedback gain vector is

$$K_{os} = [0 \quad 1] \qquad (6.2.48)$$

6.3 SLIDING MODE CONTROL OF AN UNCERTAIN LINEAR SYSTEM WITH FULL STATE FEEDBACK: BLENDING H_∞ AND SLIDING MODE METHODS

Consider the dynamics of a plant with k uncertain parameters, $\delta_1, \cdots, \delta_k$, which are expressed as

$$\dot{\mathbf{x}}(t) = A_p \, \mathbf{x}(t) + B \, \mathbf{u}(t) \tag{6.3.1}$$

where

$$A_p = A + \Delta A \tag{6.3.2}$$

Also, states $\mathbf{x}(t) \in R^n$, input $\mathbf{u}(t) \in R^m$, and the nominal system is given by known matrices A and B. Assume that the bound of the uncertain matrix ΔA is Q, i.e.,

$$\left| \Delta A \right| \leq Q \tag{6.3.3}$$

where the absolute value of the matrix implies the matrix with the absolute value of each element and the inequality sign holds good for each element of the matrices. Also, the uncertain matrix ΔA need not satisfy the so-called matching conditions, i.e.,

$$\Delta A = BD \tag{6.3.4}$$

where D is a bounded matrix (Drazenovic', 1969). The goal is to find the upper bound of δ_i for the stability of the closed-loop system.

For m inputs, m hyperplanes passing through the origin of the state space are defined by (6.2.7):

$$\mathbf{s}(t) = G\mathbf{x}(t) \tag{6.3.5}$$

To satisfy reaching conditions (6.2.9) under parametric uncertainties, the control law (6.2.10) is modified to be

$$\mathbf{u}(t) = \mathbf{u}_{eq}(t) - (GB)^{-1} \, diag(\mathbf{\psi}) \, \mathrm{sgn}(\mathbf{s}(t)) \tag{6.3.6}$$

where $\mathbf{u}_{eq}(t)$ is given by Equation 6.2.8 and $diag(\mathbf{\psi})$ is an $m \times m$ diagonal matrix with the ith diagonal element to be the ith element of an $m \times 1$ vector $\mathbf{\psi}$, which is defined as

$$\mathbf{\psi} = \left| G \right| Q \left| x(t) \right| + \mathbf{\eta} \tag{6.3.7}$$

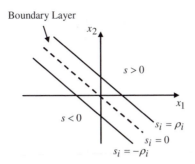

FIGURE 6.2 Boundary layer around a sliding hyperplane.

In this case,

$$\dot{\mathbf{s}} = G\Delta A\mathbf{x} - diag(\boldsymbol{\psi})\,\text{sgn}(\mathbf{s}(t)) \qquad (6.3.8)$$

Note that $\dot{s}_i$ will not be exactly equal to zero because of parametric uncertainties. Therefore, the system will try to leave the hyperplanes. But as it moves out of hyperplanes, the sgn($\cdot$) function in Equation 6.3.8 will become active, and the system will be pushed back to sliding hyperplanes. This process gives rise to chatter.

To eliminate the chattering behavior caused by the sgn($\cdot$) function, a boundary layer is introduced around each sliding hyperplane by replacing the sgn($\cdot$) function in Equation 6.3.6 by the saturation function (Asada and Slotine, 1986). Hence, the control law (6.3.6) is modified to be

$$\mathbf{u}(t) = \mathbf{u}_{eq}(t) - (GB)^{-1}\,diag(\boldsymbol{\psi})\,sat(\mathbf{s}(t)) \qquad (6.3.9)$$

The ith element of $sat(\mathbf{s}(t))$ is defined as

$$sat(s_i(t)) = \begin{cases} \text{sgn}(s_i(t)) & if \quad |s_i(t)| > \rho_i \\ s_i / \rho_i & otherwise \end{cases} \qquad (6.3.10)$$

where ρ_i is the boundary layer thickness around the ith hyperplane, as illustrated in Figure 6.2.

6.3.1 Impact of Uncertainties on System Dynamics in Sliding Modes

Having satisfied reaching conditions (6.2.9), the system is guaranteed to be on the intersection of all sliding hyperplanes for all ΔA satisfying the bound (6.3.3). However, if the matching conditions (6.3.4) are not satisfied, system dynamics is affected. To see this, the similarity transformation (6.2.11) is again used on the state equations (6.3.1). The equation (6.2.13) is modified to be

$$\dot{\mathbf{q}} = \bar{A}\mathbf{q}(t) + \Delta\bar{A}\mathbf{q}(t) + \bar{B}\mathbf{u}(t) \tag{6.3.11}$$

where

$$\Delta\bar{A} = H\Delta A H^{-1} = \begin{bmatrix} \Delta\bar{A}_{11} & \Delta\bar{A}_{12} \\ \Delta\bar{A}_{21} & \Delta\bar{A}_{22} \end{bmatrix} \tag{6.3.12}$$

When the matching condition (6.3.4) is satisfied,

$$\Delta\bar{A} = \bar{B}DH^{-1} \tag{6.3.13}$$

As the first $(n-m)$ rows of $\bar{B}$ are 0, Equation 6.3.13 yields

$$\Delta\bar{A}_{11} = 0 \quad \text{and} \quad \Delta\bar{A}_{12} = 0 \tag{6.3.14}$$

Therefore, Equation 6.2.21 still holds, and system dynamics on sliding hyperplanes chosen according to Equation 6.2.13 are insensitive to the uncertainty ΔA.

6.3.2 COMPUTATION OF SLIDING HYPERPLANE MATRIX VIA H_∞ CONTROL METHOD

When matching conditions are not satisfied,

$$\dot{\mathbf{q}}_1 = (\bar{A}_{11} + \Delta\bar{A}_{11})\mathbf{q}_1 + (\bar{A}_{12} + \Delta\bar{A}_{12})\mathbf{q}_2 \tag{6.3.15}$$

Let ΔA, which is parameterized by the k scalar uncertain parameters $\delta_1, \cdots, \delta_k$, be written as follows:

$$\Delta A = \sum_{i=1}^{k} \delta_i \hat{A}_i \tag{6.3.16}$$

where the knowledge about the structure of the uncertainty is contained in the matrix $\hat{A}_i$, which indicates how the ith uncertainty, δ_i, influences the state space model. In order to utilize the H_∞ method, the subsystem with parametric uncertainties, Equation 6.3.15, is transformed (Pai and Sinha, 1998) into the $N-\Delta-K_s$ diagram (Figure 6.3a) through the linear fractional transformation method (Chapter 5, Section 5.9.2).

Let

$$\begin{bmatrix} \hat{A}_{11i} & \hat{A}_{12i} \\ 0 & 0 \end{bmatrix} = \begin{bmatrix} L_{si} \\ W_{si} \end{bmatrix} \begin{bmatrix} R_{si} \\ Z_{si} \end{bmatrix}^* \quad \text{for each } i \tag{6.3.17}$$

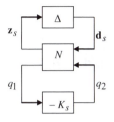 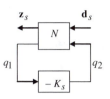

FIGURE 6.3 $N-\Delta-K_s$ diagram.

The matrices N and Δ can be expressed as follows:

$$N = \begin{bmatrix} \bar{A}_{11} & B_s & \bar{A}_{12} \\ C_s & 0 & D_s \\ I_{n-m} & 0 & 0 \end{bmatrix} \quad \text{and} \quad \begin{bmatrix} \dot{\mathbf{q}}_1(t) \\ \mathbf{z}_s(t) \\ \mathbf{y}_s(t) \end{bmatrix} = N \begin{bmatrix} \mathbf{q}_1(t) \\ \mathbf{d}_s(t) \\ \mathbf{q}_2(t) \end{bmatrix} \qquad (6.3.18)$$

$$\Delta = diag\{\delta_1 I_n, \ldots, \delta_k I_{r_k}\} \qquad (6.3.19)$$

where

$$B_s = \begin{bmatrix} L_{s1} & \cdot & \cdot & \cdot & L_{sk} \end{bmatrix} \qquad (6.3.20)$$

$$C_s = \begin{bmatrix} R_{s1} & \cdot & \cdot & \cdot & R_{sk} \end{bmatrix}^* \qquad (6.3.21)$$

$$D_s = \begin{bmatrix} Z_{s1} & \cdot & \cdot & \cdot & Z_{sk} \end{bmatrix}^* \qquad (6.3.22)$$

or

$$\dot{\mathbf{q}}_1(t) = \bar{A}_{11}\mathbf{q}_1(t) + \bar{A}_{12}\mathbf{q}_2(t) + B_s\mathbf{d}_s \qquad (6.3.23)$$

$$\mathbf{z}_s(t) = C_s\mathbf{q}_1(t) + D_s\mathbf{q}_2(t) \qquad (6.3.24)$$

The state feedback controller $K_s(s)$ is obtained to stabilize the subsystem shown in Figure 6.3b and minimize γ_s such that the cost function $\|T_{\mathbf{z}_s\mathbf{d}_s}\|_\infty < \gamma_s$, $\gamma_s > 0$. The corresponding quadratic objective function is

$$J = \frac{1}{2}\int_0^\infty [\mathbf{z}_s^T(t)\mathbf{z}_s(t) - \gamma_s^2\mathbf{d}_s^T(t)\mathbf{d}_s(t)]dt, \quad \gamma_s > 0 \qquad (6.3.25)$$

or

$$J = \frac{1}{2} \int_0^\infty [\mathbf{q}_1^T(t)Q_{11}\mathbf{q}_1(t) + 2\mathbf{q}_1^T Q_{12}\mathbf{q}_2(t) + \mathbf{q}_2^T(t)Q_{22}\mathbf{q}_2(t) - \gamma_s^2 \mathbf{d}_s^T(t)\mathbf{d}_s(t)]dt,$$

(6.3.26)

$$\gamma_s > 0$$

where

$$Q_{11} = C_s^T C_s , \quad Q_{12} = C_s^T D_s , \quad \text{and} \quad Q_{22} = D_s^T D_s$$

(6.3.27)

The optimal gain K_s is given (Lublin and Athans, 1996; Zhou et al., 1996) as

$$K_s = Q_{22}^{-1}(\bar{A}_{12}^T P_2 + Q_{12}^T)$$

(6.3.28)

where P_2 is the unique, symmetric, positive semidefinite solution of the matrix Riccati-like equation:

$$P_2(\bar{A}_{11} - \bar{A}_{12}Q_{22}^{-1}Q_{12}^T) + (\bar{A}_{11}^T - Q_{12}Q_{22}^{-1}\bar{A}_{12}^T)P_2 - P_2(\bar{A}_{12}Q_{22}^{-1}\bar{A}_{12}^T - \gamma_s^{-2}B_s B_s^T)P_2$$
$$+ (Q_{11} - Q_{12}Q_{22}^{-1}Q_{12}^T) = 0$$

(6.3.29)

Note that as $\gamma \to \infty$, Equation 6.3.29 reduces to Equation 6.2.38. Next, the sliding hyperplane matrix G is obtained from Equation 6.3.28 and Equation 6.2.19.

EXAMPLE 6.3: FLEXIBLE TETRAHEDRAL TRUSS STRUCTURE

Consider the flexible tetrahedral truss structure described in the Appendix F. For the first four controlled modes, state space equations are

$$\dot{\mathbf{x}}(t) = A\,\mathbf{x}(t) + B\,\mathbf{u}(t)$$

(6.3.30)

where

$$\mathbf{x}^T(t) = \begin{bmatrix} \psi_1 & \dot{\psi}_1 & \psi_2 & \dot{\psi}_2 & \psi_3 & \dot{\psi}_3 & \psi_4 & \dot{\psi}_4 \end{bmatrix}$$

(6.3.31)

$$A_\delta = diag\left(\begin{bmatrix} 0 & 1 \\ -\omega_{ai}^2 & -2\zeta_{ai}\omega_{ai} \end{bmatrix}\right); \quad i = 1,\dots,4$$

(6.3.32)

$$B = \begin{bmatrix} 0 & 0 & 0 \\ -0.023 & -0.067 & -0.044 \\ 0 & 0 & 0 \\ -0.112 & 0.017 & 0.069 \\ 0 & 0 & 0 \\ -0.077 & 0.271 & 0.046 \\ 0 & 0 & 0 \\ 0.189 & -0.06 & -0.249 \end{bmatrix} \qquad (6.3.33)$$

The uncertainty matrix Δ is defined by Equation 6.3.19, with $k = 8$ and $r_1 = r_2 = ... = r_n = 1$. The relations among parametric uncertainties, δ_j ($j = 1, 3, 5, 7$), actual natural frequencies, ω_{ai}, and estimated natural frequencies, ω_i, can be expressed as follows:

$$\left| \omega_{ai}^2 - \omega_i^2 \right| \le \delta_j \omega_i^2 ; \qquad (j = 2i - 1; \quad i = 1, 2, 3, 4) \qquad (6.3.34)$$

where estimated natural frequencies ω_1, ω_2, ω_3, and ω_4 are 1.342, 1.643, 2.891, and 2.958 rad/sec, respectively.

Uncertain parameters, δ_j ($j = 2, 4, 6, 8$), refer to uncertainties in $\zeta_{ai}\omega_{ai}$ defined as follows:

$$\left| 2\zeta_{ai}\omega_{ai} - 2\zeta_i\omega_i \right| \le \delta_j (2\zeta_i\omega_i); \qquad (j = 2i; \quad i = 1, 2, 3, 4) \qquad (6.3.35)$$

where the estimated damping coefficient ζ_i is assumed to be 0.005 for each mode. From Equation 6.3.16,

$$A_\delta = A + \sum_{i=1}^{8} \delta_i \, {}^i\hat{A} \qquad (6.3.36)$$

where

$$A = diag\left(\begin{bmatrix} 0 & 1 \\ -\omega_i^2 & -2\zeta_i\omega_i \end{bmatrix} \right); \qquad i = 1,\ldots,4 \qquad (6.3.37)$$

and all elements of each matrix ${}^i\hat{A}$ are zero except one, which is defined as follows:

a. For $i = 1, 3, 5, 7$

$$^i\hat{A}_{ji} = \delta_i \omega_k^2 ; \quad k = (i+1)/2, \; j = i+1 \tag{6.3.38}$$

b. For $i = 2, 4, 6, 8$

$$^i\hat{A}_{ii} = \delta_i 2\zeta_k \omega_k ; \quad k = i/2 \tag{6.3.39}$$

Using the Matlab routine, aresolv, Equation 6.3.29 is solved. The Matlab program is shown in Program 6.1. The output "wellposed" from the routine aresolv is found to be true for $\gamma_s \geq 1.475$. With $\gamma_s = 1.475$, Equation 6.3.28 and Equation 6.2.19 yield

$$G =$$

$$\begin{bmatrix} 0.2512 & -1.5928 & -0.2315 & 0.6486 & 0.1419 & -0.2934 & -0.0410 & 0.1162 \\ -1.2602 & 0.7880 & 1.1050 & 0.0023 & -0.2277 & 0.1938 & 0.2359 & 0.0109 \\ -1.1996 & 1.4711 & 0.0256 & -0.8061 & -0.2633 & 0.3103 & -0.0019 & -0.1522 \end{bmatrix}$$

MATLAB PROGRAM 6.1: UNCERTAIN TETRAHEDRAL TRUSS STRUCTURE

```
%
clear all
close all
%
freq2dm=diag([1.8 2.7 8.36 8.75]);
dampdm=0.01*2*sqrt(freq2dm);
Bf=[-0.023 -0.067 -0.044;-0.112 0.017 0.069;-0.077
0.271 0.046;0.189 -0.060 -0.249];
A=[zeros(4,4) eye(4);-freq2dm -dampdm];
B=[zeros(4,3);Bf];
Ah=[zeros(4,4) zeros(4,4);freq2dm dampdm];
%
F=null(B');
H=[F B]';
Ah=[zeros(4,4) zeros(4,4);freq2dm dampdm];
```

```
%
AA=H*A*inv(H);
AA11=AA(1:5,1:5);
AA12=AA(1:5,6:8);
for i=1:4
  Ah=zeros(8,8);
  Ah(4+i,i)=freq2dm(i,i);
AAh=H*Ah*inv(H);
AAh(6:8,1:8)=zeros(3,8);
[U,S,V]=svd(AAh);
R(:,i)=U(:,1)*S(1,1);
L(i,:)=V(:,1)';
end
%
for i=1:4
  Ah=zeros(8,8);
  Ah(4+i,i+4)=dampdm(i,i);
AAh=H*Ah*inv(H);
AAh(6:8,1:8)=zeros(3,8);
[U,S,V]=svd(AAh);
R(:,i+4)=U(:,1)*S(1,1);
L(i+4,:)=V(:,1)';
end
%
Bs=R(1:5,1:8);
Cs=L(1:8,1:5);
Ds=L(1:8,6:8);
%
Q11=Cs'*Cs;
Q12=Cs'*Ds;
Q22=Ds'*Ds;
```

```
%
AAe=AA11-AA12*inv(Q22)*Q12';
Re=AA12*inv(Q22)*AA12'-Bs*Bs'/(1.475)^2;
Qe=Q11-Q12*inv(Q22)*Q12';
%
[p1,p2,lamp,perr,wellposed,p]=aresolv(AAe,Qe,Re)
%
K=inv(Q22)*(AA12'*p+Q12')
G=[K eye(3)]*H
eig(AA11-AA12*K)
```

6.4 SLIDING MODE CONTROL OF A LINEAR SYSTEM WITH ESTIMATED STATES

Consider the following linear system:

$$\dot{\mathbf{x}} = A\mathbf{x} + B\mathbf{u} \tag{6.4.1}$$

$$\mathbf{y} = C\mathbf{x} \tag{6.4.2}$$

First, states are estimated using the Luenberger observer (Chapter 4, Section 4.2):

$$\dot{\hat{\mathbf{x}}} = A\hat{\mathbf{x}} + B\mathbf{u} + L(\mathbf{y} - C\hat{\mathbf{x}}) \tag{6.4.3}$$

Now, sliding hyperplanes are defined in the estimated state space (Kao and Sinha, 1991):

$$\mathbf{s}(t) = G\hat{\mathbf{x}}(t) \tag{6.4.4}$$

Here, the equivalent control input for $\dot{\mathbf{s}} = 0$ is

$$\mathbf{u}_{eq}(t) = -(GB)^{-1}[G(A - LC)\hat{\mathbf{x}} + GL\mathbf{y}(t)] \tag{6.4.5}$$

To satisfy the reaching conditions and eliminate chattering behavior, the control law is chosen as

$$u(t) = u_{eq}(t) - (GB)^{-1} diag(\mathbf{\eta}) sat(\mathbf{s}(t)) \tag{6.4.6}$$

where $diag(\boldsymbol{\eta})$ is a diagonal matrix with the ith diagonal element equal to a positive number η_i. The ith element of $sat(\mathbf{s}(t))$ is defined as

$$sat(s_i(t)) = \begin{cases} \mathrm{sgn}(s_i(t)) & if \quad |s_i(t)| > \rho_i \\ s_i / \rho_i & otherwise \end{cases} \quad (6.4.7)$$

where ρ_i is the boundary layer thickness around the ith hyperplane, illustrated in Figure 6.2. The system is linear inside the boundary layers. From (6.4.5) through (6.4.7),

$$\mathbf{u}(t) = -(GB)^{-1}\left[G\left(A - LC + \frac{\eta}{\mu}I_n\right)\hat{\mathbf{x}}(t) + GL\mathbf{y}(t)\right] \quad (6.4.8)$$

Without any loss of generality, η_i and ρ_i have been assumed to be identical for each hyperplane and have been set equal to η and ρ, respectively. Furthermore, inside boundary layers,

$$\dot{s}_i = -vs_i \quad (6.4.9)$$

where

$$v = \frac{\eta}{\rho} \quad (6.4.10)$$

For state estimation, $\hat{\mathbf{x}}(0)$ is usually set to zero. In this case, $\mathbf{s}(0) = 0$, and the solution of (6.4.9) indicates that $\mathbf{s}(t) = 0$. In other words, the system is always on the intersection of all sliding hyperplanes, and the resulting control input is always linear as described by (6.4.8). An important point to note here is that the control input depends on estimated states and directly on outputs as well (Sinha and Millar, 1995). This is a significant difference with the standard observer based control system where inputs only depend on estimated states. From (6.4.3) and (6.4.8), the controller transfer function is as follows:

$$\mathbf{u}(s) = -K_{es}(s)\mathbf{y}(s) \quad (6.4.11)$$

where

$$K_{es}(s) = Z_1(sI_n - A + BZ_1 + LC)^{-1}Z_2 + (GB)^{-1}GL \quad (6.4.12)$$

$$Z_1 = (GB)^{-1}G(A + vI_n - LC) \quad (6.4.13)$$

$$Z_2 = (L - \Omega L) \tag{6.4.14}$$

$$\Omega = B(GB)^{-1}G \tag{6.4.15}$$

Using (6.4.1) through (6.4.3), and (6.4.8), the dynamics of the closed-loop system is described as

$$\begin{bmatrix} \dot{\mathbf{x}} \\ \dot{\tilde{\mathbf{x}}} \end{bmatrix} = \begin{bmatrix} A - \Omega(A + \nu I_n) & \Omega(A + \nu I_n) - \Omega LC \\ 0 & A - LC \end{bmatrix} \begin{bmatrix} \mathbf{x} \\ \tilde{\mathbf{x}} \end{bmatrix} \tag{6.4.16}$$

where

$$\tilde{\mathbf{x}} = \mathbf{x} - \hat{\mathbf{x}} \tag{6.4.17}$$

Equation 6.4.16 shows that the eigenvalues of the closed-loop system are those of $A - \Omega(A + \nu I_n)$ and $A - LC$. Therefore, the sliding hyperplane matrix G and the boundary layer parameter ν, which constitute the controller parameters, can be chosen independently of the observer gain matrix L. In other words, the separation property holds, and the controller and observer can be designed independently.

Fact

The m eigenvalues of $A - \Omega(A + \nu I_n)$ are $-\nu$ and the remaining $n-m$ eigenvalues can be arbitrarily placed in the complex plane by a proper choice of G (Sinha and Miller, 1995) as the system (A,B) is controllable.

6.5 OPTIMAL SLIDING MODE GAUSSIAN (OSG) CONTROL

State Dynamics

$$\dot{\mathbf{x}}(t) = A\mathbf{x}(t) + B\mathbf{u}(t) + \mathbf{w}(t) \tag{6.5.1}$$

The vector $\mathbf{w}(t)$ is a stochastic process called *process* (or plant) *noise*. It is assumed that $\mathbf{w}(t)$ is a continuous-time Gaussian white noise vector (Appendix D). Its mathematical characterization is

$$E(\mathbf{w}(t)) = 0 \tag{6.5.2}$$

and

$$E(\mathbf{w}(t)\mathbf{w}^T(t + \tau)) = W\delta(\tau) \tag{6.5.3}$$

The matrix W is called the *intensity matrix* with the property $W = W^T \geq 0$.

Measurement Equation

$$\mathbf{y}(t) = C\mathbf{x}(t) + \mathbf{\theta}(t) \tag{6.5.4}$$

where $\mathbf{y}(t)$ is the sensor noise vector, and $\mathbf{\theta}(t)$ is a continuous-time Gaussian white noise vector. It is assumed that

$$E(\mathbf{\theta}(t)) = 0 \tag{6.5.5}$$

and

$$E(\mathbf{\theta}(t)\mathbf{\theta}^T(t + \tau)) = \Theta \delta(\tau) \tag{6.5.6}$$

where $\Theta = \Theta^T > 0$.

It is also assumed that process and measurement noise vectors are uncorrelated; i.e.,

$$E(\mathbf{\theta}(t)\mathbf{w}^T(t + \tau)) = E(\mathbf{\theta}(t))E(\mathbf{w}^T(t + \tau)) = 0 \tag{6.5.7}$$

Theorem (Sinha and Miller, 1995)

For the stochastic system (6.5.1)–(6.5.7), the matrix G chosen on the basis of Equation 6.2.37 minimizes the following objective function:

$$J = \lim_{t_f \to \infty} E\left\{ \frac{1}{t_f} \int_0^{t_f} \mathbf{x}^T(t)Q\mathbf{x}(t)dt \right\} \tag{6.5.8}$$

when states are estimated by the Kalman filter (Chapter 4, Section 4.6); i.e.,

$$\dot{\hat{\mathbf{x}}} = A\hat{\mathbf{x}} + B\mathbf{u} + L(\mathbf{y} - C\hat{\mathbf{x}}) \tag{6.5.9}$$

$$L = \Sigma C^T \Theta^{-1} \tag{6.5.10}$$

and

$$A\Sigma + \Sigma A^T + W - \Sigma C^T \Theta^{-1} C \Sigma = 0 \tag{6.5.11}$$

It should be noted here that the constraints for the minimization of the objective function (6.5.8) are

$$s(t) = G\hat{x}(t) = 0 \tag{6.5.12}$$

Proof

Using (4.8.11),

$$J = \lim_{t \to \infty} E\left\{\hat{x}^T(t)Q\hat{x}(t)\right\} + tr(\Sigma Q) \tag{6.5.13}$$

As $tr(\Sigma Q)$ is independent of the input $u(t)$, the minimization of Equation 6.5.13 is equivalent to minimizing the following objective function:

$$J_m = \lim_{t \to \infty} E\left\{\hat{x}^T(t)Q\hat{x}(t)\right\} \tag{6.5.14}$$

Define the following similarity transformation:

$$\hat{q}(t) = H\hat{x}(t) \tag{6.5.15}$$

where the matrix H is same as (6.2.12). From (6.5.9) and (6.5.15),

$$\dot{\hat{q}} = \bar{A}\hat{q} + \bar{B}u + \bar{L}v(t) \tag{6.5.16}$$

where $\bar{A}$ and $\bar{B}$ are the same as defined by (6.2.14) and (6.2.15); and

$$\bar{L} = HL \tag{6.5.17}$$

$$v(t) = y(t) - C\hat{x}(t) \tag{6.5.18}$$

Similar to Equation 6.2.16, the vector $\hat{q}$ is again partitioned as

$$\hat{q} = \begin{bmatrix} \hat{q}_1 \\ \hat{q}_2 \end{bmatrix} \tag{6.5.19}$$

The first $(n-m)$ equations of (6.5.16) can be written as

$$\dot{\hat{q}}_1 = \bar{A}_{11}\hat{q}_1 + \bar{A}_{12}\hat{q}_2 + \bar{L}_{nm}v(t) \tag{6.5.20}$$

where the matrix $\bar{L}_{nm}$ is composed of the first $(n-m)$ rows of the matrix $\bar{L}$. Similar to (6.2.35),

$$J_m = \lim_{t \to \infty} E\left\{\hat{q}_1^T(t)Q_r\hat{q}_1(t) + 2\hat{q}_1^T N\hat{q}_2 + \hat{q}_2^T R\hat{q}_2\right\} \tag{6.5.21}$$

As $\mathbf{v}(t)$ in Equation 6.5.20 is a white noise vector, the stochastic regulator theory described in Chapter 4, Section 4.7 is directly applicable here. Therefore, the following relationship will yield the optimal solution:

$$\hat{\mathbf{q}}_2(t) = -K\hat{\mathbf{q}}_1(t) \tag{6.5.22}$$

where K is given by (6.2.37). Lastly, the matrix G is given by Equation 6.2.19:

$$G = \begin{bmatrix} K & I_m \end{bmatrix} H \tag{6.5.23}$$

This completes the proof.

EXAMPLE 6.4: A THREE-DISK SYSTEM (*ECP MANUAL*)

Consider the three-disk system shown in Figure 5.11.4. Define the state vector as

$$\mathbf{x}^T(t) = \begin{bmatrix} \boldsymbol{\theta}(t) & \dot{\boldsymbol{\theta}}(t) \end{bmatrix} \tag{6.5.24}$$

The state equations are derived to be

$$\dot{\mathbf{x}} = A\mathbf{x}(t) + \mathbf{b}\tau(t) \tag{6.5.25}$$

where

$$A = \begin{bmatrix} 0 & I_3 \\ -M^{-1}C & -M^{-1}K \end{bmatrix}; \quad \mathbf{b} = \begin{bmatrix} 0 \\ M^{-1}B_f \end{bmatrix} \tag{6.5.26a,b}$$

Let the output be $y(t) = \theta_1(t)$. In this case,

$$y(t) = C\mathbf{x}(t) \tag{6.5.27}$$

where

$$C = \begin{bmatrix} 1 & 0 & 0 & 0 & 0 & 0 \end{bmatrix} \tag{6.5.28}$$

Take $W = I_6$ and $\Theta = 1$. In this case, the Kalman filter gain vector is

$$K_F = \begin{bmatrix} 0.9732 & 0.6177 & 0.7304 & -0.0264 & -0.4088 & 0.3570 \end{bmatrix}^T \tag{6.5.29}$$

To design the sliding mode control,

$$Q = C^T C \tag{6.5.30}$$

From the decomposition (6.2.34), Q_r, N, and R are computed. Because R turns out to be zero, it is arbitrarily chosen to be 10^{-7}. The OSG controller transfer function, Equation 6.4.12, is plotted in Figure 6.4. The variance of the output, $E(y^2)$, is found to be 1.0356.

Next, an LQG controller is designed by taking $Q = C^T C$ and $R = 0.01$ in Equation 4.8.8. The LQG controller transfer function, Equation 4.3.11, is also plotted in Figure 6.4. The variance of the output, $E(y^2)$, is found to be 0.9898. Therefore, both LQG and OSG controllers lead to similar levels of performance. This is also confirmed by the singular value plots of the closed-loop transfer matrices relating the 7×1 stochastic disturbance vector $[\mathbf{w}^T(t) \quad \theta(t)]^T$ to the output, $y(t)$, illustrated in Figure 6.5.

Observing locations of peaks and troughs in Figure 6.4, controller transfer functions seem to be inversions of the open loop transfer function. However, this inversion phenomenon is less severe around the natural frequency $\omega_3 = 65.39$ rad/sec. in the case of the OSG controller transfer function, in Figure 6.6. Similar results were also found for a flexible spacecraft (Sinha and Miller, 1995).

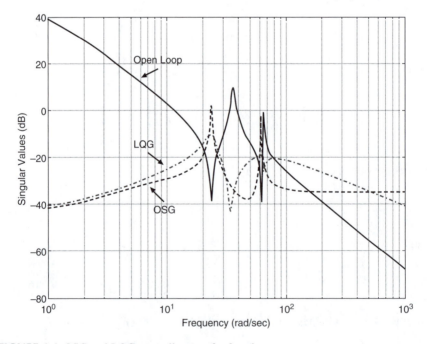

FIGURE 6.4 OSG and LQG controller transfer functions.

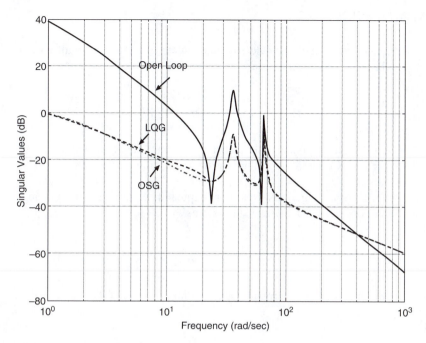

FIGURE 6.5 OSG and LQG closed-loop transfer functions.

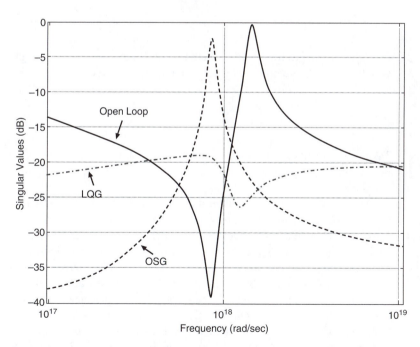

FIGURE 6.6 OSG and LQG controller transfer functions in a small frequency range.

EXERCISE PROBLEMS

P6.1   Consider the single-link flexible manipulator, Figure P3.4, with a single vibratory mode. Define the state vector as

$$\mathbf{x} = \begin{bmatrix} q_0 & \dot{q}_0 & q_1 & \dot{q}_1 \end{bmatrix}^T$$

Design a sliding hyperplane matrix to locate the eigenvalues at $-\sigma$, $-\sigma \pm j\omega_1$. Choose σ such that the damping ratio in the vibratory mode equals 0.1.

P6.2   Consider the single-link flexible manipulator, Figure P3.4, with a single vibratory mode. Define the state vector as

$$\mathbf{x} = \begin{bmatrix} q_0 & \dot{q}_0 & q_1 & \dot{q}_1 \end{bmatrix}^T$$

Design a sliding hyperplane matrix to minimize the following objective function:

$$\int_0^\infty \mathbf{x}^T \mathbf{x}\, dt$$

P6.3   Consider the single-link flexible manipulator, Figure P3.4, with a single vibratory mode. The output is $y(L,t)$. Define the state vector as

$$\mathbf{x} = \begin{bmatrix} q_0 & \dot{q}_0 & q_1 & \dot{q}_1 \end{bmatrix}^T$$

Taking $W = I$ and $\Theta = 0.5$, design an OSG controller such that $E(\mathbf{x}^T\mathbf{x})$ is minimized. Demonstrate the performance of the controller via numerical simulations.

Design an LQG control system with a comparable value of $E(y^2(t))$. Compare the OSG and LQG sensitivity and controller transfer functions.

P6.4   Consider the Problem 4.8. On the basis of the rigid and the first mode, design an OSG controller $k_{osg}(s)$ to minimize $E(y^2)$. If the weighting of equivalent input is zero, assume it to be a small number. Introduce fictitious noises $\mathbf{w}(t)$ and $\theta(t)$, Equation 4.6.1 and Equation 4.6.4, where $W = 0.1I_4$ and $\Theta = 100$. Demonstrate the performance of the controller via numerical simulations.

Design an LQG control system with a comparable value of $E(y^2(t))$. Compare the OSG and LQG sensitivity and controller transfer function.

P6.5   Consider the flexible tetrahedral truss structure (Appendix F). The controller is to be designed on the basis of the first four modes of vibration. Introduce fictitious noises $\mathbf{w}(t)$ and $\boldsymbol{\theta}(t)$, Equation 4.6.1 and Equation 4.6.4, where $W = I_8$ and $\Theta = I_3$.

   a. Design an OSG controller with collocated sensors and actuators, such that

$$J = E(\mathbf{y}^T \mathbf{y})$$

   is minimized. If the weighting of equivalent input is zero, assume it to be a small number.

   b. Design an LQG controller with collocated sensors and actuators such that

$$J = E(\mathbf{y}^T \mathbf{y} + \rho \mathbf{u}^T \mathbf{u})$$

   is minimized. Select ρ such that $E(y^T y)$ is almost equal to that for OSG control.

   c. Compare the controller transfer matrices in part (a) and part (b).

   d. Compare the closed-loop transfer matrices in part (a) and part (b).

P6.6   Consider the OSG and LQG controllers designed in P6.5. Assume that parameters are uncertain according to (6.3.34) and (6.3.35).

   a. Find the maximum range of each parametric uncertainty for robust stability using the small gain theorem.

   b. Find the maximum range of each parametric uncertainty for robust stability using μ-analysis.

References

Abate, M., Barmish, B.R., Murillo-Sanchez, C., and Tempo, R., Applications of some new tools to robust stability analysis of spark ignition engines: a case study, *IEEE Transactions on Control Systems Technology*, Vol. 2, No. 1, 22–30, 1994.

Annaswamy, A.M., Fleifil, M., Rumsey, J.W., Prasanth, R., Hathout, J.-P., and Ghoneim, A.F., Thermoacoustic instability: model-based optimal control designs and experimental validation, *IEEE Transactions on Control Systems Technology*, Vol. 8, No. 6, 905–918, 2000.

Anderson, B.D.O. and Moore, J.B., *Optimal Control: Linear Quadratic Methods*, Prentice-Hall, Englewood Cliffs, NJ, 1990.

Asada, H. and Slotine, J.J.E., *Robot Analysis and Control*, John Wiley & Sons, NY, 1986.

Astrom, K.J., Klein, R.E., and Lennartsson, A., Bicycle Dynamics and Control, *IEEE Control Systems Magazine*, Vol. 25, No. 4, 26–47, 2005.

Athans, M., A tutorial on the LQG/LTR method, *Proceedings of American Control Conference*, 1289–1296, 1986.

Athans, M., Notes for Multivariable Control Systems Course, MIT, Cambridge, MA, 1992.

Balas, M.J., Feedback control of flexible systems, *IEEE Transactions on Automatic Control*, Vol. AC-23, No. 4, 673–679, 1978.

Banavar, R. and Dominic, P., An LQG/H_∞ controller for a flexible manipulator, *IEEE Transactions on Control Systems Technology*, Vol. 3, No. 4, 409–416, 1995.

Biegler, L., Class Notes of Advanced Process Control Course, Carnegie Mellon University, Pittsburgh, PA, 1982.

Braatz, R. and Morari, M., Robust control for a noncollocated spring-mass system, *AIAA Journal of Guidance, Control and Dynamics*, Vol. 15, No. 5, 1103–1110, 1992.

Bryson, A.E., Some connections between modern and classical control concepts, *ASME Journal of Dynamic Systems, Measurement and Control*, Vol. 101, 91–98, June 1979.

Cannon, R.H. and Schmitz, E., Initial experiments on the end-point control of a flexible one-link robot, *International Journal of Robotic Research*, Vol. 3, No. 3, 1984.

Chen, C-T., *Systems and Signal Analysis*, Saunders College Publishing: Holt, Reinhart and Winston, NY, 1989.

Chiang, R.Y. and Safonov, M.G., H_∞ synthesis using a bilinear pole shifting transform, *AIAA Journal of Guidance, Control and Dynamics*, Vol. 15, No. 5, 1111–1117, 1992a.

Chiang, R.Y. and Safonov, M.G., Robust Control Toolbox — User's Guide, The MathWorks, 1992b (www.mathworks.com).

Dahleh, M., Notes on μ-analysis (Multivariable Control Systems Course), MIT, Cambridge, MA, 1992.

Dorato, P., Abdallah, C., and Cerone, V., *Linear Quadratic Control: An Introduction*, Prentice-Hall, Englewood Cliffs, NJ, 1995.

Doyle, J.C., Guaranteed margins for LQG regulators, *IEEE Transactions on Automatic Control*, Vol. 23, No. 4, 756–757, August 1978.

Doyle, J.C. and Stein, G., Multivariable feedback design: concepts for a classical/modern synthesis, *IEEE Transactions on Automatic Control*, Vol. AC-26, No. 1, February 1981.

Doyle, J.C., Lenz, K., and Packard, A., Design examples using μ-synthesis: space shuttle lateral axis FCS during reentry, *Proceedings of NATO Workshop: Modelling, Robustness and Sensitivity Reduction in Control Systems,* Curtain, R.F., Ed., 1986, pp. 127–154.

Doyle, J.C., Glover, K., Khargonekar, P.P., and Francis, B.A., State-space solutions to standard H_2 and H_∞ control problems, *IEEE Transactions on Automatic Control,* Vol. AC-34, No. 8, 831–847, 1989.

Drazenovic, B., The invariance condition in variable structure systems, *Automatica,* Vol. 5, 287–295, 1969.

ECP, Educational Control Products (ECP) Manual, http://www.ecpsystems.com/.

Friedland, B., *Control System Design — An Introduction to State Space Analysis,* McGraw-Hill, NY, 1985.

Fujita, M., Namerikawa, T., Matsumura, F., and Uchida, K., μ-synthesis of an electromagnetic suspension system, *IEEE Transaction on Automatic Control,* Vol. 40, No. 3, 530–536, 1995.

Gopal, M., *Modern Control System Theory,* John Wiley & Sons, NY, 1984.

Green, M. and Limebeer, D.J.N., *Linear Robust Control,* Prentice-Hall, Englewood Cliffs, NJ, 1995.

Gupta, N.K., Frequency-shaped cost functionals: extension of linear-quadratic-Gaussian design methods, *AIAA Journal of Guidance, Control and Dynamics,* Vol. 3, No. 6, 529–535, November–December 1980.

Hughes, R.O., Optimal control of sun tracking solar concentrators, *ASME Journal of Dynamic Systems, Measurement and Control,* Vol. 101, 157–161, June 1979.

Kailath, T., *Linear Systems,* Prentice-Hall, Englewood Cliffs, NJ, 1980.

Kalman, R.E. and Bucy, R.S., New results in linear filtering and prediction theory, *ASME Journal of Basic Engineering,* Vol. 83, 95–107 1961.

Kao, C.K., Robust Control of Vibration in Flexible Structures, Ph.D. dissertation, The Pennsylvania State University, University Park, PA, 1989.

Kao, C.K. and Sinha, A., 1991, Sliding mode control of vibration in flexible structures using estimated states, *Proceedings of American Control Conference,* Vol. 3, 1991, pp. 2467–2474.

Kar, I.N., Miyakura, T., and Seto, K., Bending and torsional vibration control of a flexible plate structure using H_∞-based robust control law, *IEEE Transactions on Control Systems Technology,* Vol. 8, No. 3, 545–553, 2000.

Kitada, H., Kondo, O., Kusachi, H., and Sasame, K., H_∞ control of molten steel level in continuous caster, *IEEE Transactions on Control Systems Technology,* Vol. 6, No. 2, 200–207, 1998.

Kuo, B.C., *Automatic Control Systems,* Prentice-Hall, Englewood Cliffs, NJ, 1995.

Kwakernaak, H. and Sivan, R., *Linear Optimal Control Systems,* John Wiley & Sons, 1972.

Lowell, J. and McKell, H.D., The stability of bicycles, *American Journal of Physics,* Vol. 50, No. 12, 1982, 1106–1112, 1982.

Lublin, L. and Athans, M., Linear quadratic regulator control, *The Control Handbook,* 1996, pp. 635–650.

Maciejowski, J.M., *Multivariable Feedback Design,* Addison-Wesley, Workingham, England, 1989.

Meditch, J.S., *Stochastic Optimal Linear Estimation and Control,* McGraw-Hill, New York, 1969.

Pai, M.C. and Sinha, A., Generating command inputs to eliminate residual vibration via direct optimization and quadratic programming techniques, Active control of vibration and noise, DE-Vol. 93, *Proceedings of the 1996 ASME International Mechanical Engineering Congress and Exposition,* Atlanta, GA, November 1996.

Pai, M.C. and Sinha, A., Active control of vibration in a flexible structure with uncertain parameters via sliding mode and H_∞ / μ techniques, Vibration and noise control, DE-Vol. 97, DSC-Vol. 65, *Proceedings of the 1998 ASME International Mechanical Engineering Congress and Exposition*, Anaheim, CA, November 1998.

Pai, M.C. and Sinha, A., Sliding mode control of vibration in a flexible structure via estimated states and H_∞ / μ techniques, *Proceedings of American Control Conference*, Chicago, IL, June 2000, pp. 1118–1123.

Pao, L. and Singhose, W.E., Robust minimum time control of flexible structures, *Automatica*, Vol. 34, No. 2, 229–236, 1998.

Papoulis, A., *Probability, Random Variables, and Stochastic Processes*, McGraw-Hill, NY, 1965.

Ray, W.H., *Advanced Process Control*, McGraw-Hill, 1981.

Reddy, B.D., *Functional Analysis and Boundary-Value Problems: An Introductory Treatment*, Longman Scientific and Technical, U.K., 1986.

Rugh, W., *Linear System Theory*, Prentice-Hall, Englewood Cliffs, NJ, 1993.

Safanov, M.G., Imaginary-axis zeros in multivariable H_∞-optimal control, *Proceedings of NATO Workshop: Modelling, Robustness and Sensitivity Reduction in Control Systems*, Curtain, R.F., Ed., 1986, pp. 71–81.

Sage, A.P. and White, C.C., *Optimum Systems Control*, Prentice-Hall, Englewood Cliffs, NJ, 1977.

Schultz, M.J. and Inman, D.J., Eigenstructure assignment and controller optimization for mechanical systems, *IEEE Transactions on Control Systems Technology*, Vol. 2, No. 2, 88–100, 1994.

Seo, B., and Chen, C.T., The relationship between the Laplace transform and the Fourier transform, *IEEE Transactions on Automatic Control*, Vol. AC-31, No. 8, 751, August 1986.

Singer, N.C. and Seering, W.P., Preshaping command inputs to reduce system vibration, *ASME Journal of Dynamic Systems, Measurement and Control*, Vol. 112, 76–82, 1990.

Singhose, W. and Pao, L., A comparison of input shaping and time-optimal flexible body control, *Control Engineering Practice*, Vol. 5, No. 4, 459–467, 1997.

Sinha, A. and Miller, D.W., Optimal sliding mode control of a flexible spacecraft under stochastic disturbances, *AIAA Journal of Guidance, Dynamics and Control*, Vol. 118, No. 3, 486–492, 1995.

Stein, G., Formal Control System Synthesis with H_2 / H_∞ Criteria, Lecture Notes for Multivariable Control Systems Course, MIT, Cambridge, MA, 1988.

Stein, G., Respect the Unstable, *IEEE Control Systems Magazine*, Vol. 23, No. 4, 12–25, 2003.

Strang, G., *Linear Algebra and Its Applications*, Harcourt Brace Jovanovich Publishers, San Diego, CA, 1988.

Strunce, R., Lin, J., Hegg, D., and Henderson, T., Actively Controlled Structures Theory, C.S. Draper Lab., Cambridge, MA, Final Report Vol. 2, R-1338, 1979.

Thompson, A.G., An active suspension with optimal linear state feedback, *Vehicle System Dynamics*, Vol. 5, 187–203, 1976.

Triantafyllou, M.S. and Hover, F.S., Loop transfer recovery, *Maneuvering and Control of Marine Vehicles*, MIT, Cambridge, MA, 2000 (www.mit.edu and www.core.org.cn).

Tsach, U., Koontz, J.W., Ignatoski, M.A., and Geselowitz, D.B., An adaptive aortic pressure observer for the Penn State electric ventricular assist device, *IEEE Transactions on Biomedical Engineering*, Vol. 37, No. 4, 374–383, April 1990.

Utkin, V.I., Variable structure systems with sliding modes: a survey, *IEEE Transactions on Automatic Control*, Vol. AC-22, 212–222, 1977.

Utkin, V.I. and Yang, K.D., Methods for constructing discontinuity planes in multidimensional variable structure systems, *Automation and Remote Control*, Vol. 31, 1466–1470, 1978.

Vidyasagar, M., *Control System Synthesis: A Factorization Approach*, MIT Press, Cambridge, MA, 1985.

Vidyasagar, M., *Nonlinear Systems Analysis*, 2nd ed., Prentice-Hall, Englewood Cliffs, NJ, 1993.

Wiberg, D.M., *State space and linear systems, Schaum's Outline Series*, McGraw-Hill, New York, 1971.

Wie, B. and Bernstein, D.S., Benchmark problems for robust control design, *AIAA Journal of Guidance, Control and Dynamics*, Vol. 15, No. 5, 1057–1059, 1992.

Yang, V., Sinha, A., and Fung, Y.T., State feedback control of longitudinal combustion instabilities, *AIAA Journal of Propulsion and Power*, Vol. 8, No. 1, 66–73, 1992.

Zhou, K. and Doyle, J.C., *Essentials of Robust Control*, Prentice-Hall, Upper Saddle River, NJ, 1998.

Zhou, K., Doyle, J.C., and Glover, K., *Robust and Optimal Control*, Prentice-Hall, Upper Saddle River, NJ, 1996.

Appendix A:
Linear Algebraic Equations, Eigenvalues/Eigenvectors, and Matrix Inversion Lemma

A.1 SYSTEM OF LINEAR ALGEBRAIC EQUATIONS (GOPAL, 1984; STRANG, 1988)

Consider the solution of the following system of linear algebraic equations:

$$Ax = \mathbf{b} \tag{A.1}$$

where A is an $m \times n$ matrix. Given an m-dimensional vector $\mathbf{b}$, the objective is to find the n-dimensional vector $\mathbf{x}$. It should be noted that m and n need not be equal.

DEFINITIONS

Linear Independence of Vectors

The vectors $\mathbf{x}_1$, $\mathbf{x}_2$,..., $\mathbf{x}_n$ are linearly independent provided the condition $\alpha_1\mathbf{x}_1 + \alpha_2\mathbf{x}_2 + ... + \alpha_n\mathbf{x}_n = 0$ is true only when scalars $\alpha_1 = \alpha_2 = ... = \alpha_n = 0$.

Range or Column Space

Range or column space consists of all combinations of column vectors of a matrix.

Rank of a Matrix

The rank of a matrix is the maximum number of linearly independent columns, which is also the dimension of the range or column space of A.

The maximum number of linearly independent columns always equals the maximum number of linearly independent rows of a matrix. Therefore, the rank of a matrix is also the maximum number of linearly independent rows.

The rank of a matrix also equals the dimension of the largest nonsingular square array.

Null Space

The null space of the matrix A consists of all vectors $\mathbf{y}$ satisfying $A\mathbf{y} = \mathbf{0}$. The dimension of the null space, also known as *nullity*, is the maximum number of independent solutions of $A\mathbf{y} = \mathbf{0}$.

Theorem A.1

Let A be an $m \times n$ matrix. Then,

$$\gamma = n - \rho \tag{A.2}$$

where ρ and γ are the rank and nullity of the matrix A.

EXAMPLE A.1

$$A = \begin{bmatrix} 1 & 3 & 5 & 7 \\ 1 & 2 & 3 & 4 \\ 2 & 5 & 8 & 11 \end{bmatrix} \tag{A.3}$$

The last row is the sum of first two rows, which are independent. Therefore,

$$rank(A) = 2 \tag{A.4}$$

From (A.2), the dimension of the null space = 2. There are two independent solutions of $A\mathbf{y} = \mathbf{0}$, which can be written as follows after ignoring the last row:

$$\begin{bmatrix} 1 & 3 & 5 & 7 \\ 1 & 2 & 3 & 4 \end{bmatrix} \begin{bmatrix} y_1 \\ y_2 \\ y_3 \\ y_4 \end{bmatrix} = \begin{bmatrix} 0 \\ 0 \end{bmatrix} \tag{A.5}$$

or

$$y_1 + 3y_2 + 5y_3 + 7y_4 = 0 \tag{A.6}$$

$$y_1 + 2y_2 + 3y_3 + 4y_4 = 0 \tag{A.7}$$

We have two equations in four unknowns. Two unknowns can be chosen arbitrarily, and then the two remaining unknowns can be determined uniquely. Equations can be expressed as

$$\begin{bmatrix} 1 & 3 \\ 1 & 2 \end{bmatrix}\begin{bmatrix} y_1 \\ y_2 \end{bmatrix} = \begin{bmatrix} -5y_3 - 7y_4 \\ -3y_3 - 4y_4 \end{bmatrix} \tag{A.8}$$

Solving for y_1 and y_2,

$$\begin{bmatrix} y_1 \\ y_2 \end{bmatrix} = \begin{bmatrix} 1 & 3 \\ 1 & 2 \end{bmatrix}^{-1}\begin{bmatrix} -5y_3 - 7y_4 \\ -3y_3 - 4y_4 \end{bmatrix} = \begin{bmatrix} y_3 + 2y_4 \\ -2y_3 - 3y_4 \end{bmatrix} \tag{A.9}$$

Choosing $y_3 = 1$ and $y_4 = 0$,

$$\mathbf{y}_{b1} = \begin{bmatrix} y_1 \\ y_2 \\ y_3 \\ y_4 \end{bmatrix} = \begin{bmatrix} 1 \\ -2 \\ 1 \\ 0 \end{bmatrix} \tag{A.10}$$

Choosing $y_3 = 0$ and $y_4 = 1$,

$$\mathbf{y}_{b2} = \begin{bmatrix} y_1 \\ y_2 \\ y_3 \\ y_4 \end{bmatrix} = \begin{bmatrix} 2 \\ -3 \\ 0 \\ 1 \end{bmatrix} \tag{A.11}$$

Solutions $\mathbf{y}_{b1}$ and $\mathbf{y}_{b2}$ are independent of each other. Therefore, any vector in the null space can be written as

$$\mathbf{y} = \alpha\mathbf{y}_{b1} + \beta\mathbf{y}_{b2} \tag{A.12}$$

where α and β are arbitrary real numbers.

Theorem A.2

The system of equations (A.1) is solvable if and only if the vector **b** belongs to the range or column space of the matrix A; i.e.,

$$rank([A \vdots \mathbf{b}]) = rank(A) \tag{A.13}$$

Theorem A.3

Consider the system of equations (A.1), which satisfies the solvability condition (A.13).

a. Irrespective of whether m is greater than or equal to or less than n, the number of independent equations will always be less than or equal to the number of unknowns. In other words, effectively, $m \leq n$.

b. When $rank(A) = n$, Equation A.1 has the unique solution

$$\mathbf{x} = A^{-1}\mathbf{b} \tag{A.14}$$

c. When $rank(A) < n$, the dimension of the null space is greater than zero (Equation A.2). In this case, there are infinite many solutions of Equation A.1, which can be written as

$$\mathbf{x} = \mathbf{x}_p + \mathbf{x}_h \tag{A.15}$$

where $\mathbf{x}_p$ is a particular solution satisfying $A\mathbf{x}_p = \mathbf{b}$, and $\mathbf{x}_h$ is any solution in the null space of the matrix A; i.e., $A\mathbf{x}_h = \mathbf{0}$. Note that there are infinite many nonzero vectors in the null space. As a result, there are infinite many solutions for $\mathbf{x}$.

Example:

$$A = \begin{bmatrix} 1 & 3 & 5 & 7 \\ 1 & 2 & 3 & 4 \\ 2 & 5 & 8 & 11 \end{bmatrix}; \quad \mathbf{b} = \begin{bmatrix} 4 \\ 3 \\ 7 \end{bmatrix} \tag{A.16}$$

The vector $\mathbf{b}$ is the sum of the first two columns of the matrix A. Therefore, $\mathbf{b}$ belongs to the column or the range space of the matrix A, and Equation A.1 is solvable. As shown in Example A.1, $rank(A) = 2$, and ignoring the last row,

$$\begin{bmatrix} 1 & 3 & 5 & 7 \\ 1 & 2 & 3 & 4 \end{bmatrix} \begin{bmatrix} x_1 \\ x_2 \\ x_3 \\ x_4 \end{bmatrix} = \begin{bmatrix} 4 \\ 3 \end{bmatrix} \tag{A.17}$$

Rewriting this equation as

$$\begin{bmatrix} 1 & 3 \\ 1 & 2 \end{bmatrix} \begin{bmatrix} x_1 \\ x_2 \end{bmatrix} = \begin{bmatrix} 4 - 5x_3 - 7x_4 \\ 3 - 3x_3 - 4x_4 \end{bmatrix} \tag{A.18}$$

Solving for x_1 and x_2,

$$\begin{bmatrix} x_1 \\ x_2 \end{bmatrix} = \begin{bmatrix} 1 & 3 \\ 1 & 2 \end{bmatrix}^{-1} \begin{bmatrix} 4 - 5x_3 - 7x_4 \\ 3 - 3x_3 - 4x_4 \end{bmatrix} = \begin{bmatrix} 1 + x_3 + 2x_4 \\ 1 - 2x_3 - 3x_4 \end{bmatrix} \tag{A.19}$$

A particular solution is generated by arbitrarily setting $x_3 = x_4 = 0$:

$$\mathbf{x}_p = \begin{bmatrix} x_1 \\ x_2 \\ x_3 \\ x_4 \end{bmatrix} = \begin{bmatrix} 1 \\ 1 \\ 0 \\ 0 \end{bmatrix} \tag{A.20}$$

The independent solutions in the null space have already been generated (A.10 and A.11). Therefore, all the infinite many solutions can be expressed as

$$\mathbf{x} = \begin{bmatrix} 1 \\ 1 \\ 0 \\ 0 \end{bmatrix} + \alpha \begin{bmatrix} 1 \\ -2 \\ 1 \\ 0 \end{bmatrix} + \beta \begin{bmatrix} 2 \\ -3 \\ 0 \\ 1 \end{bmatrix} \tag{A.21}$$

where α and β are arbitrary real numbers.

Theorem A.4

Consider the system of equations (A.1), which does not satisfy the solvability condition (A.3). In this case, $m > n$, and the following error vector $\boldsymbol{\varepsilon}$ will always be nonzero:

$$\boldsymbol{\varepsilon} = A\mathbf{x} - \mathbf{b} \neq 0 \tag{A.22}$$

Then, the solution

$$\mathbf{x} = (A^T A)^{-1} A^T \mathbf{b} \tag{A.23}$$

minimizes the following least squares error *LSE*:

$$LSE = \sum_{i=1}^{m} \varepsilon_i^2 = \boldsymbol{\varepsilon}^T \boldsymbol{\varepsilon} \tag{A.24}$$

A Systematic Test for Linear Independence of Vectors

Consider a set of vectors $\mathbf{x}_1$, $\mathbf{x}_2,\ldots,$ $\mathbf{x}_n$, with each vector having m elements. First, a *scalar* or *inner* product of any two vectors is defined as follows:

$$\left\langle \mathbf{x}_i, \mathbf{x}_j \right\rangle = \mathbf{x}_i^T \mathbf{x}_j = \mathbf{x}_j^T \mathbf{x}_i = \sum_{\ell=1}^{m} x_{i\ell} x_{j\ell} \tag{A.25}$$

To test linear independence of vectors, the following relationship is considered:

$$\alpha_1 \mathbf{x}_1 + \alpha_2 \mathbf{x}_2 + \ldots + \alpha_n \mathbf{x}_n = 0 \tag{A.26}$$

Taking the scalar product of (A.26) with the vector $\mathbf{x}_i$,

$$\alpha_1 \left\langle \mathbf{x}_i, \mathbf{x}_1 \right\rangle + \alpha_2 \left\langle \mathbf{x}_i, \mathbf{x}_2 \right\rangle + \ldots + \alpha_n \left\langle \mathbf{x}_i, \mathbf{x}_n \right\rangle = 0; \quad i = 1, 2,\ldots, n \tag{A.27}$$

Putting (A.27) in matrix form,

$$G\boldsymbol{\alpha} = 0 \tag{A.28}$$

where

$$G = \begin{bmatrix} \left\langle \mathbf{x}_1, \mathbf{x}_1 \right\rangle & \left\langle \mathbf{x}_1, \mathbf{x}_2 \right\rangle & . & . & \left\langle \mathbf{x}_1, \mathbf{x}_n \right\rangle \\ \left\langle \mathbf{x}_2, \mathbf{x}_1 \right\rangle & \left\langle \mathbf{x}_2, \mathbf{x}_2 \right\rangle & . & . & \left\langle \mathbf{x}_2, \mathbf{x}_n \right\rangle \\ . & . & . & . & . \\ . & . & . & . & . \\ \left\langle \mathbf{x}_n, \mathbf{x}_1 \right\rangle & \left\langle \mathbf{x}_n, \mathbf{x}_2 \right\rangle & . & . & \left\langle \mathbf{x}_n, \mathbf{x}_n \right\rangle \end{bmatrix} \tag{A.29}$$

and

$$\boldsymbol{\alpha} = \begin{bmatrix} \alpha_1 & \alpha_2 & . & . & \alpha_n \end{bmatrix}^T \tag{A.30}$$

Note that the matrix G is symmetric. From (A.28),

$$\boldsymbol{\alpha} = 0 \text{ if and only if } \det G \neq 0 \tag{A.31}$$

Therefore, vectors $\mathbf{x}_1$, $\mathbf{x}_2,\ldots,$ $\mathbf{x}_n$ are independent if and only if $\det G \neq 0$. detG is known as Gramian or Gram determinant.

A.2 EIGENVALUES AND EIGENVECTORS (STRANG, 1988)

The eigenvalue/eigenvector relationship for a square matrix A of order n is given by

$$A\mathbf{v} = \lambda\mathbf{v} \qquad (A.32)$$

where λ is an eigenvalue, and $\mathbf{v}$ is the corresponding eigenvector. Equation A.32 can also be written as

$$(\lambda I - A)\mathbf{v} = \mathbf{0} \qquad (A.33)$$

For nontrivial solution of $\mathbf{v}$,

$$\det(\lambda I - A) = 0 \qquad (A.34)$$

Equation A.34 is a polynomial equation of order n. Therefore, there are n roots or n eigenvalues of the matrix A. Let these eigenvalues be denoted as $\lambda_1, \lambda_2, \ldots, \lambda_n$. Then, the eigenvector $\mathbf{v}_i$ corresponding to the eigenvalue λ_i is determined by solving

$$(\lambda_i I - A)\mathbf{v}_i = \mathbf{0} \qquad (A.35)$$

It should be noted that the eigenvector $\mathbf{v}_i$ belongs to the null space of $(\lambda_i I - A)$.

A.3 MATRIX INVERSION LEMMA

$$(A + BCD)^{-1} = A^{-1} - A^{-1}B(C^{-1} + DA^{-1}B)^{-1}DA^{-1} \qquad (A.36)$$

Proof

Consider the following relationship:

$$(A + BCD)(A + BCD)^{-1} = I \qquad (A.37)$$

or

$$A(A + BCD)^{-1} + BCD(A + BCD)^{-1} = I \qquad (A.38)$$

or

$$(A + BCD)^{-1} = A^{-1} - A^{-1}BCD(A + BCD)^{-1} \qquad (A.39)$$

or

$$(A + BCD)^{-1} = A^{-1} - A^{-1}B[A(I + A^{-1}BCD)(CD)^{-1}]^{-1} \qquad \text{(A.40)}$$

or

$$(A + BCD)^{-1} = A^{-1} - A^{-1}B[A(CD)^{-1} + A^{-1}B)]^{-1} \qquad \text{(A.41)}$$

or

$$(A + BCD)^{-1} = A^{-1} - A^{-1}B(D^{-1}C^{-1} + A^{-1}B)^{-1}D^{-1}DA^{-1} \qquad \text{(A.42)}$$

or

$$(A + BCD)^{-1} = A^{-1} - A^{-1}B[D(D^{-1}C^{-1} + A^{-1}B)]^{-1}DA^{-1} \qquad \text{(A.43)}$$

or

$$(A + BCD)^{-1} = A^{-1} - A^{-1}B(C^{-1} + DA^{-1}B)^{-1}DA^{-1} \qquad \text{(A.44)}$$

This completes the proof.

Special Case

Let $A = C = D = I$. Then, the matrix inversion lemma yields

$$(I + B)^{-1} = I - B(I + B)^{-1} \qquad \text{(A.45)}$$

Appendix B: Quadratic Functions, Important Derivatives, Fourier Integrals, and Parseval's Relation

B.1 QUADRATIC FUNCTIONS

With a single variable x_1, the quadratic function is simply

$$I = \alpha x_1^2 \tag{B.1}$$

With two variables x_1 and x_2, the quadratic function is

$$I = \alpha x_1^2 + \beta x_2^2 + \gamma x_1 x_2 \tag{B.2}$$

Equation B.2 can be expressed as

$$I = \mathbf{x}^T P \mathbf{x} \tag{B.3}$$

where

$$\mathbf{x} = \begin{bmatrix} x_1 & x_2 \end{bmatrix}^T \tag{B.4}$$

and P is a nonunique matrix; for example,

$$P = \begin{bmatrix} \alpha & 0 \\ \gamma & \beta \end{bmatrix} \quad \text{or} \quad P = \begin{bmatrix} \alpha & -\gamma \\ 2\gamma & \beta \end{bmatrix} \quad \text{or} \quad P = \begin{bmatrix} \alpha & \gamma/2 \\ \gamma/2 & \beta \end{bmatrix} \tag{B.5}$$

Out of three examples of P shown in (B.5), note that one of them is a symmetric matrix.

A quadratic function in n variables can always be written as Equation B.3 where

$$\mathbf{x} = \begin{bmatrix} x_1 & x_2 & \cdot & \cdot & x_n \end{bmatrix}^T \tag{B.6}$$

and P is an $n \times n$ matrix that can always be chosen to be symmetric. Usually, the matrix P is chosen to be symmetric without any loss of generality. In fact, Equation B.3 can always be expressed as

$$I = \mathbf{x}^T P \mathbf{x} = \mathbf{x}^T \left(\frac{P + P^T}{2} \right) \mathbf{x} + \mathbf{x}^T \left(\frac{P - P^T}{2} \right) \mathbf{x} \tag{B.7}$$

The matrix $(P + P^T)/2$ is symmetric, whereas the matrix $(P - P^T)/2$ is skew symmetric. For any skew-symmetric matrix,

$$\mathbf{x}^T \left(\frac{P - P^T}{2} \right) \mathbf{x} = 0 \tag{B.8}$$

Therefore, from (B.7),

$$I = \mathbf{x}^T P \mathbf{x} = \mathbf{x}^T \left(\frac{P + P^T}{2} \right) \mathbf{x} \tag{B.9}$$

Definition B.1: Positive Definite Quadratic Function

For a positive definite quadratic function I,

$$I = \mathbf{x}^T P \mathbf{x} > 0 \quad \text{when} \quad \mathbf{x} \neq 0 \tag{B.10}$$

Definition B.2: Positive Semidefinite Quadratic Function

For a positive semidefinite quadratic function I,

$$I = \mathbf{x}^T P \mathbf{x} \geq 0 \quad \text{when} \quad \mathbf{x} \neq 0 \tag{B.11}$$

As explained earlier, the matrix P is chosen to be symmetric in (B.10) and (B.11) without any loss of generality; i.e., $P = P^T$.

Fact B.1

All eigenvalues of a symmetric matrix are real, and we can always find n orthogonal eigenvectors irrespective of whether eigenvalues are distinct or repeated (Strang, 1988).

Definition B.3: Positive Definite Matrix

For a positive definite matrix P,

$$\mathbf{x}^T P \mathbf{x} > 0 \quad \text{when} \quad \mathbf{x} \neq 0 \tag{B.12}$$

A positive definite symmetric matrix is denoted as

$$P = P^T > 0 \tag{B.13}$$

Fact B.2

All eigenvalues of a positive definite symmetric matrix $P = P^T$ are positive and nonzero numbers (Strang, 1988).

Definition B.4: Positive Semidefinite Matrix

For a positive semidefinite matrix P,

$$\mathbf{x}^T P \mathbf{x} \geq 0 \quad \text{when} \quad \mathbf{x} \neq 0 \tag{B.14}$$

A positive semidefinite symmetric matrix is denoted as

$$P = P^T \geq 0 \tag{B.15}$$

Fact B.3

All eigenvalues of a positive semidefinite symmetric matrix $P = P^T$ are zero or positive numbers (Strang, 1988).

B.2 DERIVATIVE OF A QUADRATIC FUNCTION

Fact B.5

Let

$$I = \mathbf{x}^T P \mathbf{x} \; ; \quad P = P^T \tag{B.16}$$

Then,

$$\frac{\partial I}{\partial \mathbf{x}} = \left[\frac{\partial I}{\partial x_1} \quad \frac{\partial I}{\partial x_2} \quad . \quad . \quad \frac{\partial I}{\partial x_n} \right]^T = 2P\mathbf{x} \tag{B.17}$$

Proof

Denote

$$
P = \begin{bmatrix}
p_{11} & p_{12} & \cdot & \cdot & p_{1n} \\
p_{21} & p_{22} & \cdot & \cdot & p_{2n} \\
\cdot & & \cdot & \cdot & \cdot \\
\cdot & & \cdot & \cdot & \cdot \\
p_{n1} & p_{n2} & \cdot & \cdot & p_{nn}
\end{bmatrix}
\tag{B.18}
$$

The condition $P = P^T$ is expressed as

$$
p_{ij} = p_{ji}
\tag{B.19}
$$

Equation B.16 can be written as

$$
\begin{aligned}
I = x_1 & \left[p_{11}x_1 + p_{12}x_2 + \ldots + p_{1n}x_n \right] \\
& + \quad x_2 \left[p_{21}x_1 + p_{22}x_2 + \ldots + p_{2n}x_n \right] \\
& + \quad x_3 \left[p_{31}x_1 + p_{32}x_2 + \ldots + p_{3n}x_n \right] \\
& \;\; \cdot \\
& \;\; \cdot \\
& + \quad x_n \left[p_{n1}x_1 + p_{n2}x_2 + \ldots + p_{nn}x_n \right]
\end{aligned}
\tag{B.20}
$$

Therefore,

$$
\frac{\partial I}{\partial x_1} = 2p_{11}x_1 + (p_{12}x_2 + \ldots + p_{1n}x_n) + p_{21}x_2 + \ldots p_{n1}x_n
\tag{B.21}
$$

Because of (B.19),

$$
\frac{\partial I}{\partial x_1} = 2(p_{11}x_1 + p_{12}x_2 + \ldots + p_{1n}x_n)
\tag{B.22}
$$

In general,

$$
\frac{\partial I}{\partial x_i} = 2(p_{i1}x_1 + p_{i2}x_2 + \ldots + p_{in}x_n)
\tag{B.23}
$$

Therefore,

$$
\begin{bmatrix} \dfrac{\partial I}{\partial x_1} \\[2mm] \dfrac{\partial I}{\partial x_2} \\[2mm] \cdot \\ \cdot \\ \dfrac{\partial I}{\partial x_n} \end{bmatrix} = 2 \begin{bmatrix} p_{11} & p_{12} & \cdot & \cdot & p_{1n} \\ p_{21} & p_{22} & \cdot & \cdot & p_{2n} \\ \cdot & & \cdot & \cdot & \cdot \\ \cdot & & \cdot & \cdot & \cdot \\ p_{n1} & p_{n2} & \cdot & \cdot & p_{nn} \end{bmatrix} \begin{bmatrix} x_1 \\ x_2 \\ \cdot \\ \cdot \\ x_n \end{bmatrix} \tag{B.24}
$$

or

$$
\frac{\partial I}{\partial \mathbf{x}} = 2P\mathbf{x} \tag{B.25}
$$

This completes the proof.

B.3 DERIVATIVE OF A LINEAR FUNCTION

Fact B.6

Let

$$
J = \boldsymbol{\lambda}^T \mathbf{x} \tag{B.26}
$$

where

$$
\boldsymbol{\lambda} = \begin{bmatrix} \lambda_1 & \lambda_2 & \cdot & \cdot & \lambda_n \end{bmatrix}^T \tag{B.27}
$$

Then,

$$
\frac{\partial J}{\partial \mathbf{x}} = \begin{bmatrix} \dfrac{\partial J}{\partial x_1} & \dfrac{\partial J}{\partial x_2} & \cdot & \cdot & \dfrac{\partial J}{\partial x_n} \end{bmatrix}^T = \boldsymbol{\lambda} \tag{B.28}
$$

Proof

From (B.26),

$$
J = \lambda_1 x_1 + \lambda_2 x_2 + \ldots + \lambda_n x_n \tag{B.29}
$$

Then,

$$\frac{\partial J}{\partial x_i} = \lambda_i \ ; \quad i = 1, 2, \ldots, n \tag{B.30}$$

$$\frac{\partial J}{\partial \mathbf{x}} = \left[\frac{\partial J}{\partial x_1} \quad \frac{\partial J}{\partial x_2} \quad . \quad . \quad \frac{\partial J}{\partial x_n} \right]^T = [\lambda_1 \quad \lambda_2 \quad . \quad . \quad \lambda_n]^T = \boldsymbol{\lambda} \tag{B.31}$$

This completes the proof.

B.4 FOURIER INTEGRALS AND PARSEVAL'S THEOREM

B.4.1 SCALAR SIGNAL

Consider a signal $x(t)u_s(t) \in L_2$ where the L_2 space is defined in Chapter 5, Section 5.5.1, and $u_s(t)$ is the unit step function defined as

$$u_s(t) = \begin{cases} 1 & for \quad t \geq 0 \\ 0 & for \quad t < 0 \end{cases} \tag{B.32}$$

The Fourier transform (Chen, 1989) of $x(t)u_s(t) \in L_2$ is defined as

$$x(j\omega) = \int_0^\infty x(t)e^{-j\omega t}\,dt \tag{B.33}$$

with the following property for a real function $x(t)$:

$$x(-j\omega) = x^*(j\omega) \tag{B.34}$$

where $x^*(j\omega)$ is the complex conjugate of $x(j\omega)$. Corresponding to (B.33), the inverse Fourier transform is described as

$$x(t) = \frac{1}{2\pi} \int_{-\infty}^\infty x(j\omega)e^{j\omega t}\,d\omega \tag{B.35}$$

The Laplace transform of $x(t)u_s(t)$ is defined as

$$x(s) = \int_0^\infty x(t)e^{-st}\,dt \tag{B.36}$$

Comparing (B.33) and (B.36), $x(j\omega)$ can also be obtained by replacing s in the Laplace transform $x(s)$ by $j\omega$.

EXAMPLE B.1.1

$$x(t) = (\sin bt)u_s(t) \tag{B.37}$$

where the unit step function $u_s(t)$ is defined by Equation B.32.

Note that $x(t) \notin L_2$. Therefore, its Fourier transform is not defined. However, its Laplace transform is defined and is well known to be

$$x(s) = \frac{b}{s^2 + b^2} \tag{B.38}$$

Therefore, one can substitute $s = j\omega$ in (B.38), but it will not be the Fourier transform.

Seo and Chen (1986) and Chen (1989) have used impulse functions to describe the Fourier transform of a periodic sinusoidal function: $x(t) = \sin bt$.

EXAMPLE B.1.2

$$x(t) = (e^{-at}\sin bt)u_s(t) ; \quad a > 0. \tag{B.39}$$

where the unit step function $u_s(t)$ is defined by (B.32).

The Laplace transform of the signal (B.39) is

$$x(s) = \frac{b}{(s + a)^2 + b^2} \tag{B.40}$$

Because $x(t) \in L_2$, the Fourier transform can be obtained by substituting $s = j\omega$ in (B.40):

$$x(j\omega) = \frac{b}{(j\omega + a)^2 + b^2} \tag{B.41}$$

Scalar Parseval Relation

$$\int_0^\infty |x(t)|^2\,dt = \frac{1}{2\pi}\int_{-\infty}^{+\infty} |x(j\omega)|^2\,d\omega \tag{B.42}$$

where $x(j\omega)$ is the Fourier transform of $x(t) \in L_2$.

Proof

Using (B.35),

$$\int_0^\infty |x(t)|^2\,dt = \int_0^\infty x(t)x^*(t)\,dt = \int_0^\infty x(t)\left[\frac{1}{2\pi}\int_{-\infty}^{+\infty} x(j\omega)e^{j\omega t}\,d\omega\right]^*\,dt \tag{B.43}$$

Interchanging the order of integrations on the right side of (B.43),

$$\int_0^\infty |x(t)|^2\,dt = \frac{1}{2\pi}\int_{-\infty}^{+\infty} x^*(j\omega)\left[\int_0^\infty x(t)e^{-j\omega t}\,dt\right]\,d\omega \tag{B.44}$$

Using the definition (B.33),

$$\int_0^\infty |x(t)|^2\,dt = \frac{1}{2\pi}\int_{-\infty}^{\infty} x^*(j\omega)x(j\omega)\,d\omega = \frac{1}{2\pi}\int_{-\infty}^{\infty} |x(j\omega)|^2\,d\omega \tag{B.45}$$

This completes the proof.

B.4.2 VECTOR SIGNAL

For an n-dimensional vector signal $\mathbf{x}(t)u_s(t) \in L_2$ with the unit step function $u_s(t)$ defined by (B.32), the Fourier transform pair is defined as

$$\mathbf{x}(j\omega) = \int_0^\infty \mathbf{x}(t)e^{-j\omega t}\,dt \tag{B.46}$$

and

$$\mathbf{x}(t) = \frac{1}{2\pi}\int_{-\infty}^{\infty} \mathbf{x}(j\omega)e^{j\omega t}\,d\omega \tag{B.47}$$

Multivariable Parseval Relation

For a symmetric matrix Q,

$$\int_0^\infty \mathbf{x}^H(t)Q\mathbf{x}(t)dt = \frac{1}{2\pi}\int_{-\infty}^{+\infty}\mathbf{x}^H(j\omega)Q\mathbf{x}(j\omega)d\omega \qquad (B.48)$$

where $\mathbf{x}(j\omega)$ is the Fourier transform of $\mathbf{x}(t) \in L_2$.

Proof

Using (B.47),

$$\int_0^\infty \mathbf{x}^H(t)Q\mathbf{x}(t)dt = \int_0^\infty \frac{1}{2\pi}\int_{-\infty}^{+\infty}[\mathbf{x}(j\omega)e^{j\omega t}d\omega]^H Q\mathbf{x}(t)dt \qquad (B.49)$$

Interchanging the order of integrations in (B.49),

$$\int_0^\infty \mathbf{x}^H(t)Q\mathbf{x}(t)dt = \frac{1}{2\pi}\int_{-\infty}^\infty \mathbf{x}^H(j\omega)Q\int_0^\infty e^{-j\omega t}\mathbf{x}(t)dtd\omega \qquad (B.50)$$

Using (B.46),

$$\int_0^\infty \mathbf{x}^H(t)Q\mathbf{x}(t)dt = \frac{1}{2\pi}\int_{-\infty}^\infty \mathbf{x}^H(j\omega)Q\mathbf{x}(j\omega)d\omega \qquad (B.51)$$

This completes the proof.

Appendix C:
Norms, Singular Values,
Supremum, and Infinimum

C.1 VECTOR NORMS

For a complex or real vector, the norm $\|.\|$ is a real valued function with the following properties (Vidyasagar, 1993):

i. $\|\mathbf{x}\| > 0$ for all $\mathbf{x} \neq 0$ (C.1)

ii. $\|\mathbf{x}\| = 0$ implies $\mathbf{x} = 0$ (C.2)

iii. $\|\alpha\mathbf{x}\| = |\alpha|\|\mathbf{x}\|$ for real or complex number α (C.3)

iv. $\|\mathbf{x} + \mathbf{y}\| \leq \|\mathbf{x}\| + \|\mathbf{y}\|$ for any complex or real vectors $\mathbf{x}$ and $\mathbf{y}$ (C.4)

Three commonly used norms for a complex or real vector $\mathbf{x}$ are defined as

$$\|\mathbf{x}\|_p = \left(|x_1|^p + |x_2|^p + \ldots |x_n|^p\right)^{1/p} \quad \text{where p = 1, 2, and } \infty \qquad (C.5)$$

Note that

$$\|\mathbf{x}\|_\infty = \max_i |x_i| \qquad (C.6)$$

C.2 MATRIX NORMS

For a vector norm, the induced norm of a real or complex matrix A is defined as

$$\|A\|_p = \sup_{\mathbf{x} \neq 0} \frac{\|A\mathbf{x}\|_p}{\|\mathbf{x}\|_p} = \sup_{\|\mathbf{x}\|=1} \|A\mathbf{x}\| \qquad (C.7)$$

It has been shown (Vidyasagar, 1993) that

$$\|A\|_1 = \max_j \sum_i |a_{ij}| ; \quad \text{maximum column sum} \tag{C.8}$$

$$\|A\|_\infty = \max_i \sum_j |a_{ij}| ; \quad \text{maximum row sum} \tag{C.9}$$

$$\|A\|_2 = \max_i (\lambda_i (A^H A)^{1/2}) \tag{C.10}$$

where A^H is the complex conjugate transpose of the matrix A, and $\lambda_i(A^H A)$ is the ith eigenvalue of $A^H A$.

Fact C.1

The matrix $A^H A$ is positive semidefinite.

Proof

For a vector $\mathbf{x}$, the quadratic form is

$$q = \mathbf{x}^H A^H A \mathbf{x} \tag{C.11}$$

Let

$$\mathbf{y} = A\mathbf{x} \tag{C.12}$$

Then,

$$q = \mathbf{y}^H \mathbf{y} \geq 0 \tag{C.13}$$

Because of this fact, eigenvalues of $A^H A$ are guaranteed to be nonnegative.

EXAMPLE

$$A = \begin{bmatrix} 3 & -3 & -1 \\ -1 & 3 & -1 \\ 1 & 2 & 2 \end{bmatrix} \tag{C.14}$$

Here,

$$\|A\|_1 = 8 \ , \ \|A\|_\infty = 7 \ \text{ and } \|A\|_2 = 5.33$$

C.3 SINGULAR VALUES OF A MATRIX

Singular values $\sigma_i(A)$ of a real or complex $m \times n$ matrix A are nonnegative square roots of eigenvalues of $A^H A$. The dimension of $A^H A$ is $n \times n$. Therefore, there will be n eigenvalues of $A^H A$. However, if the rank of the matrix $A^H A$ is r, $n-r$ eigenvalues of $A^H A$ will be zero. Also, note that

$$r \le p \quad \text{where} \quad p = \min(m,n) \tag{C.15}$$

Therefore, the number of singular values is chosen to be p, which is the maximum rank that a matrix A can have. Then, if there is any singular value which is zero, the matrix will not be of full rank. These singular values are usually arranged in *descending* order.

In summary, singular values of a real or a complex matrix A are given as

$$\sigma_i(A) = \sqrt{\lambda_i(A^H A)} \ ; \quad i = 1,2,\ldots,p \tag{C.16}$$

$$\sigma_1(A) \ge \sigma_2(A) \ge \ldots \ge \sigma_p(A) \tag{C.17}$$

Maximum and minimum singular values are also denoted by $\bar{\sigma}$ and $\underline{\sigma}$, respectively. Therefore,

$$\bar{\sigma}(A) = \max_{x \ne 0} \frac{\|Ax\|_2}{\|x\|_2} = \|A\|_2 \tag{C.18}$$

and

$$\underline{\sigma}(A) = \left[\max_{x \ne 0} \frac{\|A^{-1}x\|_2}{\|x\|_2} \right]^{-1} = \left[\|A^{-1}\|_2 \right]^{-1} \quad \text{if } A^{-1} \text{ exists} \tag{C.19}$$

The minimum singular value, $\underline{\sigma}(A)$, is a measure of how near the matrix A is to being singular or rank deficient. The maximum singular value, $\bar{\sigma}(A)$, is also the 2-norm of the matrix $\|A\|_2$.

C.4 SINGULAR VALUE DECOMPOSITION (SVD)

For an $m \times n$ matrix A, the SVD of the matrix A is given (Zhou et al., 1996) by

$$A = U\Sigma V^H \tag{C.20}$$

where U and V are $m \times m$ and $n \times n$ unitary matrices, respectively; i.e.,

$$UU^H = I_m \quad \text{and} \quad VV^H = I_n \tag{C.21a,b}$$

These results imply that the 2-norms of columns of U and V are unity. Furthermore, column vectors of U are orthogonal with respect to each other. Similarly, column vectors of V are orthogonal with respect to each other.

Denote

$$U = \begin{bmatrix} \mathbf{u}_1 & \mathbf{u}_2 & . & . & \mathbf{u}_m \end{bmatrix} \tag{C.22}$$

$$V = \begin{bmatrix} \mathbf{v}_1 & \mathbf{v}_2 & . & . & \mathbf{v}_n \end{bmatrix} \tag{C.23}$$

where $\mathbf{u}_i$ and $\mathbf{v}_i$ are ith column vectors of matrices U and V, respectively. The matrix Σ is an $m \times n$ matrix with the following structures:

$$\Sigma = \begin{bmatrix} \Sigma_1 & 0 \end{bmatrix} \quad \text{if } m < n \tag{C.24}$$

$$\Sigma = \begin{bmatrix} \Sigma_1 \\ 0 \end{bmatrix} \quad \text{if } m > n \tag{C.25}$$

$$\Sigma = \Sigma_1 \quad \text{if } m = n \tag{C.26}$$

where

$$\Sigma_1 = diag(\sigma_1 \quad \sigma_2 \quad . \quad . \quad \sigma_p) \tag{C.27}$$

COMPUTATION OF U AND V

Postmultiplying (C.20) by V,

$$AV = U\Sigma \tag{C.28}$$

Premultiplying (C.20) by U^H,

$$U^H A = \Sigma V^H \tag{C.29}$$

Premultiplying (C.28) by A^H,

$$A^H A V = A^H U \Sigma = V \Sigma^H \Sigma \qquad \text{(C.30)}$$

or

$$A^H A \mathbf{v}_i = \sigma_i^2 \mathbf{v}_i \qquad \text{(C.31)}$$

Therefore, columns of V are unit (right) eigenvectors of $A^H A$. They are called *right* singular vectors of the matrix A.

Postmultiplying (C.29) by A^H,

$$U^H A A^H = \Sigma V^H A^H = \Sigma \Sigma^H U^H \qquad \text{(C.32)}$$

or

$$\mathbf{u}_i^H A A^H = \sigma_i^2 \mathbf{u}_i^H \qquad \text{(C.33)}$$

Therefore, rows of U^H are unit (left) eigenvectors of AA^H. They are called *left* singular vectors of the matrix A.

SIGNIFICANCE OF SINGULAR VECTORS CORRESPONDING TO THE MAXIMUM AND MINIMUM SINGULAR VALUES

Let $\bar{v}$ and $\bar{u}$ be the right and left singular vectors corresponding to the maximum singular value, $\bar{\sigma}$, respectively. Similarly, let $\underline{\mathbf{v}}$ and $\underline{\mathbf{u}}$ be the right and left singular vectors corresponding to the minimum singular value, $\underline{\sigma}$, respectively. As singular values are arranged in descending order,

$$\underline{\sigma} = \sigma_p \quad \text{and} \quad \bar{\sigma} = \sigma_1 \qquad \text{(C.34)}$$

$$\underline{\mathbf{u}} = \mathbf{u}_p \quad \text{and} \quad \bar{\mathbf{u}} = \mathbf{u}_1 \qquad \text{(C.35)}$$

$$\underline{\mathbf{v}} = \mathbf{v}_p \quad \text{and} \quad \bar{\mathbf{v}} = \mathbf{v}_1 \qquad \text{(C.36)}$$

With these notations,

$$A\bar{\mathbf{v}} = \bar{\sigma}.\bar{\mathbf{u}} \quad \text{and} \quad A\underline{\mathbf{v}} = \underline{\sigma}\underline{\mathbf{u}} \qquad \text{(C.37a,b)}$$

If we think of a system with $\mathbf{v}_i$ and $\sigma_i \mathbf{u}_i$ as the input and output, respectively, the 2-norms of the output and input are σ_i and 1, respectively. Therefore, the vector $\bar{\mathbf{v}}(\underline{\mathbf{v}})$ represents the input direction leading to the maximum (minimum) magnitude of the output, which occurs along the $\bar{\mathbf{u}}(\underline{\mathbf{u}})$ direction (Maciejowski, 1989; Zhou et al., 1996).

C.5 PROPERTIES OF SINGULAR VALUES (CHIANG AND SAFANOV, 1992)

1. $\bar{\sigma}(A) = \max_{x \neq 0} \frac{\|A\mathbf{x}\|_2}{\|\mathbf{x}\|_2} = \|A\|_2$ (C.38)

2. $\underline{\sigma}(A) = \min_{x \neq 0} \frac{\|A\mathbf{x}\|_2}{\|\mathbf{x}\|_2}$ (C.39)

3. If A^{-1} exists, $\underline{\sigma}(A) = \dfrac{1}{\bar{\sigma}(A^{-1})}$ and $\bar{\sigma}(A) = \dfrac{1}{\underline{\sigma}(A^{-1})}$ (C.40)

4. $\bar{\sigma}(\alpha A) = |\alpha| \bar{\sigma}(A)$ for a scalar α (C.41)

5. $\bar{\sigma}(A + B) \leq \bar{\sigma}(A) + \bar{\sigma}(B)$ (C.42)

6. $\bar{\sigma}(AB) \leq \bar{\sigma}(A).\bar{\sigma}(B)$ (C.43)

7. $\max[0, (\underline{\sigma}(A) - \bar{\sigma}(B))] \leq \underline{\sigma}(A + B) \leq \underline{\sigma}(A) + \bar{\sigma}(B)$ (C.44)

8. $\underline{\sigma}(A) \leq |\lambda(A)| \leq \bar{\sigma}(A)$ where $\lambda(A)$ is an eigenvalue of A (C.45)

9. $\displaystyle\sum_{i=1}^{n} \sigma_i^2 = trace(A^H A)$ (C.46)

C.6 SUPREMUM AND INFINIMUM (REDDY, 1986)

Let a and b be two real numbers such that $b > a$. Then, we define closed and open intervals as follows:

i. closed interval $[a, b] = \{x : x \in R, a \leq x \leq b\}$ (C.47)

It is clear that

$$\text{maximum value of } x = b$$

$$\text{minimum value of } x = a$$

ii. open interval $(a,b) = \{x : x \in R, a < x < b\}$ (C.48)

Here, it is not possible to precisely define the maximum and minimum values of x. But, upper and lower bounds can be defined.

Any number greater than b is an upper bound of x. However, b is the *least upper bound* and is called the **supremum (sup)** of x. Similarly, any number smaller than a is a lower bound of x. However, a is the *greatest lower bound* and is called the **infinimum (inf)** of x.

Appendix D:
Stochastic Processes

A stochastic process $x(t)$ refers to an ensemble of functions (Papoulis, 1965), $x_1(t)$, $x_2(t)$, $x_3(t)$, ..., Figure D.1. For the purpose of contents in this book, the independent variable for these functions is restricted to be time t. Therefore, the sample number in Figure D.1 can be interpreted as time instants. At any time $t = t_i$, there will be different values of these functions, $x_1(t_i)$, $x_2(t_i)$, $x_3(t_i)$, ..., which can be plotted as a histogram. As the number of these function approaches infinity, this histogram will turn into a probability density function. In other words, there is a random variable $x(t_i)$ and the corresponding probability density function for each time t_i.

D.1 STATIONARY STOCHASTIC PROCESS

A stochastic process $x(t)$ is said to be stationary if the probability distribution function of $x(t_i)$ is independent of t_i; i.e., it is time invariant. The mean value and standard deviation of $x(t_i)$ are also time invariant.

The mean or expected value of $x(t)$ at any time t is given by

$$E(x(t)) = \int_{-\infty}^{\infty} x(t)p(x(t))dx = \text{a constant value, } \eta_x \qquad \text{(D.1)}$$

where $p(x)$ is the probability density function. For a stationary and Gaussian stochastic process, the probability density function is given by

$$p(x) = \frac{1}{\sigma\sqrt{2\pi}} e^{-\frac{(x-\eta_x)^2}{2\sigma_x^2}} \qquad \text{(D.2)}$$

where σ_x is the standard deviation; i.e.,

$$\sigma_x^2 = E((x-\eta)^2) = E(x^2) - \eta_x^2 \qquad \text{(D.3)}$$

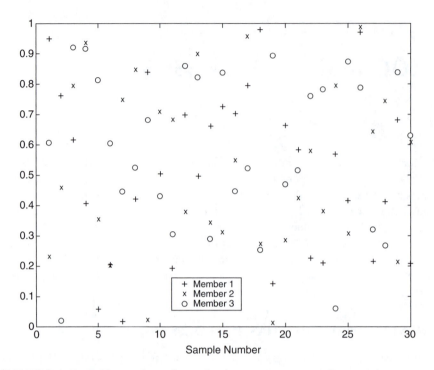

FIGURE D.1 Ensemble members of a stochastic process.

and

$$E(x^2) = \int_{-\infty}^{\infty} x^2 p(x) dx \qquad (D.4)$$

AUTOCORRELATION FUNCTION

The autocorrelation function of a stationary stochastic process $x(t)$ is defined as

$$R_{xx}(\tau) = E(x(t)x(t+\tau)) \qquad (D.5)$$

It should be noted that the autocorrelation function is only a function of τ and is independent of t. It should also be noted that

$$R_{xx}(0) = E(x^2(t)) \qquad (D.6)$$

$$R_{xx}(\tau) = R_{xx}(-\tau) \qquad (D.7)$$

Numerically, $R(\tau)$ will be computed from N observations of the stochastic process by

$$R_{xx}(\tau) = \frac{1}{N} \sum_{i=1}^{N} x_i(t)x_i(t+\tau) \tag{D.8}$$

AUTOCOVARIANCE FUNCTION

The autocovariance function of a stationary stochastic process $x(t)$ is defined as

$$C_{xx}(\tau) = E((x(t) - \eta_x)(x(t+\tau) - \eta_x)) \tag{D.9}$$

Therefore,

$$C_{xx}(\tau) = R_{xx}(\tau) - \eta_x^2 \tag{D.10}$$

and

$$C_{xx}(0) = \sigma_x^2 \tag{D.11}$$

CROSSCORRELATION FUNCTION

The crosscorrelation function for two different stationary stochastic processes $x(t)$ and $y(t)$ is defined as

$$R_{xy}(\tau) = E(x(t)y(t+\tau)) \tag{D.12}$$

It should be noted that the crosscorrelation function is only a function of τ and is independent of t. Numerically, $R_{xy}(\tau)$ will be computed from N observations of the stochastic process by

$$R_{xy}(\tau) = \frac{1}{N} \sum_{i=1}^{N} x_i(t)y_i(t+\tau) \tag{D.13}$$

CROSSCOVARIANCE FUNCTION

The crosscovariance function for two different stationary stochastic processes $x(t)$ and $y(t)$ is defined as

$$C_{xy}(\tau) = E((x(t) - \eta_x)(y(t+\tau) - \eta_y)) \tag{D.14}$$

Uncorrelated Stochastic Processes

Two stationary stochastic processes are uncorrelated if

$$R_{xy}(\tau) = E(x(t)y(t+\tau)) = E(x(t))E(y(t+\tau)) = \eta_x \eta_y \qquad (D.15)$$

Equivalently, two stationary stochastic processes are uncorrelated if

$$C_{xy}(\tau) = E((x(t) - \eta_x)(y(t+\tau) - \eta_y)) =$$
$$E((x(t) - \eta_x))E((y(t+\tau) - \eta_y)) = 0 \qquad (D.16)$$

A stationary stochastic process is called **ergodic** if all of its statistics can be determined from a single observation or a single member of the ensemble. Therefore, the mean, variance, autocorrelation function, and crosscorrelation function of ergodic processes can be computed as follows:

$$\eta_x = E(x(t)) = \lim_{T \to \infty} \frac{1}{2T} \int_{-T}^{+T} x(t)dt \qquad (D.17)$$

$$\sigma_x^2 = E((x(t) - \eta)^2) = \lim_{T \to \infty} \frac{1}{2T} \int_{-T}^{+T} (x(t) - \eta)^2 \, dt \qquad (D.18)$$

$$R_{xx}(\tau) = E(x(t)x(t+\tau)) = \lim_{T \to \infty} \frac{1}{2T} \int_{-T}^{+T} x(t)x(t+\tau)dt \qquad (D.19)$$

$$R_{xy}(\tau) = E(x(t)y(t+\tau)) = \lim_{T \to \infty} \frac{1}{2T} \int_{-T}^{+T} x(t)y(t+\tau)dt \qquad (D.20)$$

D.2 POWER SPECTRUM OR POWER SPECTRAL DENSITY (PSD)

The PSD, $S_{xx}(\omega)$, of a stationary stochastic process $x(t)$ is the Fourier transform of its autocorrelation function:

$$S_{xx}(\omega) = \int_{-\infty}^{\infty} e^{-j\omega\tau} R_{xx}(\tau)d\tau \qquad (D.21)$$

The inverse Fourier transform corresponding to Equation D.21 yields

$$R_{xx}(\tau) = \frac{1}{2\pi} \int_{-\infty}^{\infty} e^{j\omega\tau} S_{xx}(\omega) d\omega \qquad (D.22)$$

D.3 WHITE NOISE: A SPECIAL STATIONARY STOCHASTIC PROCESS

The autocorrelation function of the white noise, $v(t)$, is given by

$$R_{vv}(\tau) = P\delta(\tau) \qquad (D.23)$$

where P is a constant, and $\delta(\tau)$ is a delta-Dirac function. Equation D.23 indicates that

$$R_{vv}(\tau) = 0 \quad \text{when} \quad \tau \neq 0 \qquad (D.24)$$

Therefore, there is no correlation between $v(t_1)$ and $v(t_2)$ where $t_1 \neq t_2$. Because $R_{vv}(0)$ is infinite, the standard deviation of the process is infinite. The PSD of the white noise, $v(t)$, is given by

$$S_{vv}(\omega) = \int_{-\infty}^{\infty} e^{-j\omega\tau} P\delta(\tau) d\tau = P \qquad (D.25)$$

That is, the PSD of white noise is a constant for each frequency, and consequently, white noise will have infinite power. Thus, white noise is a mathematical fiction.

GENERATION OF APPROXIMATE GAUSSIAN WHITE NOISE IN MATLAB

Gaussian random numbers with the zero mean and the standard deviation σ can be generated as follows:

$$w(k) = \sigma r(k); \quad k = 1, 2, 3, \ldots \qquad (D.26)$$

where random numbers $r(k)$ wth zero mean and unity standard deviation are generated using the Matlab routine *randn*. Then, a time-domain signal is obtained by considering this sequence of numbers $w(k)$ as a discrete-time signal with a constant sampling interval T. The continuous-time signal $v(t)$ obtained via a zero-order hold as shown in Figure D.2 can be viewed as a Gaussian white noise with the power spectral density $\sigma^2 T$ (Meditch, 1969).

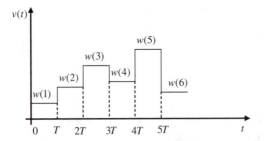

FIGURE D.2 Continuous-time white noise $v(t)$ from random numbers $w(k)$.

FIGURE D.3 A SISO system.

D.4 RESPONSE OF A SISO LINEAR AND TIME-INVARIANT SYSTEM SUBJECTED TO A STATIONARY STOCHASTIC PROCESS

Consider the linear system, Figure D.3, with the transfer function $G(s)$ where the input $x(t)$ is a stationary stochastic process with PSD equal to $S_{xx}(\omega)$. The output $y(t)$ will also be a stochastic process. For a stable system, the output will become a stationary stochastic process in the statistical steady state.

In the statistical steady state,

$$S_{yy}(\omega) = \left|G(j\omega)\right|^2 S_{xx}(\omega) \tag{D.27}$$

where $S_{yy}(\omega)$ is the PSD of the output $y(t)$. When the input is a white noise with unit PSD, i.e., $S_{xx}(\omega) = 1$,

$$S_{yy}(\omega) = \left|G(j\omega)\right|^2 \tag{D.28}$$

D.5 VECTOR STATIONARY STOCHASTIC PROCESSES

Define the vector

$$\mathbf{v}(t) = \begin{bmatrix} v_1(t) & v_2(t) & . & . & v_n(t) \end{bmatrix}^T \tag{D.29}$$

where $v_1(t), v_2(t), \ldots, v_n(t)$ are n scalar stochastic processes. Then, the following quantities are defined (Athans, 1992; Kwakernaak and Sivan, 1972):

Mean

$$\boldsymbol{\eta} = E(\mathbf{v}(t)) = \begin{bmatrix} E(v_1(t)) & E(v_2(t)) & . & . & E(v_n(t)) \end{bmatrix}^T =$$

$$\begin{bmatrix} \eta_1 & \eta_2 & . & . & \eta_n \end{bmatrix}^T$$

(D.30)

Autocorrelation Matrix

$$R(\tau) = E(\mathbf{v}(t)(\mathbf{v}(t+\tau))^T)$$

(D.31)

Autocovariance Matrix

$$C(\tau) = E((\mathbf{v}(t) - \boldsymbol{\eta})(\mathbf{v}(t+\tau) - \boldsymbol{\eta})^T)$$

(D.32)

Variance Matrix

$$\Xi = C(0) = E((\mathbf{v}(t) - \boldsymbol{\eta})(\mathbf{v}(t) - \boldsymbol{\eta})^T)$$

(D.33)

PSD Matrix

$$S(\omega) = \int_{-\infty}^{\infty} e^{-j\omega\tau} R(\tau) d\tau$$

(D.34)

$$R(\tau) = \frac{1}{2\pi} \int_{-\infty}^{\infty} e^{j\omega\tau} S(\omega) d\omega$$

(D.35)

Appendix E: Optimization of a Scalar Function with Constraints in the Form of a Symmetric Real Matrix Equal to Zero (Zhou et al., 1996)

Definition E.1

Suppose that an n-dimensional real vector $\mathbf{x}_o$ is a local minimum of the scalar function $f(\mathbf{x})$ subject to constraints $U(\mathbf{x}) = 0$ where $U(\mathbf{x})$ is a symmetric real matrix. Then, $\mathbf{x}_o$ is a *regular point* of the constraints if for a symmetric matrix P, $\nabla trace(U(\mathbf{x}_o)P) = 0$ has a unique solution $P = 0$.

Definition E.2

Let $\mathbf{x} = \begin{bmatrix} x_1 & x_2 & \cdot & \cdot & x_n \end{bmatrix}^T$. Then, the gradient $\nabla f(\mathbf{x})$ is defined as

$$\nabla f(\mathbf{x}) = \begin{bmatrix} \dfrac{\partial f}{\partial x_1} & \dfrac{\partial f}{\partial x_2} & \cdot & \cdot & \dfrac{\partial f}{\partial x_n} \end{bmatrix}^T \tag{E.1}$$

Definition E.3

Given an $m \times n$ matrix X, define the following vector:

$$\mathbf{x} = VecX = \begin{bmatrix} x_{11} & \cdot & x_{m1} & x_{12} & \cdot & x_{m2} & \cdot & x_{1n} & \cdot & x_{mn} \end{bmatrix}^T$$

Then, for a scalar function $f(\mathbf{x})$, the gradients with respect to the matrix X and $\mathbf{x} = VecX$ are defined as

$$\frac{\partial f(\mathbf{x})}{\partial X} = \begin{bmatrix} \dfrac{\partial f(\mathbf{x})}{\partial x_{11}} & \dfrac{\partial f(\mathbf{x})}{\partial x_{12}} & \cdot & \cdot & \dfrac{\partial f(\mathbf{x})}{\partial x_{1n}} \\ \cdot & \cdot & \cdot & \cdot & \cdot \\ \cdot & \cdot & \cdot & \cdot & \cdot \\ \dfrac{\partial f(\mathbf{x})}{\partial x_{m1}} & \dfrac{\partial f(\mathbf{x})}{\partial x_{m2}} & \cdot & \cdot & \dfrac{\partial f(\mathbf{x})}{\partial x_{mn}} \end{bmatrix} \qquad (E.2)$$

$$\nabla f(\mathbf{x}) =$$

$$\begin{bmatrix} \dfrac{\partial f(\mathbf{x})}{x_{11}} & \cdot & \dfrac{\partial f(\mathbf{x})}{x_{m1}} & \dfrac{\partial f(\mathbf{x})}{x_{12}} & \cdot & \dfrac{\partial f(\mathbf{x})}{x_{m2}} & \cdot & \dfrac{\partial f(\mathbf{x})}{x_{1n}} & \cdot & \dfrac{\partial f(\mathbf{x})}{x_{mn}} \end{bmatrix} \qquad (E.3)$$

It should be noted that

$$\frac{\partial f(\mathbf{x})}{\partial X} = 0 \qquad (E.4)$$

is equivalent to

$$\nabla f(\mathbf{x}) = 0 \qquad (E.5)$$

Fact E.1

Suppose that an n-dimensional real vector $\mathbf{x}_o$ is a local minimum of the scalar function $f(\mathbf{x})$ subject to constraints $U(\mathbf{x}) = 0$ where $U(\mathbf{x})$ is a symmetric real matrix. Also, assume that $\mathbf{x}_o$ is a *regular point* of the constraints. Then, there exists a unique multiplier symmetric matrix P such that if we define

$$g(\mathbf{x}, P) = f(\mathbf{x}) + trace(U(\mathbf{x})P) \qquad (E.6)$$

then at the optimal solution $\mathbf{x}_o, P_o,$

$$\nabla g(\mathbf{x}_0, P_o) = \nabla f(\mathbf{x}_0) + \nabla trace(U(\mathbf{x}_0)P_o) = 0 \qquad (E.7)$$

Note that the gradient here is defined with respect to the following vector:

$$\begin{bmatrix} \mathbf{x} \\ vecP \end{bmatrix} \qquad (E.9)$$

Fact E.2

$$\frac{\partial}{\partial X} tr(AXB) = A^T B^T \tag{E.10}$$

$$\frac{\partial}{\partial X} tr(AX^T B) = BA \tag{E.11}$$

$$\frac{\partial}{\partial X} tr(AXBX) = A^T X^T B^T + B^T X^T A^T \tag{E.12}$$

$$\frac{\partial}{\partial X} tr(AXBX^T) = A^T XB^T + AXB \tag{E.13}$$

Appendix F:
A Flexible Tetrahedral
Truss Structure

The tetrahedral flexible structure, Figure F.1, has been developed at the C. S. Draper Laboratory (Strunce et al., 1979). The structure is supported on the ground by three right-angled bipods on which actuators are mounted. The finite element model is represented by

$$M\ddot{\mathbf{p}} + K\mathbf{p}(t) = B_f \mathbf{f}(t) \tag{F.1}$$

where M and K are mass and stiffness matrices, respectively. The vector $\mathbf{p}(t)$ contains displacements in x, y, and z directions of nodes 1-4. The 3×1 force vector $\mathbf{f}(t)$ is composed of the forces provided by actuators mounted on elements 1 -3.

The modal matrix Φ and the diagonal eigenvalue matrix Λ are defined as follows:

$$K\Phi = M\Phi\Lambda \tag{F.2}$$

where

$$\Lambda = diag[\omega_1^2 \quad \omega_2^2 \quad . \quad . \quad \omega_{12}^2] \tag{F.3}$$

and

$$\Phi^T M \Phi = I \tag{F.4}$$

The modal matrix is provided in the report (Strunce et al., 1979) and the thesis (Kao, 1989). The natural frequencies are listed in Table F.1.

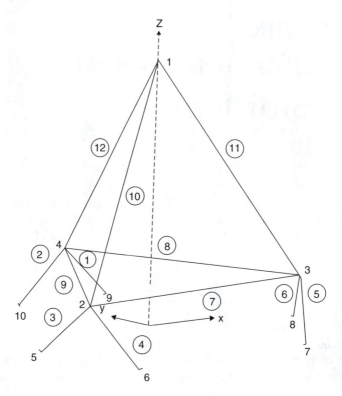

FIGURE F.1 A flexible tetrahedral truss structure.

TABLE F.1
Natural Frequencies (rad/sec)

ω_1	ω_2	ω_3	ω_4	ω_5	ω_6	ω_7	ω_8	ω_9	ω_{10}	ω_{11}	ω_{12}
1.342	1.643	3.391	4.207	4.658	4.754	8.538	4.754	8.538	9.252	10.296	12.923

Define modal coordinate vector $\boldsymbol{\psi}$ as

$$\mathbf{p}(t) = \Phi \boldsymbol{\psi}(t) \tag{F.5}$$

Substituting (F.5) into (F.1) and premultiplying both sides by Φ^T,

$$\ddot{\boldsymbol{\psi}} + \Lambda \boldsymbol{\psi}(t) = B_\psi \mathbf{f}(t) \tag{F.6}$$

where

$$B_\psi = \Phi^T B_f = \begin{bmatrix} -0.023 & -0.067 & -0.044 \\ -0.112 & 0.017 & 0.069 \\ -0.077 & 0.271 & 0.046 \\ 0.189 & -0.060 & -0.249 \\ 0.156 & -0.049 & 0.351 \\ -0.289 & 0.289 & -0.289 \\ -0.320 & -0.369 & -0.049 \\ 0.365 & 0.299 & -0.069 \\ -0.229 & 0.250 & 0.231 \\ 0.167 & -0.150 & -0.317 \\ -0.145 & 0.146 & -0.220 \\ 0.025 & -0.013 & 0.114 \end{bmatrix} \tag{F.7}$$

$$\boldsymbol{\psi}^T = [\psi_1 \quad \psi_2 \quad . \quad . \quad \psi_{12}] \tag{F.8}$$

Appendix G:
Space Shuttle Dynamics during Reentry (Doyle et al., 1986)

The 4-state rigid body aircraft model (Figure G.1) has state variables $\mathbf{x}(t)$ and measurements $\mathbf{y}_{meas}$:

$$\mathbf{x} = \begin{bmatrix} \beta \\ p \\ r \\ \phi \end{bmatrix} = \begin{bmatrix} sideslip & angle \\ roll & rate \\ yaw & rate \\ bank & angle \end{bmatrix} ; \quad \mathbf{y}_{meas} = \begin{bmatrix} p \\ r \\ \eta_y \\ \phi \end{bmatrix} \tag{G.1a,b}$$

where η_y is the lateral acceleration. In Figure G.1, α and V are the angle of attack and the velocity of the shuttle, respectively. Also,

$$\mathbf{c}_{fm} = \begin{bmatrix} c_y \\ c_n \\ c_l \end{bmatrix} = \begin{bmatrix} side & force \\ yawing & moment \\ rolling & moment \end{bmatrix} \text{ and } \mathbf{c}_{an} = \begin{bmatrix} c_\beta \\ c_\alpha \\ c_l \end{bmatrix} = \begin{bmatrix} sideslip & angle \\ aileron & angle \\ rudder & angle \end{bmatrix} \tag{G.2a,b}$$

Then,

$$\mathbf{c}_{fm} = C_{aer}\mathbf{c}_{an} \tag{G.3}$$

$$C_{aer} = \begin{bmatrix} c_{y\beta} & c_{ya} & c_{yr} \\ c_{n\beta} & c_{na} & c_{nr} \\ c_{l\beta} & c_{la} & c_{lr} \end{bmatrix} \tag{G.4}$$

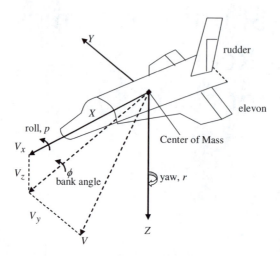

FIGURE G.1 Space shuttle variables (Doyle et al., 1986).

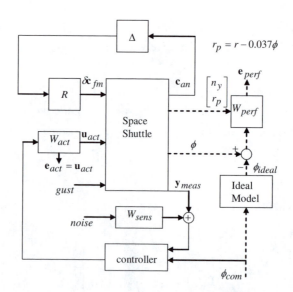

FIGURE G.2 Space shuttle control system (Doyle et al., 1986).

The elements of the aerodynamic coefficients matrix C_{aer} are uncertain. Let

$$C_{aer} = \bar{C}_{aer} + \delta C_{aer} \tag{G.5}$$

where

$$\delta C_{aer} = \begin{bmatrix} r_{y\beta}\delta_{y\beta} & r_{ya}\delta_{ya} & r_{yr}\delta_{yr} \\ r_{n\beta}\delta_{y\beta} & r_{na}\delta_{na} & r_{nr}\delta_{nr} \\ r_{l\beta}\delta_{l\beta} & r_{la}\delta_{la} & r_{lr}\delta_{lr} \end{bmatrix} = R\Delta \tag{G.6}$$

$$R = \begin{bmatrix} r_{y\beta} & 0 & 0 & r_{ya} & 0 & 0 & r_{yr} & 0 & 0 \\ 0 & r_{n\beta} & 0 & 0 & r_{na} & 0 & 0 & r_{nr} & 0 \\ 0 & 0 & r_{l\beta} & 0 & 0 & r_{la} & 0 & 0 & r_{lr} \end{bmatrix} \tag{G.7}$$

$$\Delta^T = \begin{bmatrix} \delta_{y\beta} & 0 & 0 & \delta_{ya} & 0 & 0 & \delta_{yr} & 0 & 0 \\ 0 & \delta_{n\beta} & 0 & 0 & \delta_{na} & 0 & 0 & \delta_{nr} & 0 \\ 0 & 0 & \delta_{l\beta} & 0 & 0 & \delta_{la} & 0 & 0 & \delta_{lr} \end{bmatrix} \tag{G.8}$$

Let

$$\mathbf{c}_{fm} = \bar{\mathbf{c}}_{fm} + \delta\mathbf{c}_{fm} \tag{G.9}$$

where $\bar{\mathbf{c}}_{fm}$ is the nominal value.
    Then,

$$\delta\mathbf{c}_{fm} = \delta C_{aer}\mathbf{c}_{an} = R\Delta\mathbf{c}_{an} \tag{G.10}$$

STATE SPACE MODEL

$$\dot{\mathbf{x}} = A\mathbf{x}(t) + B\mathbf{u}(t) \tag{G.11}$$

$$\mathbf{y}(t) = C\mathbf{x}(t) + D\mathbf{u}(t) \tag{G.12}$$

where

$$\mathbf{y} = \begin{bmatrix} \mathbf{c}_{an} \\ \mathbf{y}_{meas} \end{bmatrix}; \quad \mathbf{u} = \begin{bmatrix} \delta\mathbf{c}_{fm} \\ \mathbf{u}_{act} \\ gust \end{bmatrix}; \quad and \quad \mathbf{u}_{act} = \begin{bmatrix} u_e \\ u_r \end{bmatrix} \tag{G.13}$$

$$A = \begin{bmatrix} -9.460e-02 & 1.409e-01 & -9.900e-01 & 3.637e-02 \\ -3.595e+00 & -4.284e-01 & 2.809e-01 & 0 \\ 3.950e-01 & -1.263e-02 & -8.142e-02 & 0 \\ 0 & 1 & -1.405e-01 & 0 \end{bmatrix} \tag{G.14}$$

$$B =$$

$$\begin{bmatrix} 1.275e-05 & 0 & 0 & -1.240e-02 & 1.023e-02 & -1.086e-04 \\ 0 & -3.114e-05 & -3.117e-03 & 6.571e+00 & 1.256e+00 & -4.126e-03 \\ 0 & -1.905e-04 & -6.443e-05 & 3.783e-01 & -2.560e-01 & 4.533e-04 \\ 0 & 0 & 0 & 0 & 0 & 0 \end{bmatrix}$$

$$\tag{G.15}$$

$$C = \begin{bmatrix} 1 & 0 & 0 & 0 \\ 0 & 0 & 0 & 0 \\ 0 & 0 & 0 & 0 \\ 0 & 1 & 0 & 0 \\ 0 & 0 & 1 & 0 \\ -6.804e+01 & -1.744 & -4.058 & -3.720e-05 \\ 0 & 0 & 0 & 1 \end{bmatrix} \tag{G.16}$$

$$D =$$

$$\begin{bmatrix} 0 & 0 & 0 & 0 & 0 & 1.148e-03 \\ 0 & 0 & 0 & 1 & 0 & 0 \\ 0 & 0 & 0 & 0 & 1 & 0 \\ 0 & 0 & 0 & 0 & 0 & 0 \\ 0 & 0 & 0 & 0 & 0 & 0 \\ 1.111e-02 & -1.111e-02 & -1.111e-02 & 2.667e+01 & -2.952 & -7.810e-02 \\ 0 & 0 & 0 & 0 & 0 & 0 \end{bmatrix}$$

$$\tag{G.17}$$

Each measurement is corrupted by the white noise passing through a filter. The weighting matrix W_{sens} is described as follows:

$$W_{sens} =$$

$$diag\left[\frac{0.003(1+100s)}{(1+2s)}\quad\frac{0.003(1+100s)}{(1+2s)}\quad\frac{0.25(1+20s)}{(1+0.1s)}\quad\frac{0.007(1+100s)}{(1+0.5s)}\right]$$

(G.18)

As p and r are both measured with similar gyroscopes, their sensor noise weights are assumed to be identical.

The actuators are represented by the elevon $\mathbf{u}_e$ and the rudder $\mathbf{u}_r$. The actuator weights are

$$W_{act} = diag\left[W_e \quad W_r\right]$$

(G.19)

where

$$W_e = \frac{1}{1+1.44(s/14)+(s/14)^2}\frac{1-1.732(s/173)+(s/173)^2}{1+1.732(s/173)+(s/173)^2}$$

(G.20a)

$$W_r = \frac{1}{1+1.5(s/21)+(s/21)^2}\frac{1-1.732(s/173)+(s/173)^2}{1+1.732(s/173)+(s/173)^2}$$

(G.20b)

The performance weights are

$$W_{perf} = diag\left[\frac{0.8(1+s)}{(1+10s)}\quad\frac{500(1+s)}{(1+100s)}\quad\frac{250(1+s)}{(1+100s)}\right]$$

(G.21)

The dynamics of the ideal model is given by

$$G_{ideal}(s) = \frac{1}{1+1.4(s/1.2)+(s/1.2)^2}$$

(G.22)

Index